# Treatise on Materials Science and Technology

## VOLUME 1

# TREATISE ON MATERIALS SCIENCE AND TECHNOLOGY

EDITED BY

## HERBERT HERMAN

*Department of Materials Science*
*State University of New York at Stony Brook*
*Stony Brook, New York*

## VOLUME 1

 1972

ACADEMIC PRESS   New York and London

ACADEMIC PRESS, INC.
111 Fifth Avenue, New York, New York 10003

*United Kingdom Edition published by*
ACADEMIC PRESS, INC. (LONDON) LTD.
24/28 Oval Road, London NW1

LIBRARY OF CONGRESS CATALOG CARD NUMBER: 77-182672

PRINTED IN THE UNITED STATES OF AMERICA

# Contents

## On the Energetics, Kinetics, and Topography of Interfaces

### W. A. Tiller

## Fracture of Composites

### A. S. Argon

## Theory of Elastic Wave Propagation in Composite Materials

### V. K. Tewary and R. Bullough

## Substitutional–Interstitial Interactions in bcc Alloys

### D. F. Hasson and R. J. Arsenault

## The Dynamics of Microstructural Change

### R. T. DeHoff

## Studies in Chemical Vapor Deposition

*R. W. Haskell and J. G. Byrne*

# List of Contributors

Numbers in parentheses indicate the pages on which the authors' contributions begin.

A. S. ARGON (79), Massachusetts Institute of Technology, Cambridge, Massachusetts

R. J. ARSENAULT (179), Engineering Materials Group, College of Engineering, University of Maryland, College Park, Maryland

R. BULLOUGH (115), Theoretical Physics Division, A.E.R.E., Harwell, United Kingdom

J. G. BYRNE (293), Division of Materials Science and Engineering, University of Utah, Salt Lake City, Utah

R. T. DEHOFF (247), Department of Metallurgical and Materials Engineering, University of Florida, Gainesville, Florida

R. W. HASKELL (293), Division of Materials Science and Engineering, University of Utah, Salt Lake City, Utah

D. F. HASSON (179), Engineering Materials Group, College of Engineering, University of Maryland, College Park, Maryland

V. K. TEWARY (115), Theoretical Physics Division, A.E.R.E., Harwell, United Kingdom

W. A. TILLER (1), Department of Materials Science, Stanford University, Stanford, California

# *Preface*

Materials limitations are often the major deterrents to the achievement of new technological advances. In modern engineering systems, materials scientists and engineers must continually strive to develop materials which can withstand extreme conditions of environment and maintain their required properties. In the last decade we have seen the emergence of new types of materials, literally designed and processed with a specific use in mind. Many of these materials and the advanced techniques which were developed to produce them, came directly or indirectly from basic scientific research.

Clearly, the relationship between utility and fundamental materials science no longer needs justification. This is exemplified in such areas as composite materials, high-strength alloys, electronic materials, and advanced fabricating and processing techniques. It is this association between the science and technology of materials on which we intend to focus in this treatise.

The topics to be covered in this Treatise on Materials Science and Technology will include the fundamental properties and characterization of materials, ranging from simple solids to complex heterophase systems. This treatise is aimed at the professional scientist and engineer, as well as at graduate students in materials science and associated fields.

Volume 1 includes an article by W. A. Tiller on an atomistic view of interfaces, with emphasis on mechanisms and topography. There are two articles on the mechanical properties of composite materials: A. S. Argon discusses fracture of a wide range of composites, and V. K. Tewary and R. Bullough examine the theory of the propagation of elastic waves through composites. The complex nature of substitutional–interstitial interactions in bcc alloys is reviewed by D. F. Hasson and R. J. Arsenault. In his contribution, R. T. DeHoff develops a range of ideas on the dynamics of quantitative metallography. Both the science and technology of chemical vapor deposition are outlined in an article by R. W. Haskell and J. G. Byrne.

Forthcoming volumes will examine fatigue and other aspects of mechanical properties of a variety of materials, phase transformations in metal alloys and both crystalline and noncrystalline ceramics, a wide range of topics in the fields of ceramics and polymers, and numerous other topics.

The editor would like to express his sincere appreciation to the members of the Editorial Advisory Board who have given so generously of their time and advice.

H. HERMAN

# On the Energetics, Kinetics, and Topography of Interfaces

W. A. TILLER

*Department of Materials Science*
*Stanford University*
*Stanford, California*

## I. Introduction

The present scientific understanding of surfaces is largely thermodynamic in nature (Mullins, 1963) and, although it is capable of giving one a reliable procedure for categorizing experimental behavior, it gives little insight into the detailed mechanisms which lead to this behavior. In order to progress further in this area, it is necessary to follow a path of study aimed at revealing the individual factors which determine the interfacial energy and the interface configuration between any two materials.

To illustrate the various distinguishable contributions involved in the interfacial energy, let us consider the formation of an $\alpha/\beta$ interface from an $\alpha$ crystal and a $\beta$ crystal (each are of unit cross-sectional area) by the procedure illustrated in Fig. 1. We make imaginary cuts through the cross section of each crystal and separate the halves of each crystal without allowing either atomic or electronic relaxation. This requires work $W_{\alpha\alpha} + W_{\beta\beta}$. We next bring the $\alpha$ and $\beta$ half-crystals together along the cut surfaces gaining work $W_{\alpha\beta}$. Next, we let the free electrons relax their positions which does work $2W_e$. Finally, we let the atom positions relax which does work $2W_a$. Since we end up with two unit areas of $\alpha/\beta$ interface of specific interfacial free energy $\gamma_{\alpha\beta}$, the energy balance in the process requires that

$$\gamma_{\alpha\beta} = \{[(W_{\alpha\alpha} + W_{\beta\beta})/2] - W_{\alpha\beta}\} - W_a - W_e \tag{1a}$$

In Eq. (1a), all of the work terms are positive. We will see that the excess free energy associated with the interface $\gamma$ can be treated as arising from five different sources: (1) a short range quasi-chemical term $\gamma_c$ due to the typical molecular interactions at the interface, (2) a longer range electrostatic term $\gamma_e$ associated with the equalization of the electrochemical potential of the free electrons, (3) a short-range transition structure term $\gamma_t$ associated with the atomic diffuseness of the interface, (4) a long-range strain energy term $\gamma_d$ associated with atomic displacements (dislocations) at the interface, and (5) a chemical adsorption term $\gamma_a$ associated with the redistribution of different chemical species in the interface field due to (1)–(4); i.e.,

$$\gamma = \gamma_c + \gamma_e + \gamma_t + \gamma_d + \gamma_a \tag{1b}$$

In Eq. (1b), $\gamma_d$ is always positive whereas $\gamma_e$, $\gamma_a$, and $\gamma_t$ are always negative. However, $\gamma_c$ may be either positive or negative depending upon the relative interaction energies of the AA, BB, and AB bonds (because of special electron

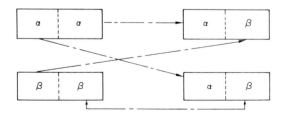

**Fig. 1.** Illustration of $\alpha/\beta$ interface formation process from single crystal blocks of $\alpha$ and $\beta$.

effects it is most likely to be positive). In reality, these contributions are not completely independent of each other; however, we can consider them so, under somewhat idealized conditions and can evaluate the magnitude of each contribution. In this article, attention will be restricted to $\gamma_c$, $\gamma_e$, and $\gamma_t$.

It has been our habit in the past to think of $\gamma$ as a single and indivisible entity that is unchanging in any measurable way as a particular material surface is utilized in its many potential applications: processing (phase transformations, foams, emulsions, aerosols, catalysis, coagulation), protection (coatings, paints), adhesion (joining, brazing), corrosion and degradation, or for its unique properties (composites, films, membranes). This practice has contributed greatly to the slowness of progress in the field. We are now capable of proceeding to the next level of discernment and modeling wherein we use the approach of Eq. (1b) not just for general computational convenience but to indicate that each of the components will be altered in magnitude by different amounts depending upon the particular application in which the surface is involved.

We have amassed a reasonable amount of data concerning the excess free energy of the liquid–vapor surface $\gamma_L$ and the solid–vapor surface $\gamma_S$. Only a small amount of data exists concerning the excess free energy of interfaces $\gamma_{\alpha\beta}$ and practically no data exists on the individual components of $\gamma_{\alpha\beta}$ identified in Eq. (1b). Here, we are specifically interested in $\gamma_{SL}$.

The long-standing nucleation data of Turnbull (1950) provides us with most of our information on $\gamma_{SL}$. He found that the ratio of $\gamma_{SL}$ per atom to the latent heat of fusion $\Delta H$ per atom was approximately 0.5 for most metals and approximately 0.35 for nonmetals. An illustration of his data is given in Table I. Using contact angle techniques, Miller and Chadwick (1967) measured the ratio $\gamma_{SL}/\gamma_{GB}$, where GB refers to high angle grain boundary, on the alloy systems Al–Sn and Zn–Sn. They extrapolated this data to a zero concentration of tin and found that $\gamma_{SL}/\gamma_{GB} \to 0.47$ in both cases. However, using data on $\gamma_{SL}$ and $\gamma_{GB}$ for pure systems from other sources, this ration of 0.47 is not found for these systems and, for many systems, one finds the ratio to be greater than 0.5. Obviously, we have a discrepancy here that needs to be cleared up in the future. Using an electron microscope technique to study dihedral angles at grain boundaries, Glicksman and Vold (1968) found for bismuth, $\gamma_{SL} = 61.3$ ergs/cm$^2 \approx 1.1\gamma_{SL}$ (Turnbull). Finally, Morris (see Chadwick, 1970) studied the lead solid–liquid interface in a temperature gradient and used 100 ppm of antimony for demarcation of the interface contour in the vicinity of the grain boundary. Comparing this with the predicted shape by Bolling and Tiller (1960), he found $\gamma_{SL} \approx 3\gamma_{SL}$ (Turnbull) for lead. He also found a strong anisotropy of $\gamma_{SL}(\gamma_{(111)} \sim 0.8\gamma_{avg})$. We begin to see, from the foregoing, that the measured $\gamma_{SL}$ appears to depend upon the technique used for its measurement [which may indicate the influence of the technique on the different components of $\gamma$ in Eq. (1b)]. Obviously, we are in need of more experimental data on $\gamma_{SL}$ gathered very carefully via a number of different techniques.

## TABLE I

RELATIONSHIP BETWEEN MAXIMUM SUPERCOOLING, SOLID–LIQUID INTERFACIAL ENERGY AND HEAT OF FUSION

| Metal | Interfacial energy $\gamma_{SL}$ (ergs/cm²) | $\gamma g$ (cal/mole) | $\gamma g / \Delta H$ | $\Delta T$ (deg) |
|---|---|---|---|---|
| Mercury | 24.4 | 296 | 0.53 | 77 |
| Gallium | 55.9 | 581 | 0.44 | 76 |
| Tin | 54.5 | 720 | 0.42 | 118 |
| Bismuth | 54.4 | 825 | 0.33 | 90 |
| Lead | 33.3 | 479 | 0.39 | 80 |
| Antimony | 101 | 1430 | 0.30 | 135 |
| Germanium | 181 | 2120 | 0.35 | 227 |
| Silver | 126 | 1240 | 0.46 | 227 |
| Gold | 132 | 1320 | 0.44 | 230 |
| Copper | 177 | 1360 | 0.44 | 236 |
| Manganese | 206 | 1660 | 0.48 | 308 |
| Nickel | 255 | 1860 | 0.44 | 319 |
| Cobalt | 234 | 1800 | 0.49 | 330 |
| Iron | 204 | 1580 | 0.45 | 295 |
| Palladium | 209 | 1850 | 0.45 | 332 |
| Platinum | 240 | 2140 | 0.45 | 370 |

## II. Energetics: $\gamma_c$

Though all molecular and atomic forces ultimately find their root in the mutual behavior of the constituent parts of the atoms, i.e., the nuclei and the electrons and, although theoretically they all may be derived from the fundamental wave equations describing the atoms, it is still convenient here, as in other branches of physics, to treat the various forms of mutual interactions of atoms as different forces acting independently. Such a procedure, which deliberately ignores the theoretical knowledge that all forces are already included in the wavefunction describing the mutual behavior of atoms, is permitted only because of the good empirical results obtained with it. To properly deal with this topic area, it is necessary to survey such forces as the exchange forces leading to covalent bonds, the Coulomb forces between ions and dipoles, the electrostatic polarization of atoms or molecules by ions, the nonpolar van der Waals' (dispersion) forces, the repulsion forces, and the energy states of the free electrons. A detailed understanding of these forces is not only important for an assessment of $\gamma_c$, but is vital for the understanding of molecular adsorption to interfaces, crystal habit modification, and modification of layer kinematics. In this section, we shall discuss each of these items briefly. Next, we shall consider the various methods of calculation available to us, and, finally, we shall consider some relevant examples to illuminate the teachings of the earlier discussion.

## A. Basic Components

### 1. REPULSIVE FORCES

Whatever may be the detailed nature of the attraction forces between atoms and molecules, when they approach each other repulsion forces will check the attraction forces and, at a certain distance (equilibrium distance), these forces will balance each other. The repulsion forces arise from the interpenetration of the electron clouds of the atoms (Pauli principle) and will be greater the more electrons there are in the outer shells of the atoms. The potential energy $E_R$ arising from these forces has been given two simple forms

$$E_R = +b\,e^{-r/\rho} \tag{2a}$$

$$= +(b/r)n \tag{2b}$$

where $b$, $\rho$, and $n$ are constants and where $r$ is the distance between atoms.

### 2. DIPOLE–DIPOLE FORCES

The elementary theory of dipole–dipole interaction utilizes the configuration given in Fig. 2 and the familiar electrostatic inverse square relationship for

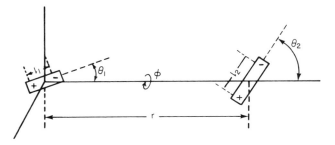

**Fig. 2.** Construction for general calculation of force and energy of interaction of two dipoles $\mu_1$ and $\mu_2$.

point charges (subject to the condition $r \gg l$ which often does not hold in real situations). For fixed orientations, the mutual potential energy of the two dipoles is given by

$$E_A = -(\mu_1 \mu_2/r^3)[2\cos\theta_1 \cos\theta_2 - \sin\theta_1 \sin\theta_2 \cos(\phi_1 - \phi_2)] \qquad (3a)$$

where $\mu_1$ and $\mu_2$ are the permanent dipole moments of the two molecules and where $\theta$ and $\phi$ are the angles, in polar coordinates, giving the orientations of the dipoles. If the dipoles are free to rotate relative to each other, they will take up the head-to-tail (lowest energy) configuration. Then Eq. (3a) reduces to

$$E_A = -2\mu_1 \mu_2 r^{-3}. \qquad (3b)$$

These formulas apply only to dipoles in fixed orientation. Two dipoles may be free to rotate, relative to each other, provided the chemical binding of other groups in the respective molecules or the crystal allows it and provided the dipole attraction energy is less than the average thermal energy $kT$ ($kT = 4.1 \times 10^{-14}$ ergs at $T = 25°$ C). If a dipole is not restricted by chemical bonds to the other molecular segments and if the $E_A$ is small, the mutual orientation of two dipoles may be destroyed in the collisions occurring between molecules and the average interaction energy becomes

$$E_A = -(2\mu_1{}^2 \mu_2{}^2/3kT)r^{-6} \qquad (3c)$$

In Table II, $E_A$ is given for some dipole–dipole interactions considering only the above-mentioned orientation polarization in the most favored orientation (here, $\mu$ is given in debyes, i.e., units of $1 \times 10^{-18}$ esu-cm). From Table II, we

## TABLE II

BIMOLECULAR ENERGIES OF INTERACTION FOR SOME DIPOLES[a]

| | | | Assumed intermolecular distance $a°$ | Energy (ergs) | |
|---|---|---|---|---|---|
| | $\mu_1$ | $\mu_2$ | | $2\mu_1{}^2\mu_2{}^2/3kTr^6$ Eq. (3c) | $2\mu_1\mu_2/r^3$ Eq. (3b) |
| $CHI_3$–$CHI_3$ | 0.9 | 0.9 | 5.0 | $7 \times 10^{-16}$ | |
| $CHCl_3$–$CHCl_3$ | 1.1 | 1.1 | 4.5 | $6 \times 10^{-15}$ | |
| $CH_3Cl$–$CH_3Cl$ | 1.94 | 1.94 | 4.0 | | $1.2 \times 10^{-13}$ |
| $CH_3Cl$–$H_2O$ | 1.94 | 1.85 | 3.5 | | $1.7 \times 10^{-13}$ |
| $CH_3NO_2$–$CH_3NO_2$ | 3.1 | 3.1 | 4.5 | | $2.1 \times 10^{-13}$ |
| $H_2O$–$H_2O$ | 1.85 | 1.85 | 3.0 | | $2.5 \times 10^{-13}$ |
| $CH_3NO_2$–$CH_3Cl$ | 3.1 | 1.94 | 4.2 | | $3.1 \times 10^{-13}$ |

[a] At 25° C ($kT = 4.1 \times 10^{-14}$ ergs).

note that water is expected to exhibit a surface with aligned dipoles (the protons are found to be immersed in the liquid and the oriented surface layer is 6 to 8 molecules thick; inside this layer, the $H_2O$ molecules are randomly oriented).

## 3. DIRECT COULOMB FORCES

If we idealize the surface of an ionic compound (e.g., alkali halide crystal) as a two-dimensional network of ions and we assume a negative ion adsorbed just on top of one of the positive ions of the surface, one finds that the attraction of this negative adsorbed ion by the surface is far weaker than it would be if the positive ion with which it has direct contact were on its own (i.e., $Z_1 Z_2/r^2$). Denoting $r_c$ as the shortest interionic distance in the crystal and setting the equilibrium distance of the adsorbed ion equal to $r_c$, we find that the electrostatic field emanating from the surface falls off very rapidly with distance (the field is negligible at $4r_c-5r_c$) and that the interaction energy is given by

$$E_A = -(aZ_sZ_i/r_c) \qquad (4)$$

where $Z_s$ is the charge of the surface ions and $a$ depends upon the location of the ion, with charge $Z_i$, on the surface ($a = 0.066, 0.090, 0.247$, and $0.874$ for face-middle, crystal-edge, crystal-corner and half-crystal positions, respectively). The electrostatic forces will be extremely strong at all places where the normal periodicity of the constituent ions of the crystal is interrupted. Crystal or lattice disturbances at the surface are, for this reason, active spots for adsorption and catalysis.

## 4. INDUCTION FORCES

If a molecule, having a permanent dipole moment $\mu$, is located in an external electric field $F$, the energy of interaction is given by

$$E_A = -\mu \cdot F \qquad (5a)$$

The external field described by $F$ may be produced by the solid per se (ionic solids) or it may be produced by charge distributions induced in the solids by the presence of the moments of molecules (important for conductors). If a nonpolar molecule, of polarizability $\alpha$, is located in the external field $F$, an induced dipole moment of magnitude $\mu = \frac{1}{2}\alpha F$ is created in the direction of the field and the interaction energy is again given by Eq. (5a).

An electric charge polarizes a metal in such a way that we describe it as if an electric charge of opposite sign were induced at a distance below the surface equal to the distance between the actual inducing charge and the metal surface (image charge). In this approximation, we are treating the metal as an ideally polarizable structure. For an adsorbent of more restricted polarizability, a similar phenomenon happens; however, the shifting of the electrons due to the polarization is confined to bound electrons. An ion of charge $Z_i$ at a distance $r$ from an atom of polarizability $\alpha$ induces a dipole of moment $\alpha Z_i/r^2$. The interaction energy of the ion and the atom due to this electrostatic polarization is

$$E_A = -(\alpha Z_i^2 r^{-4}/2) \tag{5b}$$

Equation (5b) gives the contribution toward the adsorption energy of an ion, on a condensed phase, by the polarization of only the nearest center of attraction. The dielectric constant and the molar polarization of a dielectric by an electric charge take full account of all the interactions so that the interaction energy, due to this electrostatic induction of a dielectric by an adsorbed ion, may be given as

$$E_A = -\frac{Z_i^2}{4r}\left(\frac{k-1}{k+2}\right) \tag{5c}$$

where $K$ is the dielectric constant of the material. In Table III, comparative values of $E_A$ from Eq. (4) (Section II, A, 3) and $E_A$ from Eq. (5c) (Section II, A, 4) are given for a variety of situations.

### TABLE III

ENERGY CONTRIBUTIONS BY VARIOUS FORCES DUE TO MONOVALENT ION
AT A DISTANCE OF 2.81 Å

| Surface | Ergs/ion ($\times 10^{12}$) | Kcal/mole |
|---|---|---|
| NaCl (smooth cubic face) | | |
| $E_A{}^b$ | 0.54 | 7.8 |
| $E_A{}^c$ | ~1.0 | ~15 |
| NaCl, cubic face | | |
| (H-crystal position) | | |
| $E_A{}^b + E_A{}^c$ | 7.15 | 103 |
| Metal Surface ($E_A{}^b$) | 2.04 | 29.5 |
| Dielectric Surface $E_A{}^c$ | | |
| ($K = 1.5$) | 0.60 | 8.6 |
| ($K = 10$) | 1.53 | 22 |
| ($K = 110$) | 1.98 | 28.5 |

[a] If the polarizability of the dielectric is only due to electronic shifts and not to displacement of ions or dipoles, $K = n^2$ where $n$ is the index of refraction [TiO$_2$ in the form of Rutile has $K = 114$ ($n^2 = 7.35$)].
[b] From Eq. (4), Sect. II, A, 3.
[c] From Eq. (5c), Sect. II, A, 4.

## 5. Nonpolar Dispersion Forces

The largest contribution to van der Waals' forces is from the London dispersion forces arising from mutually induced in-phase components in the oscillations of the electron cloud. These forces predominate in the case of nonpolar molecules interacting with covalent or metallic solids. In the quantum-mechanical ground state of an atom, the electrons still possess considerable kinetic energy. Thus, the instantaneous charge distribution over the atoms will possess electric moments even though, on the average, the distribution might be spherically symmetric. The instantaneous moments of neighboring molecules then interact to produce a nonvanishing force of attraction. For a single pair of atoms, the potential energy of interaction is found to be

$$E_A = -Cr^{-6} - C^1 r^{-8} - C^{11} r^{-10} \dots \tag{6}$$

where $C$, $C^1$, $C^{11}$, etc., are constants associated with instantaneous dipole–dipole, dipole–quadrupole, quadrupole–quadrupole, etc., interactions, respectively. In most normal cases, the second and third terms are neglected leaving only the dipole–dipole term.

When simple atoms in their ground state are approximated by simple isotropic harmonic oscillators of frequency $v_1$ and $v_2$ (obtained from the optical dispersion curve of the atoms), London obtained the result

$$C = \tfrac{3}{2}\alpha_1 \alpha_2 [hv_1 v_2/(v_1 + v_2)] \tag{7}$$

where $h$ is Plank's constant. The above equation holds at short interaction distances; at large separation distances, however, a relativistic effect enters the picture to change things somewhat. At a distance $r$ from the fluctuating instantaneous dipole of molecule $a$, the electric field $F$ will fluctuate synchronously if $r$ is small and so will the dipole that is induced in molecule $b$. But electromagnetic disturbances are propogated at the speed of light $c$ thus, if $r$ is large enough, the fluctuations of the field at $r$ will be retarded behind the fluctuations of the dipole at $a$ and the interaction energy will be reduced. This retardation effect causes the attraction energy to fall off as $r^{-7}$ at large distances instead of as $r^{-6}$ which holds for small distances (small with respect to the wavelengths $\lambda$, corresponding to the transitions between the ground and excited states of the atoms).

Lifshitz (1956) has applied the retardation concept to the interaction of two bulk phases in a very novel way. From this point of view, the interaction of two macroscopic objects is regarded as occurring through the medium of the fluctuating electromagnetic field which is always present in the interior of any absorbing medium and also extends beyond its boundaries (as a result of statistical fluctuations in the position and motion of charges in a body, spontaneous local electric and magnetic moments occur in it). This electromagnetic field is present partially in the form of traveling waves radiated by the body and partially of standing waves which are damped exponentially as they move away from the surface of the body. The field does not vanish even at absolute zero, at which point it is associated with the zero point vibrations of the radiation field. Experiments carried out with $r \sim 0.1$ to $1.0\,\mu$ separation completely confirm the details of this analysis. The analysis, which is too detailed to be presented here, indicate that $E_A$ should be considerably greater for two metals of high conductivity than for two metals of low conductivity. In the case where the conductivities of two media are decidedly different, the magnitude of $E_A$ is governed largely by the conductivity of the poorer material. This approach promises to be of significant interest for the interaction of two metals separated by a thin layer of oxide such as one finds in brazing applications.

## 6. Exchange Forces

If an atom comes near the surface of a solid, there may be an exchange of electrons resulting in a covalent and/or an ionic bond. The situation is vastly more complicated and less well understood than the other forces already discussed. Thus, we shall not penetrate too deeply. In an ionic bond, one imagines that the energy of the solid-ad-atom complex is lowered by transferring an electron from the solid to the ad-atom or vice versa. In Fig. 3, this situation is sketched for an ad-atom with ionization potential $I$ (the minimum energy required to remove an electron from an atom) and affinity $A$ (the energy binding an extra electron to an otherwise neutral atom). Nearby is the surface of a metal having work function $\psi$ and Fermi energy $E_F$. If $A > \psi$, the system lowers its energy by transferring an electron from the metal to the ad-atom, whereas if $I < \psi$, an electron is transferred from the ad-atom to the metal. Image potential effects increase $A$ and decrease $I$ by roughly $e^2/4x$ relative to their free atom values. A large dipole moment accompanies the ionic bonding corresponding to an electronic charge transferred through a distance of order $x$. Thus, the ionic bond corresponds to a 100% probability for the ad-atom being non-neutral.

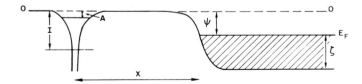

**Fig. 3.** Schematic illustration of affinity $A$ and ionization $I$ levels in an adsorbed molecule at a metal surface, the metal having a work function $\psi$, Fermi level $E_F$, and Fermi energy $\zeta$.

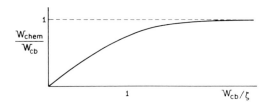

**Fig. 4.** Schematic illustration of the chemisorption energy $W_{chem}$, divided by the binding energy of the ad-atom–metal atom "molecule" $W_{cb}$ plotted as a function of the ratio $W_{cb}$ to the Fermi energy $\zeta$.

Gomer and Schrieffer (1969) very recently proposed the concept of an induced covalent bond to metals, which appears to have great merit. They imitate the Heitler–London scheme for forming a diatomic molecule, i.e., one atom of the molecule is the ad-atom and the other is the metal sample. The problem is to form an unpaired spin in the metal adjacent to the ad-atom which requires the formation of an electron wave packet costing energy of about one half the electron bandwidth in the metal. If the energy reduction associated with the covalent bond between this spin and the ad-atom is larger than the energy required to form the wave packet, a covalent chemisorption bond is formed. In practice, a covalent bond can always be formed even in $W_{cb} \ll E_F$, although the bond will have a net energy of order $W_{cb}(W_{cb}/E_F)$. The qualitative results are sketched in Fig. 4. The two limiting regimes are $W_{cb}/E_F \ll 1$, when a weak covalent bond is formed between the ad-atom and the weakly spin-polarized metal ($W_{chem}/W_{cb} \sim W_{cb}/E_F$) and when a very strong metal-ad-atom bond is formed, breaking the metal atom-metal bond (a weak covalent bonding of the molecule to the metal surface occurs and $W_{chem}/W_{cb} \sim 1$).

## 7. FREE ELECTRON ENERGY STATES

A contribution to $\gamma_c$ is expected to arise from the change in the kinetic energy of the electrons due to the different volume available for them at the surface and due to the penetration of the surface by the electron wave. A positive contribution to $\gamma_c$ arises from the increase in the kinetic energy of the electrons imposed by the steep potential barrier presented by the newly created surface. In a metal of infinite extent, the wavenumbers $k$ of the free electrons form a continuum. However, the standing electron waves should have nodes at a free surface. Thus, for a metal of finite thickness $L$ the wavenumbers of the free electrons are quantized with values proportional to $L^{-1}$ so that the Fermi energy should vary as $L^{-2}$ and $\gamma_c$ increase. The surface penetration term could be considered as a relaxation effect and thus would contribute a negative contribution to $\gamma_c$.

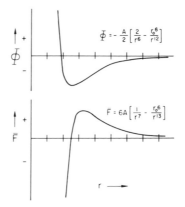

**Fig. 5.** Energy and force of interaction between two molecules according to the Lennard–Jones potential.

## 8. Surface Entropy

This is too large a subject to be dealt with here. People have generally swept the excess entropy of the surface $S$ under the proverbial rug. We shall do likewise with this contribution $S_e$; however, let us note that it needs some very careful attention in the future.

## B. *Methods of Calculation*

### 1. Atom Summation†

For a molecular system, a typical energy–distance curve for a pair of atoms will resemble Fig. 5, which has been drawn to the shape of a Lennard–Jones potential $\Phi$, i.e.,

$$\Phi = -\frac{A}{2}\left[\frac{2}{r^6} - \frac{r^6}{r^{12}}\right] \tag{8a}$$

where $A$ is constant and $r_0$ is the intermolecular distance at equilibrium. The net force $\bar{F}$ between the atoms is given by

$$\bar{F} = -\frac{d\Phi}{dr} = 6A\left[\frac{1}{r^7} - \frac{r_0^6}{r^{13}}\right] \tag{8b}$$

---

†Even in a metal, the free electrons cluster around the ion cores giving much the same electron configuration as in the bound electron case. The energy levels of these electrons are different than in the bound electron case and, even though clustered around ion cores, they move freely from cluster to cluster. Thus, although $\alpha$ might be selected as the free atom value without great error, the $v_i$ are quite different for the condensed phase so that one should not use Eq. 7 for condensed phases.

If the interaction energy of two atoms at a distance $r$ is given by $\Phi(r)$ then, for a perfect infinite crystal, the total lattice energy per atom is given by

$$E_L = \tfrac{1}{2} \sum_i M_i \Phi(r_i) \tag{9a}$$

where $r_i$ is an interatomic distance, $M_i$ is the number of neighbors of a given atom at this distance and the sum extends over all possible $r_i$. The method of calculation is therefore simply to list, for each structure, the possible values of $r_i$ and their associated multiplicities $M_i$ and then evaluate the lattice sum for particular potential functions $\Phi(r)$. This corresponds to writing Eq. (9a) in the form

$$E_L = \sum_i \Phi_i^s \tag{9b}$$

where

$$\Phi_i^s = \tfrac{1}{2} M_i \Phi(r_i) \tag{9c}$$

represents the contribution to the lattice energy from the $i$th shell of neighbors.

One may consider Morse potentials of the form

$$\frac{\Phi(r)}{\Phi_0} = \left\{ 1 - \exp\left[ -\alpha\left( \frac{r - r_0}{r_0} \right) \right] \right\}^2 - 1 \tag{9d}$$

or Mie potentials of the form

$$\Phi(r)/\Phi_0 = (m - n)^{-1} \{ n(r_0/r)^m - m(r_0/r)^n \} \tag{9e}$$

where, in all potentials, $\Phi_0$ corresponds to the maximum interaction energy which occurs at a separation $r_0$. For a particular material, one can use experimental values of the sublimation energy, lattice parameter, and compressibility to deduce values of either $\Phi_0$, $\alpha$, $r_0$ in Eq. (9d) or $m$, $n$, and $r_0$ in Eq. (9c). Thus, Eqs. (9d) and (9e) become useful fitting functions for surface energy calculations.

With such pairwise interactions, it is convenient to say that each atom is linked to each other atom by a "bond" and to associate an energy with such a bond. When an infinite crystal is divided into two parts, a number of these bonds will be broken (they will increase to infinite length), and the energy associated with each new surface will be computed as one half the energy associated with breaking these bonds. To evaluate the energy associated with a two-phase interface, one must consider three types of bonds. Two are associated with the splitting of each of the condensed phases into two parts and the third is associated with the joining, in a particular configuration, of these two different phases to form the $\alpha/\beta$ interfaces illustrated in Fig. 1. Nicholas

(1968) has shown that for a crystal divided along a plane of a particular orientation specified by the normal vector $\bar{h}$ with magnitude $h$, the component of surface energy $\gamma_c(\bar{h})$ is given by

$$\gamma_c(\bar{h}) = \sum_j \mu_j(\bar{h}) E_j \tag{10a}$$

where

$$\mu_j(\bar{h}) = (1/\Omega h)\,\bar{h} \cdot \sum_k \bar{u}_k \tag{10b}$$

is the "multiplicity" factor associated with a broken bond from the $j$th shell and

$$E_j = -\tfrac{1}{2}\Phi(u_j) \tag{10c}$$

is the energy contribution from breaking a bond in the $j$th shell at a distance $u_j$. In Eqs. (10), $\Omega$ is the volume of the crystal per atom, the sum over $k$ covers all bonds, with positive $\bar{h} \cdot \bar{u}_k$, that have a particular magnitude $u_j$ and the sum over $j$ covers all possible interatomic distances in the crystal. Physically, this means first summing (with $k$) over all bonds linking an atom to the atoms in a particular shell of neighbors and then summing (with $j$) over all shells.

Nicholas (1968) has applied this approach to fcc and bcc crystal lattices using up to 300 shells of neighbors for a variety of Morse potentials and up to 500 shells of neighbors (22–23 lattice distances) for a variety of Mie potentials to obtain the requisite accuracy. He plotted relative energy contour maps on stereographic triangles for these two systems showing (a) that the fcc system exhibited an anisotropy of less than 15% and exhibited a plateau in the middle portion of the diagram, and (b) that the bcc system tends to one or two plateaus with a ridge joined to the (110) and a similar degree of anisotropy. He then proceeded to calculate the situation wherein (a) only first nearest neighbor bonds are considered and (b) where both first and second nearest neighbors bonds are considered (with $E_2/E_1 = \varepsilon$ of arbitrary magnitude). A most interesting result, which is illustrated in Fig. 6, was forthcoming. Case (a) differs considerably from the exact calculations and produces a much greater anistropy (30%–40%); however, case (b) bears a close resemblance to the exact calculation but with a slightly higher anisotropy for the bcc case. Thus, although it is common procedure in bond-type calculations to take account of nearest neighbors only, a significantly more reliable result is obtained by the inclusion of second neighbors provided $E_1$ and $\varepsilon$ are treated as adjustable parameters. It is completely profitless to calculate these values from a given potential function; however, when only relative values are needed, a good estimate is obtained by taking $\varepsilon$ between $\tfrac{1}{3}$ and $\tfrac{1}{2}$ for fcc and between $\tfrac{3}{4}$ and 1 for bcc crystals.

One reason why a high degree of accuracy is needed in these calculations is that small changes in $\gamma$ cause large changes in the equilibrium shape of the

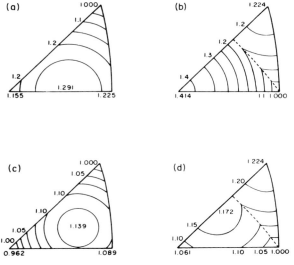

**Fig. 6.** Theoretical contour plots of surface energy (1) when only nearest neighbor interactions are considered in (a) fcc and (b) bcc crystals and (2) when nearest and second-nearest neighbor interactions are considered in (c) fcc crystals with $\varepsilon = 0.5$ and (d) bcc crystals with $\varepsilon = 1$. The dashed lines show where corners occur in the contours.

crystal. This may be illustrated by considering the variation of the anisotropy parameter $\lambda = \gamma_{max}/\gamma_{min}$ over the stereographic triangle; the resulting features of the equilibrium form:

$\lambda \sim 1.01 - 1.02$;    no flat surfaces (nearly spherical form)
$\lambda \sim 1.04 - 1.1$;    flat areas surrounded by curved regions
$\lambda \sim 1.15 - 1.25$;    polyhedral form with rounded corners
$\lambda \gtrsim 1.3$;    polyhedron

## 2. Integration Approximation

Instead of carrying out elaborate summations, integration procedures have often been used to simplify the calculations. The method has been described by Fowler and Guggenhiem (1936) and essentially involves the approximation of smearing the atoms throughout each condensed phase so that they contain a uniform density of matter. Following this procedure, one finds that the energy $\Delta E$ required to separate two phases from a distance $Z$ to an infinite distance is given by

$$\Delta E_{\alpha\beta}(Z) = -2\pi \int_Z^\infty \rho_\alpha \, d_j \int_j^\infty df \int_f^\infty \rho_\beta \, \Phi_{\alpha\beta}(r) \, dr \qquad (11a)$$

where $\rho_\alpha$ and $\rho_\beta$ are the densities of the two phases and $\Phi_{\alpha\beta}$ is assumed to be a function only of $r$, the distance between two differential volumes of mass, and independent of Z. However, even without free electron redistribution or atom relaxation, we might expect $\Phi_{\alpha\beta}$ to be a function of $Z$ especially if the exchange forces and the overlap of the free electron waves are important contributions to $\gamma_c$ (the latter certainly is). However, if we assume $\Phi_{\alpha\beta}$ to be given by a simple potential of the form

$$\Phi_{\alpha\beta} = 4\varepsilon[(\sigma/r)^m - (\sigma/r)^n] \tag{11b}$$

then

$$\Delta E_{\alpha\beta}(Z) = 8\pi\varepsilon\sigma^4 \rho_\alpha \rho_\beta \left[ \frac{(\sigma/Z)^{m-4}}{(m-2)(m-3)(m-4)} - \frac{(\sigma/Z)^{n-4}}{(n-2)(n-3)(n-4)} \right] \tag{11c}$$

The force of adhesion $F$ between these two phases can be readily found from the relationship $F = -d\Delta E/dZ$ and the normal equilibrium conditions.

From Eq. (11c) we note that for a 6–12 type potential, $\Delta E$ varies as $Z^{-2}$ (except for very small $Z$). Thus, for $\alpha$ and $\beta$ chosen as the same substance, $\Delta E$ is approximately twice $\gamma_c$ and we find (neglecting the repulsive contribution)

$$\gamma_c \approx \pi\rho^2 C/24Z^2 \tag{12}$$

with $C$ given by Eq. (7) or by some paramaterization procedure. Even if $\phi_{\alpha\beta}$ is independent of $Z$, as assumed to obtain Eq. (11a), such an integration procedure is only allowed if $Z$ is much greater than the interatomic distance, otherwise the smeared atom model is not correct. It is the use of this approximation that eliminates any possibility of predicting an orientation dependence in $\gamma_c$. Using this procedure instead of the pairwise summation procedure underestimates the magnitude of $\Delta E$ and thus $\gamma_c$ by a factor of 2–5. However, the procedure does show us the $Z^{-2}$ dependence of $\gamma_c$ ($Z^{-3}$ dependence for a single atom in the vicinity of a surface) which should certainly hold at distances greater than several atomic distances. Further, it does allow us to simply calculate the order of magnitude of $\Delta E_{\alpha\beta}$ and $\gamma_{\alpha\beta}^c$. As an exercise, one can estimate the magnitude of the adhesive forces between copper and iron using a 6–12 potential and find $F_{max}$ to be between 5 and 23 tons/cm$^2$ for appropriate choices of $E$, $r$, and the equilibrium separation $Z_0$. The fact that the adhesive strength may assume such large values is in general agreement with the observations that cold-welded joints may be stronger than either component (the tensile strength of pure Fe $\sim$ 8 tons/cm$^2$).

## 3. SUMMATION OF PLANES APPROXIMATION

A convenient method of summation of the interaction forces is to use a model of a solid having all the interacting volume elements in the solid located

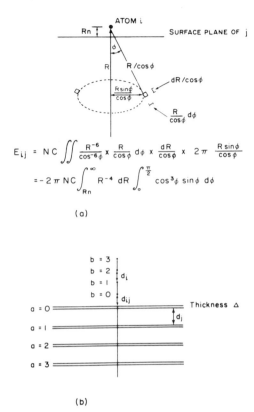

(a)

(b)

**Fig. 7.** (a) Model for integration of interactions of atom $i$ with all volume elements of bulk phase $j$. (b) Model for summation of interactions of all atoms in phase $i$ with all atoms (in planes) in phase $j$.

in thin parallel sheets of uniform density separated by the interelement distance $d_j$. This model is simple for calculation purposes and is an especially good approximation for systems where the volume elements of the other phase are mobile. The results of this calculation give attractive energies about three times greater than that found with Eq. (11a).

A similar but slightly different model has been developed by Fowkes (1967). This is illustrated in Fig. 7 and its usefulness discussed by considering just the attractive forces between a bulk solid and a single atom (denoted by $i$) situated at a distance $R_n$ from the solid (denoted by $J$). The atoms of solid $J$ are assumed to be smeared into uniform sheets of thickness $\Delta$ separated by the interplanar spacing $d_j$. These sheets, therefore, have a density $\rho_j d_j / \Delta$. The attractive energy of atom $i$ for each sheet is

$$E_{ij} = -\pi(\rho_j d_j C_{ij}/6\Delta)[R^{-3}+(R+\Delta)^{-3}] \tag{13a}$$

where $C_{ij}$ is given by Eq. (7). This result could be expanded in powers of $\Delta/R$ and, looking at the small value limit,

$$E_{ij} = (\pi/2)\rho_j d_j C_{ij} R^{-4} \qquad \text{for} \quad \Delta/R \ll 1 \tag{13b}$$

When the interaction is summed over all layers of $J$, one obtains for the energy of attraction (setting $R_n = d_j$)

$$E_{iJ} = -(\pi\rho_j C_{ij}/2d_j^3) \sum_{a=1}^{\infty} a^{-4} \tag{13c}$$

where the summation of $a^{-4}$ for thousands of layers ($\sim$micron) is 1.082. Since one always finds, because of the equilibrium force balance, that the repulsive forces tend to be some fraction of the attractive forces, they can be readily accounted for by a coefficient $f$ in the above formula ($f \sim 2/3$) so that the equilibrium binding energy (for $d_i = d_j$) would be in this case

$$\Phi_{iJ} = -(\pi\rho_j C_{ij} f_{ij}/2d_j^3) \sum_{a=1}^{\infty} a^{-4} \tag{13d}$$

One can readily extend this procedure to the case of $d_i \neq d_j$ and to the case of an interface between phases $I$ and $J$ (see Fig. 7b)

$$\Phi_{IJ} = -\frac{\pi}{2}\rho_i \rho_j f_{ij} C_{ij} d_i d_j \sum_{a=0}^{\infty} \sum_{b=0}^{\infty} (d_{ij}+ad_j+bd_i)^{-4} \tag{14a}$$

where $d_{ij}$ is the equilibrium separation distance at the interface. For the special case of $d_i = d_j = d$, we find

$$\Phi_{IJ} = -0.6\pi\rho_i \rho_j f_{ij} C_{ij} d^{-2} \tag{14b}$$

which leads to a value of $\gamma_c$ about a factor of 4 larger than that given by Eq. (12).

Using this model, we still see the $Z^{-2}$ dependence but also see an orientation dependence of $\Phi_{IJ}$ via the $d_i$ and the $d_j$. One path for increasing the accuracy of the method would be to use atomic pairwise summation procedures for the two contacting planes at the interface, i.e., $a = 1$ and $b = 1$, and use the sheet summation procedure for the remainder of the volume.

## C. Examples

1. We can use the understanding of this section to illustrate, in a straightforward way, that the excess energy of a layer edge on the interface may be appreciably reduced below that for the same area of surface locate in an infinite plane of that same orientation. Consider Fig. 8, the hatched region is the solid. Region 1-6 experiences a decreased $\gamma_c$ due to the interaction of the

volume 1-2. We expect this energy change to decrease as we move from the
1 2 3 contact point to the left. Region 2-4 experiences an increase in $\gamma_c$ identical
with the decrease for the region 1-6 provided we ascribe the missing atoms in
volume 6 to the 2-6 interface and provided the step height is large enough.
Region 2-6 appears to have an increased interaction associated with the 1-2
volume interaction. However, this interaction has already been counted in the
1-6 segment. Region 2-6 also has a normal loss of 2-5 volume interaction;
however, this has already been counted as a portion of the 2-4 interfacial
energy. Thus, we find that the excess free energy of the 2-6 edge interface is
due only to the loss of bonds between the volumes 2 and 6 rather than to the
loss of bonds between volumes 2 and $6+1+5$ so that

$$\frac{\gamma_{edge}^{c}}{\gamma_{\infty}^{c}} = \frac{(\sum_j \mu_j \Phi_j)_{2-6}}{(\sum_j \mu_j \Phi_j)_{2-6} + (\sum_j \mu_j \Phi_j)_{2-1} + (\sum_j \mu_j \Phi_j)_{2-5}} \tag{15}$$

2. In the area of nucleation, it has become standard practice to use the
macroscopic value of $\gamma$ in our calculations of embryo population and critical
embryo size for the nucleation event. This procedure is not too unreasonable
when the dispersion forces dominate the magnitude of $\gamma$. However, when the
dipole forces dominate and when dipole alignment occurs in the macroscopic
surface case, this approximation can lead to serious error. This point has
been nicely demonstrated by Abraham (1969) who showed that the interfacial
energy of small spheres of water could increase, by misalignment of the
dipoles, by as much as 50% of the bulk surface value. The relative situation is
illustrated in Fig. 9 where the dipole configuration is given for both the flat
surface and a sphere. For the sphere, each dipole is rotated by angle $\psi$ relative
to an immediate neighbor. This deviation from the parallel orientation involves
an increase in the potential energy of the surface molecule by the amount
$\frac{1}{2}\mu^2(1-\cos\psi)n/a^3$, where $\mu$ is the magnitude of the dipole moment of a
molecule, $n$ is the number of surface nearest neighbors, and $a$ is the distance

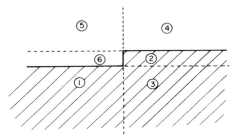

**Fig. 8.** Schematic illustration of layer edge to calculate the lowering of the 2–6 interfacial
energy at the edge compared to that for a macroscopic segment of surface.

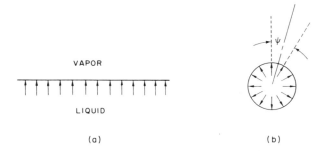

VAPOR

LIQUID

(a)                                              (b)

**Fig. 9.** Illustration of a polar molecule with dipole alignment at either (a) a flat surface or (b) a spherical surface.

between nearest neighbors. The total increase of $\gamma_c$ for this curved surface with respect to planar surface is then

$$\Delta\gamma_c = \tfrac{1}{2}\left[\frac{\mu^2(1-\cos\psi)\,nN}{2a^3}\right] \tag{16}$$

where $N$ is the total numbers of molecules in the surface of the droplet and the factor of $\tfrac{1}{2}$ arises because each dipole of a dipole pair shares one half the interaction energy.

As we shall see in the next example, the dipole–dipole contribution constitutes about 60% of the total interfacial energy of water. Thus, the effect is significant, not only for the vapor–liquid condensation case, but also for the water–ice formation case. If we consider a strongly polar crystal with a layer edge as in Fig. 8, the dipole orientation effect at an interface might cause $\gamma_{\text{edge}}^c$ to become greater than $\gamma_\infty^c$.

3. From the solubility work of Hildebrand and Scott (1962) one finds that the excess energy of interaction between dissimilar materials in a solution is largely determined by the geometrical mean of the dispersion force interaction. This result has been utilized by Fowkes (1967) to rationalize that at an interface between two materials interacting via dispersion forces, the chemical component of the interfacial tension is given by

$$\gamma_{\alpha\beta}^c = \gamma_\alpha{}^c + \gamma_\beta{}^c - 2(\gamma_\alpha^{c_d}\gamma_\beta^{c_d})^{1/2} \tag{17}$$

where $c_d$ refers to the dispersion component of the chemical part. He then selected eight saturated hydrocarbons as reference liquids, because they are known to interact only via London dispersion forces, and measured the interfacial tension (excess free energy) between these liquids and mercury plus that between these liquids and water. He measured $\gamma_{\alpha\text{Hg}}$, $\gamma_\alpha$, and $\gamma_{\text{Hg}}$ by standard techniques and, since $\gamma_\alpha \approx \gamma_\alpha{}^c = \gamma_\alpha^{c_d}$, he found $\gamma_{\text{Hg}}^{c_d} = 200 \pm 7$ dyn/cm² at 20° C for the eight reference liquids. In a similar fashion, he found $\gamma_{\text{H}_2\text{O}}^{c_d} = 21.8 \pm 0.7$

dyn/cm$^2$ at 20° C. Presuming that we can make the following simplifying approximations for the Hg and the H$_2$O interfaces:

$$\gamma_{Hg} \approx \gamma_{Hg}^{Cd} + \gamma_{Hg}^{Ce} \tag{18a}$$

$$\gamma_{H_2O} \approx \gamma_{H_2O}^{Cd} + \gamma_{H_2O}^{CD} \tag{18b}$$

where $c_e$ refers to the free-electron contribution to $\gamma_c$ and $c_D$ refers to the dipole interaction portion of $\gamma_c$, this gives $\gamma_{Hg}^{Ce} \approx 284$ dyn/cm$^2$ and $\gamma_{H_2O}^{CD} \approx 51.0$ dyn/cm$^2$ (both at 20° C). Fowkes then proceeded to investigate the $\gamma_{Hg/H_2O}$ interface. He assumed that the free electrons in the mercury did not interact significantly with the water and that the aligned dipoles in the water did not interact significantly with the mercury (which is surprising in view of image charge concepts) and calculated $\gamma_{Hg/H_2O} = 424.8 \pm 4.4$ dyn/cm$^2$ at 20° C. This result is in remarkable agreement with the experimental value $\gamma_{Hg/H_2O} = 426$–$427$ dyn/cm$^2$.

It is this author's opinion that extension of these concepts and techniques to the study of the various components of $\gamma$ for other systems would be extremely worthwhile.

4. In this final example, we consider a rationalization for the strong surface active nature of sulphur in cast iron. We know that it strongly adsorbs to the basal plane of graphite which causes the graphite particles to develop a flake form with the basal plane of the graphite forming the large flat surfaces. Removal of sulphur allows the particle morphology to change. This, and an abundance of other data, strongly support the surface active nature of S for C and indicate the importance of its effect upon processes that proceed via kink or ledge reactions. This type of action, associated with a minor constituent, is very important in a number of examples found in the crystallization area and is thus worthy of deeper consideration.

The bond structure of the graphite surface is illustrated in Fig. 10a. It consists of hexagonal layers of molecules which are separated by a distance so large (3.40 Å) that there are no covalent bonds between them. The four

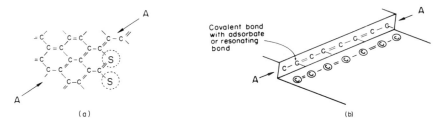

**Fig. 10.** Organization of carbon atoms with sulfur impurities in graphite for (a) the basal plane and (b) a layer edge on the basal plane.

**TABLE IV**

POLARIZABILITIES PARALLEL TO, $\alpha^{\parallel}$, AND PERPENDICULAR TO, $\alpha^{\perp}$, THE BOND CHAIN

| Bond chain | $\alpha^{\parallel}$ (in $10^{-25}$ cm$^3$) | $\alpha^{\perp}$ (in $10^{-25}$ cm$^3$) |
|---|---|---|
| C—C | 18.8 | 0.2 |
| (C—C) Aromatic | 22.5 | 4.8 |
| C=C | 28.6 | 10.6 |
| C≡C | 35.4 | 12.7 |
| C—Cl | 36.7 | 20.8 |
| C—Br | 50.4 | 28.8 |
| C=O | 19.9 | 7.5 |
| C=S | 75.7 | 27.7 |
| C≡N | 31.0 | 14.0 |
| C—H | 7.9 | 5.8 |
| N—H | 5.8 | 8.4 |
| S—H | 23.0 | 17.2 |
| O$_2$ | 24.3 | 11.9 |
| N$_2$ | 24.3 | 14.3 |
| CO | 26.0 | 16.2 |
| Cl$_2$ | 60.0 | 36.2 |
| HCl | 31.3 | 23.9 |
| HBr | 42.3 | 33.2 |
| H$_2$ | 65.8 | 48.9 |
| SO$_2$ | 29.0 | 14.7 |

valences of each carbon atom are thought to be used in forming bonds with its three neighbors in the plane. We would expect ledges on the surface to lie along the line A-A of Fig. 10a so that the ledge structure would be that of Fig. 10b and only one half a bond per ledge carbon atom is available for covalent bonding with adsorbed species. We can think of the S atom adsorbing at a ledge site, as indicated in Fig. 10a, and forming a single covalent bond or a double resonating bond with a ledge C atom.

Little to no information is available on the actual process of chemisorption of S on graphite basal planes and none at all for adsorption on ledges. Considering the formation of either carbon monosulphide or carbon disulphide bonds, we find that the interatomic distance is favorable but that the C–S bond is not particularly favored energetically compared to the C–C bond, whereas the CS$_2$ molecule is thermodynamically stable at low temperatures and is a slight possibility. Since C and S have about the same electronegativity, a strong ionic bond is not likely to be formed and we must turn to a consideration of the London dispersion forces for the explanation of the favorable binding.

Recalling Eq. (7), the relevant parameter upon which to focus attention is the polarizability $\alpha$. To appreciate this situation, we note that, since C–X often forms chainlike molecules, we must refer to polarizabilities both parallel and perpendicular to the bond between atoms in the chain by the symbols $\alpha^{\parallel}$ and $\alpha^{\perp}$, respectively. In Table IV, a list of $\alpha^{\parallel}$ and $\alpha^{\perp}$ are given for several different organic and inorganic bonding types. From this table, we see that the polarizability of the C=S bond strongly favors adhesion of S to C via London dispersion forces. For a complete analysis, one would need to consider the interaction of the S with Fe and Fe with C; however, these factors are of second order compared to the polarizability factor under discussion.

## III. Energetics: $\gamma_e$

### A. The Macropotential

When we think of the electrical potential inside a solid, we have come to think of the periodic field associated with the molecules and have somewhat lost sight of the more primitive viewpoint of a macroscopic continuum. To regain this perspective, let us consider an insulator as a dielectric between the plates of a charged condenser as in Fig. 11. Then, by way of example, we have high potential values on the right side of this insulator and low values on the left. Hence, because of its negative charge, an electron has a low electrostatic

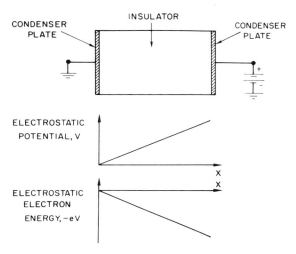

*Fig. 11.* Spatial distribution of electrostatic potential and electron energy in a charged condenser.

energy on the right side and a high one on the left. To obtain the total energy of the electron $E_T$ in the insulator, we must add an electric energy term $-eV$ due to the macropotential $V$ to the lattice energy term $E_L$ arising from the binding forces exerted by the crystal lattice on the free electron, i.e.,

$$E_T = E_L - eV \qquad (19)$$

Even if we remove the condenser in Fig. 11, we will find that a macropotential $\phi$ still exists in the insulator arising from the presence of macroscopic surfaces, space charges, double layers, etc.

The origin of the macropotential or interface potential can be thought to reside in three simultaneously acting contributions (that are not independent). These are called the distribution, polarization, and adsorption potentials.

## 1. DISTRIBUTION POTENTIAL

This potential arises as a result of the redistribution of charged particles that are soluble in both phases at the interface between the phases. In metals, the prime example is the redistribution of free electrons; a second example is that of a metal immersed in an electrolyte containing ions of this metal.

Metals consist of a space lattice of fixed positively charged ions and electrons free to move. Electrolytes contain no free electrons but contain ions in a state of association with solvent molecules. When a metal dips into an electrolyte, as illustrated in Fig. 12, the charged particles passing across the phase boundary are generally the positive metallic ions. The mechanism whereby the potential difference between a metal such as copper and a solution containing copper ions is set up is the passage of Cu ions from the metal to the solution or vice versa. The phase boundary acts somewhat as a membrane permeable to

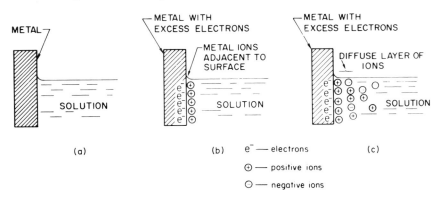

**Fig. 12.** Representation of electrostatic double layer at a metal–solution interface (a) at the instant of immersion; (b) first stage of formation of double layer; (c) complete double layer.

positively charged ions only. For such a case, if the Cu metal electrode is insulated at its exposed end (projecting out of the solution), electrons cannot pass to it or leave from it. Positive ions are deposited on the metal from the solution and leave the metal for the solution. If, initially, the rate of deposition of ions from the solution is less than the rate of passage from the metal into the solution, the metal acquires a negative charge and the solution a positive charge. The process of dissolution continues until a double layer of such strength is set up that the energy level of the hydrated ions in the solution is raised and that of the positive ions in the metal lowered. When these levels are equalized, net ion transfer ceases, i.e., transfer occurs until the electrochemical potentials of the ions in the metal and in the solution become equal. The double layer consists, of course, of an accumulation of positive ions in the solution near the surface of the metal and of electrons near the surface inside the metal (see Fig. 12).

## 2. POLARIZATION POTENTIAL

This potential occurs as a result of an orientation of neutral molecules having a dipole moment, permanent or induced, at the surface. Most molecules contain such dipoles and, as discussed in the previous section, their presence is one of the principle reasons for the orientation of molecules at surfaces; the oriented row of dipoles is a double layer which is not diffuse. If we consider a metal–vacuum interface, several sources of polarization potential exist. (1) Within the metal, the electrons are in vigorous random thermal motion. The electrons moving toward the surface of the metal (even if their energy is insufficient to leave the metal) will move somewhat beyond the lattice of fixed positive ions before they turn back. Hence, a thin negative skin is found beyond the positive skin of the ions (a double layer). (2) At the metal–vacuum boundary, the ions of the uppermost atomic layer are, by comparison with the bulk of the metal, subject only from one side to the "proper" forces of a complete lattice. Therefore, the ion cores of the outermost atomic layers take up displaced positions relative to the configuration in the bulk of the metal. Since quasi-neutrality prevails in the bulk of the metal, a dipole double layer forms at the surface of the metal. (3) Rather than ion displacement occurring at the surface, ion deformation may occur to produce surface dipoles. (4) Finally, we are all aware of the adsorption of monatomic impurity layers on metallic surfaces. Here, the adsorbed foreign atoms are pulled apart to form dipoles, which accounts for their adhesion as well as for their double-layer effect.

## 3. ADSORPTION POTENTIAL

This potential arises when neither the positively nor the negatively charged particles can leave either of the phases in appreciable amounts, but where the particles of one sign tend to be adsorbed at the interface to a greater degree than those of the other sign. For example, at the air–water interface, either the cations or the anions may be adsorbed in the greater quantity. In this case, there is negligible transference of charged particles between air and water, but an electrical double layer due to dipole orientation is present at the interface region and adsorption occurs as a response to this potential distribution (only because of defects in the oriented dipole layer can a field gradient exist adjacent to the layer and produce ion attraction. If it were a perfectly planar dipole layer, a constant potential would exist everywhere in the fluid). Another example is that of the mercury–water (plus solute) interface. In this case, one type of ion (e.g., anions) are attracted to the $H_2O$–Hg interface more than the other type and are thus adsorbed at the interface.

## 4. ADDITIONAL DEFINITIONS

As briefly discussed earlier, for each phase one may distinguish an inner or Galvani potential $\phi$ and an outer or Volta potential $V$. The inner potential $\phi$ is the electrostatic potential in the interior of the phase with respect to an infinitely distant point in the charge free vacuum. By this definition, electrical energy $q\phi$ is required to transport a charge $q$ from infinity to the interior of the phase. The Volta or outer potential $V$ is the energy $qV$ required to transport a charge $q$ from infinity to the outside surface of the phase at a distance of $10^{-4}$ cm from the surface. In the absence of dipole layers or any other excess charges of either sign on the surface, the Volta potential is zero, i.e., $V = 0$.

The Galvani potential (inner) is different from the Volta potential (outer) only when a potential difference exists between the interior and the surface of the phase. This potential difference is called the surface potential $\chi$. These three electrical potentials are related by the following equation,

$$\phi = V + \chi \tag{20}$$

The Volta potential is a well-defined electrical quantity and can be measured in different ways. The Galvani potential of a phase cannot be measured and even the Galvani potential difference between two phases is not accessible to unambiguous measurement. For such a measurement, it would be necessary to determine the electrostatic work required to bring a small charge from the bulk of one phase to the other. However, the amount of work involved in transporting an electron may be quite different from the work involved in the transport of an ion. To find the work done against the potential difference, the total work in the transport process has to be diminished by the difference in chemical potential $\Delta\mu$ of the transported component.

In Fig. 13a the energy band picture of an electron in two solids that are isolated from each other is shown relative to the zero level of electron energy

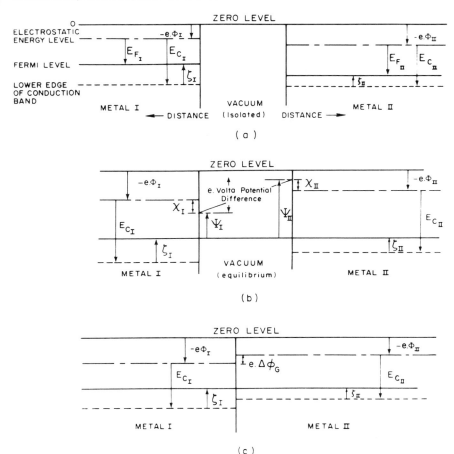

**Fig. 13.** Energy level diagram for two solids illustrating the macropotential condition in (a) the electrically isolated and separated condition, (b) the electronic equilibrium but separated condition, and (c) the intimate contact condition.

(an electron at infinity in a perfect vacuum): $\phi$ is the macropotential so that $-e\phi < 0$ is the electrostatic binding energy; $E_C$ is the chemical binding of the electron in the lattice at the lower edge of the conduction band; $E_F$ is the chemical binding energy at the Fermi level and $\zeta = E_F - E_C > 0$ is called the Fermi energy. From Fig. 13a, we note that the electrochemical potential (Fermi level) of the electrons is different in metal II than in metal I. Let us allow the metals to come to electronic equilibrium (by thermionic emission, say). The electron transfer creates an additional distribution potential between the two phases and continues until the electrostatic potential shift brings the two Fermi levels into alignment. Then, there is no further driving force for electron transfer. From Fig. 13b, we note that the surface potential (double layer jump) $\chi$ differs in sign for the two phases so that the Volta potential difference will be different from the Galvani potential difference ($\Delta\phi_G = \phi_{II} + \phi_I$). From this figure, we can also note the definition of the work function $\psi$ of a material. It is the difference in electron energy between a point just outside the surface to a point just inside the surface, i.e., on the energy diagram, it is the distance from the Fermi level to the electrostatic potential immediately in front of the metal surface. When the two solids are brought into intimate contact, most, but not all, of the Volta potential difference is likely to disappear so that some interface potential barrier should still exist.

### B. The Formula for $\gamma_e$

One useful method for the evaluation of $\gamma_e$ is to consider the electron redistribution that occurs on contact between two phases. If we take $\delta n$ electrons from the phase with higher Fermi level and place them in the phase with the lower Fermi level, this causes a decrease in free energy of the system by an amount $-|\delta n \Delta E|$, where $\Delta E$ is the difference in Fermi level ($E_F - e\phi$) of the two materials. Unit area of interface is assumed. Now let these electrons redistribute to build an equilibrium space charge at the interface with a potential change $\delta\phi$; this causes an increase in energy of $|\delta n \cdot e\delta\phi|$. Allow this process to continue until $\Delta n$ electrons have been transferred and the potential difference $\Delta\phi$ generated which is sufficient to bring about Fermi level alignment. Thus, we have

$$\gamma_e = -|\Delta n \, \Delta E| + \int_0^\sigma \delta\phi dq \tag{21a}$$

where $\sigma$ is the charge per square centimeter of interface ($\sigma = e\Delta n$). In the final state, the initial energy difference $\Delta E$ must exactly outweigh the electrostatic energy change so that

$$\Delta E + e\Delta\phi = 0 \tag{21b}$$

and Eq. (21a) becomes

$$\gamma_e = -\sigma\Delta\phi + \int_0^\sigma \delta\phi dq \tag{21c}$$

in which, obviously, the first term is always larger than the second term. By partial integration both terms can be summarized into a single term so that we finally obtain

$$\gamma_e = -\int_0^{\Delta\phi} \sigma d\phi \tag{21d}$$

where $\Delta\phi$ is given from Eq. (21b). Referring to Fig. 13, we find

$$\eta_{i_o}^e = -e\phi_{i_o} + E_{C_i} + \zeta_i; \qquad i = \text{I, II} \tag{22a}$$

and

$$\Delta E = \eta_{\text{II}_o}^e - \eta_{\text{I}_o}^e \tag{22b}$$

where the subscript o refers to the isolated state (Fig. 13a) and $\eta_0^e$ refers to the electrochemical potential of the electrons in the isolated state (Fermi level). Thus, we see that $\Delta\phi$ is equal to the Galvani potential difference $\Delta\phi_G$ only under very special circumstances.

An alternate procedure for obtaining a quantitative expression for $\gamma_e$ is to consider the Gibbs adsorption equation, i.e.,

$$d\gamma = -\sum_i \Gamma_i \, d\mu_i - \sigma \, d\phi \tag{23a}$$

where $\Gamma_i$ is the surface excess of the $i$th chemical constituent, $d\mu_i$ is the change in chemical potential of this $i$th constituent during the adsorption process, and $\sigma$ is the surface charge density of electrons associated with the change in electrostatic potential $d\phi$ (here, $\Gamma_e = \sigma/-e$ and $d\mu_e = -e \, d\phi$). For a pure system or a system where no atomic redistribution is allowed

$$\gamma = \gamma_0 - \int_0^{\Delta\phi} \sigma \, d\phi \tag{23b}$$

where $\gamma_0$ is the surface tension in the absence of any potential difference so that $\gamma_e$ is again given by Eq. (21d). This equation is known as the Lippman equation.

If the electric surface potential is so small that $\sigma$ is linearly related to $\Delta\phi$, the simple condenser model can be applied and Eq. (21d) simplifies to

$$\gamma_e = -\tfrac{1}{2}\sigma\Delta\phi \tag{24a}$$

Thus, in this case, the electric work to be done is exactly half the chemical work gained in the process. For large values of potential change, the charge increases more rapidly with increasing potential [for a single double layer in an ionic system, it increases roughly according to $\exp(e\Delta\phi/kT)$] and, accordingly, the value of $\gamma_e$ is generally greater than that given by Eq. (24a), i.e.,

$$-\tfrac{1}{2}\sigma\Delta\phi \leqslant \gamma_e < 0 \tag{24b}$$

Referring back to Eq. (23b), we note that the excess surface free energy shows a maximum for $\sigma = 0$ and, on both sides of this zero point of charge, $\gamma$ is lowered.

### 1. HOMOGENEOUS CONDENSER MODEL

For simplicity, we may think of the charge distribution in the double layer which gives rise to the required potential distribution, illustrated in Fig. 14, as being concentrated into sheets at the centroid of the respective charge distributions. This leads to a homogeneous condenser model as an approximation (same dielectric constant in both phases). For such a model, simple electrostatics gives

$$\sigma = \varepsilon \Delta\phi / 4\pi d \tag{25}$$

where $d$ is the separation between the charges and $\varepsilon$ is the dielectric constant. The next step is to set up and solve Poisson's equation for the particular charge statistics under consideration. This has been carried out by Tiller and Takahashi (1969) for both Boltzmann and Fermi–Dirac statistics. For Boltzmann statistics, they obtained

$$(d\phi/dx)^2 = [(8\pi nkT/\varepsilon)\, e^{e\phi/kT} + e^{-e\phi/kT} - 2] \tag{26}$$

where $n$ is the number of positive and negative ions per unit volume and $\phi$ is the potential change relative to the bulk medium potential. Since $\sigma$ and $(d\phi/dx)_0$ are related via Gauss' law $[\sigma = (\varepsilon/4\pi)(d\phi/dx)]$, we have
(a) *for the case of* $e\phi/kT \ll 1$

$$d\phi/dx = (8\pi ne^2/\varepsilon kT)^{1/2}\, \phi \tag{27a}$$

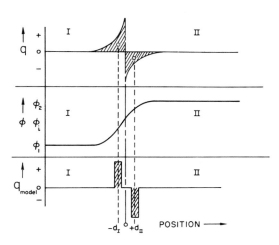

**Fig. 14.** Illustration of charge $q$ and potential $\phi$ distribution across an interface between two phases of similar electrical characteristics plus the condenser approximation for the charge.

so that

$$d = (\varepsilon kT/8\pi ne^2)^{1/2} \tag{27b}$$

which is called the Debye length and
(b) *for the case of $e\phi/kT \gg 1$,*

$$d\phi/dx = (8\pi nkT/\varepsilon)^{\frac{1}{2}} e^{e\phi/2kT} \tag{27c}$$

and, if Eq. (25) is used,

$$d \approx (\varepsilon/8\pi nkT)^{1/2} \Delta\phi \, e^{-e\Delta\phi/2kT} \tag{27d}$$

However, for this case, Eq. (25) does not hold, so that this value of $d$ may be appreciably in error.

For metallic systems using an effective dielectric constant $\varepsilon^*$ and Fermi–Dirac statistics, Tiller and Takahashi (1969) showed that with $e\phi/E_F \ll 1$,

$$\frac{d\phi}{dx} = \frac{4\pi e}{h} \left(\frac{2m}{\varepsilon^*}\right)^{1/2} \left(\frac{3n}{\pi}\right)^{1/6} \phi \tag{28a}$$

and

$$d = \left[\frac{\varepsilon^* h^2}{32\pi^2 \, me^2} \left(\frac{\pi}{3n}\right)^{1/3}\right]^{1/2} \tag{28b}$$

Alternatively,

$$\sigma = C^* n^{1/6} \varepsilon^{*1/2} \phi \tag{28c}$$

where $C^* = 2.65 \times 10^3$ c.g.s. units. This result is considerably different from that using Boltzmann statistics since $\sigma$ is relatively insensitive to small variations in carrier density $n$ ($n^{1/6}$ instead of $n^{1/2}$).

## 2. Inhomogeneous Condenser Model

For the general case where both $n$ and $\varepsilon$ (or $\varepsilon^*$) change at the interface, the model and variables involved are illustrated in Fig. 15 (two condensers of equal charge density $\sigma$, in series). Here, $n_\alpha$ and $n_\beta$ are the free carrier densities,

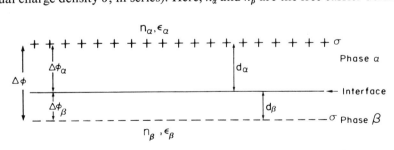

**Fig. 15.** Schematic illustration of an inhomogeneous condenser model for a two-phase interface.

$\varepsilon_\alpha$ and $\varepsilon_\beta$ are the dielectric constants, $\Delta\phi_\alpha$ and $\Delta\phi_\beta$ are the potential jumps in materials $\alpha$ and $\beta$, respectively, $\Delta\phi$ is the total potential difference, and $\sigma$ is the surface charge density of each of the electrical double layers. The general conditions to be satisfied are

$$\Delta\phi = \Delta\phi_\alpha + \Delta\phi_\beta \qquad (29a)$$

$$\sigma = \sigma_\alpha = \sigma_\beta \qquad (29b)$$

and

$$\varepsilon_\alpha (d\phi_\alpha/dx)_0 = \varepsilon_\beta (d\phi_\beta/dx)_0 \qquad (29c)$$

or

$$\frac{\Delta\phi_\alpha}{\Delta\phi_\beta} = \frac{(\varepsilon_\beta/d_\beta)}{(\varepsilon_\alpha/d_\alpha)} \qquad (29d)$$

where Eqs. (25)–(28) are used to relate $\sigma_i$, $d_i$, and $\Delta\phi_i$. For this condenser model, we have

$$\sigma = \Delta\phi/4\pi\left[(d_\alpha/\varepsilon_\alpha)+(d_\beta/\varepsilon_\beta)\right] \qquad (30)$$

We are now in a position to draw some qualitative conclusions concerning $\gamma_e$ from the preceding equations. From Eqs. (27) and (28) we note that the larger the dielectric constant and the smaller the number of free carriers in the bulk material, the larger is $d$. For nonmetals, $d/\varepsilon$ is generally large, whereas for metals, $d/\varepsilon$ is generally small. Thus, the charge $\sigma$ needed to produce a given potential difference $\Delta\phi$ is small for nonconductors and much larger for conductors, so that $\gamma_e$ should be small in magnitude at the interface between two nonconductors and much larger at an interface between two conductors. At a conductor–nonconductor interface, Eq. (29d) shows us that most of $\Delta\phi$ will be developed in the nonconductor, and Eq. (30) shows us that $\sigma$ will be small in this case so that $\gamma_e$ will be small in magnitude. Tiller and Takahashi (1969) have used the Drude free electron theory (Garbuny, 1965) to calculate values for $\varepsilon^* \sim 10^3$–$10^4$ for most metals, whereas an alternate approximation for $\varepsilon^*$ (Tiller, 1969) finds $\varepsilon^* \sim 10^2$–$10^3$ for most metals. This is one of the greatest areas of doubt concerning the quantitative accuracy of our $\gamma_e$ calculation. Much more effort needs to be directed towards finding reliable estimates of $\varepsilon^*$.

To evaluate $\gamma_e$, we need to obtain an expression for $\Delta\phi$ in specific instances. Some hope for this exists if we consider a solid–liquid interface and use free electron theory to calculate the change in Fermi energy $\zeta$ due to the volume change on melting. We will assume that in the isolated condition the energy level of the bottom of the conduction band for both liquid and solid are equal so that

$$\Delta\phi_{SL} = \frac{\Delta\zeta_{SL}}{e} \qquad (31a)$$

i.e., we are assuming that $-e\phi_{I_0} + E_{C_I} = -e\phi_{II_0} + E_{C_{II}}$. Since the Fermi energy relative to the bottom of the conduction band is given by

$$\zeta_F \approx \zeta_F(\text{at } 0° \text{ K}) = (h^2/2m)(3n/8\pi)^{2/3} \tag{31b}$$

where $n = n_{eff} N/V$, $n_{eff}$ is the effective number of free electrons per atom, $N$ is Avogadro's number, and $V$ is the molar volume. Letting $\Delta V$ be the volume change on melting, the number of free electrons per unit volume in the liquid becomes

$$n' = \frac{n_{eff} \cdot N}{V[1 + (\Delta V/V)]} \approx n[1 - (\Delta V/V)] \tag{31c}$$

since $\Delta V/V \ll 1$ and provided that $n_{eff}$ does not change on melting. Thus, using Eqs. (31b) and (31c),

$$\zeta_F(L) \approx \zeta_F(S)[1 - \tfrac{2}{3}(\Delta V/V)] \tag{31d}$$

which, from Eq. (31a), yields $\Delta\phi_{SL} \sim 0.1$ volt for the solid–liquid interface. In Table V values of $\Delta\phi_{SL}$ and $|\gamma_e|$, as well as the experimental values of $\gamma_{SL}$, are given for a number of metals. These values may be somewhat in error for several reasons: our choice of $\varepsilon^*$, our use of the condenser approximation, the Eq. (31a) approximation, and the neglect of exchange and correlation energies of the electrons in the metal [i.e., the use of Eq. (31b)]. In Table V, the Drude value of $\varepsilon^*$ was used yielding values of $d \sim 10$ atom spacings. It is interesting to note that $|\gamma_e|$ is large compared to $\gamma_{SL}$ and that $\gamma_{SL}$ is found to be small when $|\gamma_e|$ is calculated to be large (since $\gamma_{SL}$ contains $\gamma_e$ which is negative). We might expect $\Delta\phi_{SL}$ to be in error by as much as a factor of 2 and $|\gamma_e|$ to be in error by as much as a factor of 3.

**TABLE V**

CALCULATED VALUES OF $\Delta\phi_{SL}$ AND $|\gamma_e|$

| Element | $e\Delta\phi_{SL}$ (eV) | $|\gamma_e|$ (ergs/cm$^2$) | $\gamma_{SL}$[a] (ergs/cm$^2$) |
|---------|--------|--------|--------|
| Li | 0.047 | 10.1 | — |
| Al | 0.258 | 323.0 | 95 |
| Cu | 0.191 | 154.0 | 181 |
| Ag | 0.161 | 164.0 | 128 |
| Au | 0.194 | 142.0 | 135 |
| Fe | 0.186 | 17.6 | 208 |
| Co | 0.165 | 23.0 | 239 |
| Zn | 0.256 | 185.0 | — |

[a] Data in this column from Holloman and Turnbull (1950).

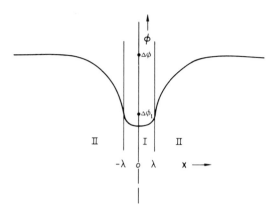

**Fig. 16.** Space charge distribution between two materials, I and II, when the thickness of I is much less than its Debye length.

### 3. Size Effects

If we consider a thin slab of material I immersed in material II, as illustrated in Fig. 16, the calculations of the previous sections will satisfy provided the slab is thicker than $2d_I$. However, as the slab decreases in thickness below twice the Debye length $2d_I$, the space charge distribution must change its characteristics and $\gamma_e$ must consequently change in magnitude. As shown by Tiller and Takahashi (1969), the following happens: (a) an increasingly larger fraction of the total potential difference $\Delta\phi$ (assumed constant) is developed in the external phase (phase II), and (b) both the charge density $\sigma$ and excess energy $\gamma_e$ become more characteristic of phase II as a homogeneous capacitor. This indicates that the adhesion between a metal and a submerged nonmetallic film may be excellent when the film is very thin and that it may strongly decrease as the film thickness becomes larger than $2d_I$. Likewise, the ability of a film to serve as a suitable site for heterogeneous nucleation may be affected by its thickness. Even if it is a semiconductor or insulator, it may become a good catalyst if it is significantly thinner than its Debye length. Using similar reasoning, it can be readily appreciated that this effect will be operative in homogeneous nucleation when the critical embryo radius is smaller than the Debye length of the solid. This will make $\gamma$ a function of embryo radius ($\gamma$ will increase with increase of radius) and thus will change the critical embryo size (increase $r^*$) and the free energy of formation of the critical embryo (decrease $\Delta G^*$) for materials like Ge and Si.

### 4. Solute Redistribution Effect

The presence of the interface electric field causes dipole formation in the ion cores of the constituent atoms or molecules and an interaction occurs between the interface field $\xi$ and the dipole moment of the polarized ion cores. The force $F$ on each dipole tends to pull it in the direction in which the field increases most rapidly. The magnitude of the force is given by

$$F = \frac{\alpha}{2} \frac{\partial}{\partial Z} [\xi^2(Z)] \qquad (32a)$$

where $\alpha$ is the polarizability of the ion core under consideration. For a binary alloy system, it is the relative force between the solute $F_s$ and the solvent $F_S$ that will cause the redistribution of the two species. Thus,

$$\Delta F(Z) = F_s - F_S = \frac{(\alpha_s - \alpha_S)}{2} \frac{\partial}{\partial Z} [\xi^2(Z)] \qquad (32b)$$

For $\xi(Z)$, we may use an exponential solution; however, this presumes that the solute and solvent atoms occupy the same volume, contribute the same number of free electrons/atoms to the alloy and do not cause a change in the dielectric constant of the surface region. If this is not the case (which is most likely), the whole space charge distribution will be altered and the surface tension changed in a subtle way.

One means of determining the equilibrium solute distribution in the interface region is the following. First, equilibrium demands constant electrochemical potential of each species throughout the entire system. Defining $\Delta\mu(Z)$ as the change in electrochemical potential due to replacing an S atom in the interface region with an s atom, we require that

$$\mu_j^s(\infty) + \Delta\mu_j^s(Z) + kT\ln a_j^s(Z) = \mu_j^s(\infty) + kT\ln a_j^s(\infty) \qquad (33a)$$

where $\mu_j^s(\infty)$ is the standard electrochemical potential of the solute far from the interface in phase $j$, and $a_j^s$ is the activity of the solute in phase $j$ in the absence of any fields. Thus, from Eq. (33a), we have

$$\frac{N_j^s(Z)}{N_j^s(\infty)} \approx \frac{a_j^s(Z)}{a_j^s(\infty)} = \exp -(\Delta\mu_j^s(Z)/kT); \qquad j = 1, 2 \qquad (33b)$$

Neglecting any change in $\xi(Z)$ due to the redistribution, we find that $\Delta\mu_j^s$ is given by

$$\Delta\mu_j^s(Z) = -\tfrac{1}{2}(\alpha_s - \alpha_S)\xi_j^2(Z); \qquad j = 1, 2 \qquad (33c)$$

and the lowering of free energy of the system due to this dipole formation $\Delta\gamma$ by

$$\Delta\gamma = \int_{-\infty}^{\infty} \Delta\mu_j^s(Z)[N_j^s(Z) - N_j^s(\infty)]\,dZ$$

$$-\tfrac{1}{2}\alpha_s \int_{-\infty}^{\infty} \xi_j^2(Z)[N_j^S(Z) - N_j^S(\infty)]\,dZ \qquad (34)$$

If $\xi(Z)$ changes from $\xi^0(Z)$ to $\xi'(Z)$ due to the redistribution and thus to a change in $\varepsilon$, we find $\Delta\mu_j{}^s$ is given by

$$\Delta\mu_j{}^s(Z) = -\tfrac{1}{2}\{\alpha_s[\xi_j'(Z)]^2 - \alpha_s[\xi_j{}^0(Z)]^2\} \tag{35}$$

We note that the magnitude of this solute redistribution will increase the lower is the temperature, the larger is $\xi_j(\Delta\phi$ and $\varepsilon_j$ and $n_j)$, and the larger is $(\alpha_s - \alpha_S)$.

In general, one would expect atoms with many electrons to have a larger polarizability $\alpha$ than those with few electrons, since the electrons in the outer electronic shells are not so strongly bound to the nucleus as those in the inner shells. Following this line of reasoning, positive ions should have smaller electronic polarizabilities than neutral atoms, whereas negative ions should have larger electronic polarizabilities than natural atoms. This picture is borne out in practice for isolated atoms. The picture for a particular alloy system is not complete, however, since we generally do not know the exact ionization state of a solute atom in a particular solvent.

In metallic systems, the electronic polarizability of solute atoms will probably be much larger than in either the gas phase or nonconducting solutions. This is because of the screening charge surrounding the solute atom which exhibits charge oscillations out to radii of about 25 Å. These shells of oscillating charge are so loosely bound to the nucleus that they are easily distorted by the field. No data is presently available on the magnitudes of these polarizabilities.

## C. Examples

### 1. APPLICATION TO HETEROGENEOUS NUCLEATION

The formal theory of heterogeneous nucleation incorporates the catalytic potency of the substrate in terms of the familiar "wetting angle" description. The critical size embryo of solid is presumed to make some contact angle $\theta$ with the substrate, as illustrated in Fig. 17. Agreement with experiment is satisfactory in that a certain contact angle $\theta$ may be determined that characterizes the nucleation process. However, this is really a thermodynamic definition and no clear insight is obtained into what determines $\theta$ and how it varies with

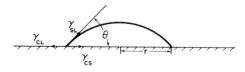

**Fig. 17.** Schematic illustration of a solid embryo forming from a liquid at a substrate surface with a wetting angle $\theta$.

lattice disregistry between the substrate and the stable phase, topography of the substrate surface, chemical nature of the substrate, adsorbed impurities on the substrate surface, etc.

Turnbull and Vonnegut (1952) attempted to rectify this situation by developing a crystallographic theory of nucleation catalysis which predicted that the order of the catalytic potency between different catalysts should be identical with the order of the reciprocal of the disregistry between the catalyst and the forming crystal on low index planes of similar atomic arrangement. Experimental data for the nucleation of $NH_4I$ on mica and for the nucleation of ice on AgI and quartz appear to be in reasonable agreement with their theoretical considerations.

Experimental data pointing out the incompleteness of the crystallographic theory has recently come from several sources. Bradshaw *et al.* (1958), studying the nucleation of gold droplets by various substrates, found that the compounds WC, ZrC, TiC, and TiN were much more potent catalysts than the oxides, which had little effectiveness even though the disregistries were similar. Further, even among these intermetallic compounds, dominance of the disregistry effect did not appear to coincide with the observations. Glicksman and Childs (1962), studying the nucleation of bulk tin with various substrates, found that metallic substrates appeared to be more potent catalysts for tin than nonmetallic substrates. When the substrates composed of different chemical constituents were compared, the effect of lattice mismatch appeared to be appreciably masked by other factors.

Very recently, an extensive investigation of the undercooling required for the nucleation of the second phase by the primary phase in 60 binary eutectic alloys was carried out by Sundquist and Mondolfo (1961). Although Wagner's experiments (1962) raise a serious question concerning certain quantitative results of this investigation, the qualitative conclusions are probably still valid and these are most startling. The most significant finding of this study was that in a system of $\alpha$ and $\beta$ solid, $\alpha$ usually nucleates $\beta$ at low undercoolings ($\Delta T \gtrsim 10°$ C), whereas $\beta$ does not nucleate $\alpha$ except possibly at undercoolings approaching homogeneous nucleation ($\Delta T \sim 100°$ C). They also found that the metals investigated could be placed in a series of the form

$$I(Tl, Pb): II(Ag\{Au, Cu, Ni, Co, Fe\}): III(Ge, Sn, Zn\{Bi, Sb\}).$$

In this series, any element of group II can readily nucleate any element of group I or I–II compounds and has no observable nucleating effect upon any element of group III or II–III compounds. Within any group, the exact location of an element was not so well established.

From the foregoing it should be clear that we possess an inadequate understanding of what constitutes a suitable substrate for nucleation catalysis. Let us see how the $\gamma_e$ values between the substrate and liquid and the substrate and solid can be arranged to favor embryo formation, i.e., $\theta \to 0$. By minimizing the free energy of the embryo, we are led to the following relationship for the wetting angle $\theta$, i.e.,

$$\cos \theta = (\gamma_{CL} - \gamma_{CS})/\gamma_{SL} \tag{36a}$$

$$= (\gamma_{CL}^0 - \gamma_{CS}^0)/\gamma_{SL} + \delta_{\gamma_e} \tag{36b}$$

where C refers to the catalyst and $\delta_{\gamma_e} = (\gamma_{CL}^e - \gamma_{CS}^e)/\gamma_{SL}$. Tiller and Takahashi (1969) have rationalized that $\pi/2 \lesssim \theta \lesssim \pi$ if $\delta_{\gamma_e} \sim 0$ and that one needs to select a situation such that $\delta_{\gamma_e} \gtrsim 2$ for a substrate to lead to a low value of $\theta$ and become a good nucleating agent.

Defining the capacitance $K_{\alpha\beta} = \sigma/\Delta\phi$ in Eq. (30), the expression for $\delta_{\gamma_e}$ (using the condenser approximation) is

$$\delta_{\gamma_e} = (1/2\gamma_{SL})[K_{CS}\Delta\phi_{CS}^2 - K_{CL}\Delta\phi_{CL}^2] \tag{37a}$$

neglecting any size effects. If we assume that $K_{\alpha\beta}$ is largely determined by the properties of the substrate so that $K_{CS} \approx K_{CL} \approx K$, we have

$$\delta_{\gamma_e} \approx (K/2\gamma_{SL})[2\phi_C - (\phi_S + \phi_L)][\phi_L - \phi_S] \tag{37b}$$

under the assumptions that hold for Eq. (31a). From Eqs. (31a) and (31d), we find that $\Delta\phi_{LS} = \phi_L - \phi_S$ is positive for close-packed or bcc metals which expand upon melting and is negative for those open-structured metals such as bismuth and gallium which expand upon freezing. Thus, for the case of $\Delta\phi_{LS} > 0$, a good nucleating agent must have a macropotential (in the contact situation) which is greater than that of the average of the solid and liquid; for the case of $\Delta\phi_{LS} < 0$, a good nucleating agent must have a macropotential (in the contact situation) which is smaller than that of the average of the solid and liquid. Further, in order for the magnitude of $\delta_{\gamma_e}$ to be large, $K$ must be large, i.e., the substrate must approach a metallic like character with respect to $(d/\varepsilon)$. Finally, we note that if the catalyst is a good nucleating agent for the solid ($\phi_C > \phi_S$, say, and $\delta_{\gamma_e} > 0$), then this solid is a poor nucleating agent for the catalyst ($\phi_C < \phi_S$) since $\delta_{\gamma_e}$ changes sign for these two situations.

We are at a loss to apply the above criterion to a real situation because it has not been possible to measure the macropotential $\phi$ of a phase or even to obtain a reliable measurement of the Galvani potential difference between two phases. The closest one can come to an operational situation, at the moment, is to relate the macropotential difference for two materials to the work function difference. The work function $\psi$, was defined earlier and is illustrated in Fig. 13b.

Having no other course of action open to them, Tiller and Takahashi (1969) made such an approximation to rationalize the data of Sundquist and Mondolfo (1961). In Fig. 18, their $\Delta T_E/\Delta T_{max}$ data are plotted vs. $\Delta\psi = \psi_\alpha - \psi_\beta$, where

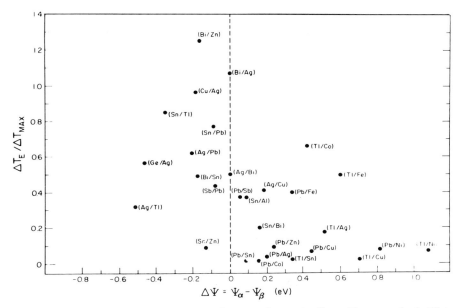

**Fig. 18.** Plot of $\Delta T_E/\Delta T_{\max}$ vs. $\psi_\alpha - \psi_\beta$ for several eutectic alloys. The parenthesis $(\beta/\alpha)$ means $\beta$ nucleated by $\alpha$.

$\Delta T_E$ is the supercooling below the eutectic temperature, $\Delta T_{\max}$ is the maximum supercooling of the major constituent in the nucleated solid, $\psi_\alpha$ is the work function of the major constituent in the primary phase, and $\psi_\beta$ is the work function of the major constituent in the nucleated solid. Here, the mean work function (averaged over various crystal faces) compiled by Samsonov (1964) was used. It should be noted that when the $\alpha$ phase nucleates the $\beta$ phase, i.e., $\Delta T_E/\Delta T_{\max} < 1$, $\Delta\psi = \psi_\alpha - \psi_\beta$ is positive, whereas when the $\alpha$ phase does not nucleate the $\beta$ phase, $\Delta\psi$ is negative. This striking contrast is exactly in agreement with the predictions following from Eq. (37b) provided we are justified in replacing $e\Delta\phi_{ij}$ by $\Delta\psi_{ij}$. It appears at this point that we were justified in making this approximation.

From Eq. (37b) it can also be concluded that nonconducting substrates would be poor nucleating agents as far as the electrostatic contribution is concerned. This seems to be in accord with the existing experimental data such as (i) Pound and LaMer's experiment (1952) showing that tin droplets coated with oxide could be greatly undercooled, (ii) Wagner's experiment (1962) on the freezing of alloys on Ge substrates showing that Ge is a very poor catalyst for nucleation, (iii) Glicksman and Childs' experiments (1962) on the solidification of molten tin showing that oxides and graphite were ineffective nucleating catalysts compared to metals, and (iv) Bradshaw *et al.*'s experiment

(1958) on gold droplets showing that oxides were ineffective nucleating agents compared to carbides for gold. In all these experiments, it has been found that nonconductive materials are ineffective nucleating agents for metals. On the other hand, conductive carbides and nitrides (Plenum Press, 1964), were found to be effective nucleating catalysts for certain metals, i.e., TiN, TiC, WC, and ZrC for the nucleation of gold (Bradsham *et al.*, 1958) and TiC for the nucleation of tin (Glicksman and Childs, 1962), whereas SiC, which is a semiconductor, was a poor nucleation catalyst for tin (Glicksman and Childs, 1962). The definite correlations between the existing data and the predictions of this section seem to indicate that the electrostatic contribution is indeed a major factor controlling the effectiveness of nucleating agents.

## 2. APPLICATION TO PARTICLE AGGLOMERATION

In ionic systems containing colloidal particles, the stability of the dispersion relative to coagulation is a very important consideration (Verwey and Overbeek, 1948). From the foregoing we see that even in good conductors analogous forces are operative, and it is of value to mention the force balance that leads to either a stable dispersion or a particle coarsening via agglomeration.

The force of attraction between two particles, separated by a distance $h$, in a solution is the van der Waals dispersion force $Q$ discussed in the previous section ($Q = A/h^3$ for small $h$ and $Q = B/h^4$ for large $h$, where $A$ and $B$ depend upon the materials of the particles and the medium). The force of repulsion

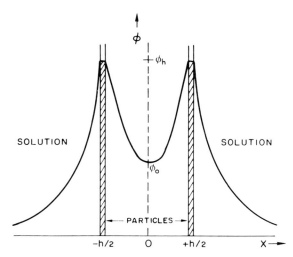

**Fig. 19.** Schematic representation of the electrical potential $\phi$ between two infinitely large parallel plates at a distance $h$ apart.

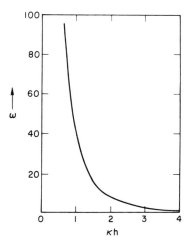

**Fig. 20.** Plot of the functional dependence of the quantity $\omega$ on the variable $\kappa h$.

arises because of the overlap of charge distributions in the electrical double layer surrounding the particles. With fixed potential at the surface of the particles, the electrostatic potential at the midpoint between them $\phi_0$ increases as the particles come closer together. This is illustrated in Fig. 19 for two lath-shaped particles. Equilibrium of such a system requires that at every point of the solution, the gradient of the hydrostatic pressure and the force on the space charge balance each other, i.e.,

$$dP - \rho d\phi = 0 \qquad (38a)$$

where $\rho$ is the volumetric charge density. Using Poisson's equation for $\rho$, we find

$$P - \varepsilon/8\pi (d\phi/dx)^2 = \text{constant} \qquad (38b)$$

as the equation describing the pressure distribution. It is the pressure difference between the fluid between the particles $p_0$ and that of the reservoir $p_\infty$ that tries to drive the particles apart, i.e., $\Delta P = p_0 - p_\infty$. Defining $\omega = \Delta P/(CZ)kT$, where $C$ is the total amount of charge of one sign per cubic centimeter and $Z$ is the valence of the clustering ion, $\omega$ is plotted versus $\varkappa h$ in Fig. 20 (here $\varkappa = 1/d$, the inverse Debye length). For various limits, the following approximations hold

$$\Delta P = (\pi/2)\,\varepsilon(kT/eZ)^2\,h^{-2}; \qquad \varkappa h \ll 1 \qquad (39a)$$

$$\Delta P = 64(C/Z)kTe^{-\varkappa h}; \qquad \varkappa h \gg 1 \qquad (39b)$$

Denoting $R = P - Q$ as the resulting repulsion between the particles, a plot of $R$ vs. $h$, such as illustrated in Fig. 21, shows three possibilities. We find that the force of attraction dominates at both large and small distances and that only the curve 3 case is capable of producing a stable dispersion. A first approximation to the stability limit occurs at a particle separation $h^*$ such that

$$\left(\frac{\partial R}{\partial h}\right)_{h^*} = 0 \quad \text{and} \quad (R)_{h^*} = 0 \tag{40}$$

which leads to a stability condition of the form

$$m = (Z/C') = a(A^2/\varepsilon^3)[(eZ)^6/(kT)^5] \tag{41}$$

where $a \sim (1 \text{ to } 2) \times 10^{-2}$ and $Z$ is the valence of the dominant ion. Thus, if the solution concentration exceeds a value $C'$, the coagulation process occurs. Because of the strong dependence on $Z$, we note that an extremely small concentration of a hexavalent ion can have the same coagulating effect as a much larger ($\sim 10^4$) concentration of a monovalent ion. Actually, one should use $R_{h^*} = \delta \sim 25\,kT$ for complete stability because thermal motion of the particles allows them to tunnel through small activation barriers.

## IV. Diffuseness: $\gamma_t$

In this section we will not treat all aspects of atomic relaxation, but will restrict our attention to the roughening of the interface, i.e., no discussion of chemical redistribution. We shall concern ourselves primarily with the transition in structure at the interface between two phases.

The driving force for either surface or interface roughening is the creation

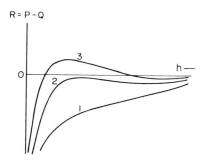

**Fig. 21.** Plot of the resultant repulsion force $R = P - Q$ between the two parallel plates as a function of separation distance $h$. Only curve 3 prevents coalescence.

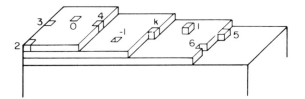

***Fig. 22.*** Representation of the different, energetically unique, surface sites for molecules.

of new, unique energy states that provide configurational entropy to the system at the cost of an increase in enthalpy for each of these new states. The minimum free energy condition yields the population of the different states and the energy decrease due to this roughening.

To better appreciate the different states of molecular attachment between one condensed phase and another, let us first consider the surface structure of a solid in contact with its vapor. We shall regard the crystal as a simple cubic lattice for the sake of illustration, the sites of which are occupied by molecules. Depicting the molecules as cubes, typical situations for surface molecules will be as illustrated in Fig. 22. Molecules may be in the surface (0), adsorbed on a surface site (1), in a corner (2), in an edge (3), in a step (4), at a kink in a step ($k$), or on a step (5). Removal of a molecule from the surface creates a vacant site in the surface ($-1$). Removal of a molecule from a step causes a vacant site in the step (6). The position $k$ has a special position in the theory of crystal growth; it is referred to as the "half-crystal" position. According to their position, molecules will be bound to the crystal with different energies.

Burton *et al.* (1949) extended the work of Onsager (1944) and quantitatively described the roughening of the solid–vapor surface illustrated in Fig. 22. The driving force for the roughening is the creation of new unique states that provide configurational entropy to the system at the cost of an increase in enthalpy for each of these new states. The minimum free energy condition yields the population of the different states. Jackson (1958) applied these same considerations to the solid–liquid interface. He described the roughening of a two-level model (see below) and related the excess enthalpy of a solid–liquid bond to some fraction of the latent heat of fusion. Cahn and Hilliard (1958) described this interfacial relaxation in thermodynamic terms calculating the interface diffuseness via that path. Cahn (1960) proceeded to calculate the change in crystal face energy and crystal edge energy in terms of its equilibrium roughness and generated a clear picture of the importance of this interface diffuseness on the molecular attachment kinetics for a phase transformation. Cahn and Kikuchi (1961) next applied these considerations to the evaluation of the energy and entropy of domain walls in ordered structures as a function of orientation and composition. Next, Mullins (1959) and Mutaftschiev and Kern (1964, 1965) made contributions to the three-level model and, most recently, Temkin (1966) developed the $N$-level model for a simple cubic lattice predicting the same general features as Cahn (1960) did.

The foregoing all deal with lattice models for the condensed phase even if it is a liquid. Thus, although they are intended to apply to the solid–liquid interface, they really apply to the analysis of domain walls (except the thermodynamic analyses) and give only a qualitative rather than a quantitative picture of what is happening at a solid–liquid interface. We will see later what is in error with these approaches and the directions we must travel to correct them. However, for pedagogical reasons, we shall illustrate the important approaches and features of interface relaxation following the foregoing authors.

### A. Solid–Vapor Case

#### 1. ROUGHNESS AT STEPS

At $T = 0$, steps tend to be as straight and as smooth as possible. At higher temperatures, thermal fluctuations will produce roughness in the form of kinks. Kinks may be positive forward jumps of the edge or negative backward drops, and they may be single or multiple units of the lattice distance (see Fig. 23a). The energy to form a kink of amount $m$ in a step on the (100) surface of a simple cubic (S.C.) crystal is $m[(\phi_1/2)+\phi_2]-\phi_2$ if we consider only first and second nearest neighbors. For simplicity we shall neglect the effects of neighbors higher than the first (despite our earlier criticism of this practice; see page 14). The roughness $r$ of the step is defined as the sum of all the amounts of positive and negative jumps in a length of step divided by the number of sites in that length, and analysis (Dunning, 1966) shows

$$r = [\sinh \phi_1/2kT]^{-1} \qquad (42)$$

which is plotted in Fig. 23b as $r$ vs. $\eta_1$, $[\eta_1 = \exp(-\phi_1/2kT)]$. The partial excess Helmholtz free energy associated with unit length of step $f^* = f_c + f_d$ is given by

$$f^* = (\phi_1/2a_0) + (kT/a_0)\log[\tanh(\phi_1/4kT)] \qquad (43)$$

where $a_0$ is the lattice spacing. In this expression, the electronic and chemical redistribution effects have been neglected and all differences in vibrational and rotational frequencies have been omitted for simplicity.

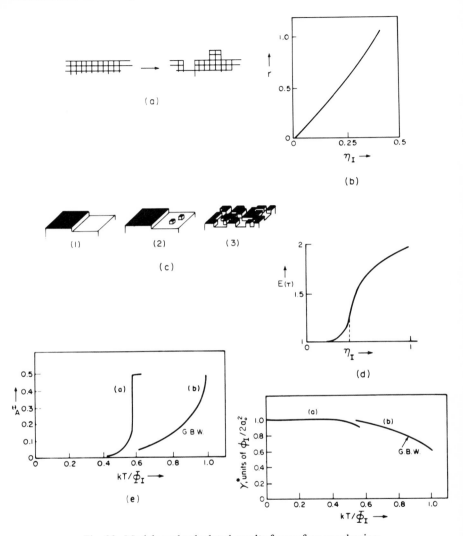

***Fig. 23.*** Models and calculated results for surface roughening.

## 2. SURFACE ROUGHNESS

At low temperatures there will be small concentrations of self-adsorbed molecules and of vacant sites on the surface (see Fig. 22). As the temperature increases, more and more molecules jump out of the surface leaving more and more vacancies and, with increasing surface concentrations, clusters of two, three, or more molecules and vacancies appear on the surface. At still higher temperatures, if the crystal has not melted, more than three levels are involved. This progressive roughening can be viewed as a cooperative disordering of the surface. The exact solution of the eigenvalue problem for the two-dimensional Ising model was found by Onsager and is presented here.

Since the Onsager theory is restricted to "two levels," the flat (100) surface of the S.C. crystal at $T = 0$ is considered as being half covered by a monomolecular half layer as illustrated in Fig. 23c(1). Neglecting the energy of the step, the potential energy of the surface at $T = 0$ is $\phi_1/2a_0^2$ per unit area on a first nearest neighbor (N.N.) bond picture. At a slightly higher temperature [Fig. 23c(2)], molecules in the half layer on the left have evaporated onto the lower half surface to the right forming self-adsorbed molecules and leaving vacancies behind. The energy is increased by $4(\phi_1/2)$, four "half bonds," for each singlet vacancy or adsorbed molecule present. When the surface coverage of vacancies is $\theta_V$ on the left and of adsorbed molecules $\theta_A$ on the right, the energy of the surface $E$ becomes

$$E(0+) = \phi_1/2a_0^2 + \theta_A(4\phi_1/2a_0^2) \tag{44a}$$

on both the left and the right. This is so because the energy of a vacancy is also $4\phi_1/2$ and $\theta_V = \phi_A$.

At higher temperatures, the surface energy is given by

$$E(T) = (\phi_1/2a_0^2)\{2 - \tfrac{1}{2}[1 + (2/\pi)(1 - k_1^2)^{\frac{1}{2}} K(k_1)] \coth H\} \tag{44b}$$

where

$$H = \phi_1/2kT = \ln \eta_1$$

$$k_1 = 2 \sinh H/\cosh^2 H \tag{44c}$$

$$K(k_1) = \int_0^{\pi/2} (1 - k_1^2 \sin^2 \omega)^{-\frac{1}{2}} d\omega$$

Here, $E(T) - \phi_1/2a_0^2$ is the configurational energy (neglecting contributions from altered vibrational modes). A plot of $E(T)$ vs. $\eta_1$ is shown in Fig. 23d. We see that the curve has a vertical tangent at the transition temperature $T_c$ given by

$$\sinh H_c = 1 \tag{45a}$$

or

$$T_c \sim 0.57\phi_1/k \tag{45b}$$

The surface coverage $\theta_A$ is given by

$$\theta_A = \tfrac{1}{2} - \tfrac{1}{2}(1 - \cosh^4 H)^{1/8} \tag{46}$$

and its dependence on temperature is shown in Fig. 23c, curve (a). At the transition temperature, $\theta_A = 0.5$, and the surface concentration of adsorbed molecules and of vacancies is the same on the left and the right parts of the surface [Fig. 23c(3)] and the step disappears.

With increasing temperature, the partial excess Helmholtz free energy $f$ which is equal to the partial surface tension, $\gamma^* = \gamma_c + \gamma_d$, begins to decrease slowly as the roughness increases [see Fig. 23f, curve (a)]. At the transition point, $\gamma^*(0) - \gamma^*(T_c) = 0.1103\phi_1/2a_0{}^2$, a decrease of about 10% from the value of $\gamma_c(0) = \phi_1/2a_0{}^2$.

For inert-gas crystals, it is found from experimental data that the melting point $T_M \sim 0.7\phi_1/k$ which suggests that $T_c$ calculated by Eq. (45) is close to the melting point [for the (100) face]. For the (111) face of an fcc lattice, theory gives $T_c \sim 0.91\phi_1/k$ so that the transition temperature is probably above the melting point for this surface. For the (001) face of an fcc crystal, the coverage of self-adsorbed molecules is about 1% for $T = 0.75\, T_c \sim 0.6 T_M$.

If we had used the Gorsky–Bragg–Williams (G.B.W.) approximation in the two-level problem, the analysis would be greatly simplified, but the predictions of $\theta_A$ and $\gamma^*$ are appreciably different as can be seen by curves (b) in Fig. 23e and Fig. 23f, respectively. In this approximation, the coverage $\theta_A \sim 0.5$ (at $T_{0.5} = \phi_1/k$) occurs at almost twice the temperature found by Onsager's exact calculation. This type of analysis is sensitive to such simplifying approximations as the G.B.W. approximation and the nearest neighbor interaction only approximation and must be kept in mind when we evaluate the quantitative accuracy of the following.

## B. Solid–Liquid Case

In the case where the crystal is surrounded by its melt, we can see that the binding energy of the molecule to the half-crystal position relative to the bulk liquid is just the latent heat of fusion $\Delta H_f$, i.e.,

$$\Delta\phi_k \approx \Delta H_f$$

The molecule in this position can, on the simple bonding model, be thought of as being half bonded to the solid and half bonded to the liquid. A subsequent molecule from the liquid which joins the solid at an adjacent kink site does not change the free energy of the system if $T_i = T_E$ (perhaps due to a change in equilibrium diffuseness); however, a latent heat of $\Delta H'$ is liberated. In an analogous fashion, we can define binding energies for the other eight states in Fig. 22 and can determine the change in enthalpy caused by moving a molecule from the kink position to one of these other states. This result would be approximately given in terms of the change in number of first, second, and third nearest neighbor bonds made with the solid in the two states. In general, the change in internal energy from one state to the next $\Delta E_{ij}$ will be given by

$$\Delta E_{ij} = \Delta n_1^{ij} \Delta \phi_1 + \Delta n_2^{ij} \Delta \phi_2 + \Delta n_3^{ij} \Delta \phi_3 = a_{ij} \Delta H' \tag{47}$$

where $\Delta \phi_1$, $\Delta \phi_2$, and $\Delta \phi_3$ represent the differences in internal energy per bond between molecules in the interface relative to the bulk liquid for 1st, 2nd, and 3rd N.N. bonds, respectively, and the $\Delta n_{ij}$ represent the change in the number of bonds of a given type between the state $i$ and the state $j$. Since $\phi_3$ and $\phi_2$ can be related to $\phi_1$ by a suitable molecular force law, the 1st, 2nd, and 3rd N.N. bond contributions $\Delta \phi$ can be related to $\Delta H'$. Thus, the enthalpy of a molecule occupying any of the unique states identified in Fig. 22 may be specified in terms of the number of bonds it makes with molecules in the solid and with molecules in the liquid relative to the state of the bulk liquid. In a similar manner, an entropy of fusion can be identified with each state as a fraction of the total entropy of fusion per molecule.

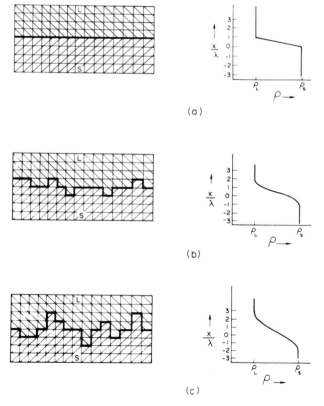

**Fig. 24.** Representation of surface energy-bond model showing population of 0, $\pm 1$, $\pm 2$ states and the density variation $\rho$ in the transition region between phases.

## 3. The Surface Energy–Bond Model

In this section, we shall first consider the structure of an infinite surface between a pure solid and its melt for the case where the interface contains no layer edges. We shall initially restrict our attention to a simple cubic solid with a similarly structured liquid at the equilibrium melting temperature $(T_i = T_E)$ and will consider the interface in Fig. 24 as the degree of diffuseness changes. The interface illustrated in Fig. 24a exhibits zero diffuseness and consists of "solid" molecules in state 0 (Fig. 22), whereas in Fig. 24b and 24c, the "solid" molecules are in $\pm 1$ states and $\pm 1$, $\pm 2$ states, respectively. We shall first deduce the excess free energy per site $f_c$ in Fig. 24a and will then determine the lowering of the free energy $f_d$ due to roughening for the cases of Fig. 24b and Fig. 24c (3-level and 5-level models, respectively) and will first consider only singlet and doublet formation (neglect of triplets, etc.). We will then proceed to evaluate the effects of (i) departure of $T_i$ from $T_E$, (ii) interaction potentials, i.e., only 1st N.N. vs. 1st, 2nd, 3rd N.N. interactions, (iii) orientation of interface, (iv) different crystal structures, (v) layer-edge diffuseness, and (vi) and $N$-level interface. We shall assume that at $T_i = T_E$, every molecule in the system whether tagged as part of the solid or the liquid has the same chemical potential $\mu$ which can be calculated from the bulk properties, i.e., they have the proper free volume, vibrational frequency, etc. In addition, for this model, the molecules at the interface are assumed to have an extra free energy contribution proportional to the number of solid–liquid bonds associated with the particular molecular state. Thus, as the molecules at the interface change state during the transition from the 1-level case of Fig. 24a to the 5-level case of Fig. 24c, the free energy of the system increases as a result of the increase in the number of S/L bonds but decreases more due to the increased configurational entropy (as a result of the number of possible arrangements of the different molecular states among the n possible interface sites).

On the nearest neighbor picture, each molecule of a simple cubic solid in a (100) interface makes one S/L bond for the 1-level model. The excess energy $\varepsilon$ of this double bond over the sum of a solid–solid bond plus a liquid–liquid bond is equal to the excess surface free energy per molecule $f_c$. On the basis of a 1st, 2nd, and 3rd N.N. bond only picture, in which the bond strengths are related by a 6-12 potential one 1st, four 2nd, and four 3rd N.N. bonds with excess energies $\varepsilon_1$, $\varepsilon_2$, and $\varepsilon_3$, respectively, need be considered. Because of the atomic spacings in the S.C. system, $\varepsilon_2 \approx \varepsilon_1/8$ and $\varepsilon_3 \approx \varepsilon_1/27$ so that

$$f_c \approx \varepsilon_1 + 4\varepsilon_2 + 4\varepsilon_3 = 1.64\varepsilon_1 \tag{48}$$

and $\varepsilon_1$ can be evaluated if $f_c$ is known [$f_c$ might be somewhat larger if we had followed the procedure of Nicholas (1968)]. By evaluating the excess energy of a half crystal site, one finds $\varepsilon_1 = \Delta H_f/3.9$ (provided all of the heat of fusion is associated with molecular changes).[†]

For the 3-level model, let $X_0$, $X_1$, and $X_{-1}$ be the fraction of molecular states of type 0, 1, and $-1$, respectively, per unit area of interface. Therefore,

$$X_0 + X_1 + X_{-1} = 1 \tag{49}$$

On the 1st N.N. bond only picture, each $X_1$ and $X_{-1}$ singlet state generates four additional S/L bonds over the 1-level model of extra energy $4\varepsilon_1$. This means that each state is equally probable and that $X_1 = X_{-1} = X = \frac{1}{2}(1 - X_0)$. Considering also that the formation of $(1, 1)$ or $(-1, -1)$ neighbors results in a decrease of one S/L bond each, the excess energy $\Delta u$ over that of the 1-level model is given by

$$\Delta u = 8\varepsilon_1 X(1 - X) \tag{50}$$

In calculating $\Delta u$, only that portion arising from bonds parallel to the surface need be considered since the number of bonds normal to the surface is unaltered by the presence of states of type $+1$ and $-1$. The average entropy change per site is simply determined by the number of arrangements $\omega$ of the $n$-sites for fixed $X$ which is given by $\Delta S = (k/n)\ln \omega$. Thus, we have

$$\Delta S = k/n \ln \{n!/(nX)!^2 [n(1-2X)]!\} \tag{51a}$$

$$= k[2X \ln X + (1-2X)\ln(1-2X)] \tag{51b}$$

and the total free energy $\Delta F(X)$ change due to roughening is given by

$$\Delta F(X) = \Delta u - T_E \Delta S \tag{52}$$

The optimum degree of diffuseness is given by minimizing $\Delta F$ with respect to variations of $X$ (neglecting free electron effects). This optimum value of $X, X = X^*$, is given by

$$\frac{X^*}{1 - 2X^*} = e^{-v_1(1-2X^*)} \tag{53}$$

[†]Even in a pure system, the total free energy is comprised from a portion associated with the atomic cores and a portion associated with the free electrons in the system. In different phases of the same material, the density and chemical potential of the free electrons will be different. Thus, when we consider a phase transition, the total enthalpy change $\Delta H$ is comprised of a portion due to rearrangement of the atomic cores and a portion associated with electronic changes. A similar situation exists for the total entropy change. To evaluate $\varepsilon_1$, we must first evaluate that portion of the total enthalpy change associated with the atomic core rearrangement.

where $v_1 = 4\varepsilon_1/kT_E$ and the change in the specific interfacial free energy is given by

$$\frac{\gamma_t}{kT_E} = \frac{\Delta F(X^*)}{kT_E} = 2v_1 X^*(1-X^*) + 2X^* \ln X^* + (1-2X^*)\ln(1-2X^*)$$

(54)

A plot of $X^*$ vs. $\varepsilon_1/kT_E$ is given in Fig. 25 ($\xi = 4$ for this case). Here, we can see that $\varepsilon_1/kT_E \gtrsim 0.25$ leads to a completely roughened condition and that the diffuseness decreases rapidly for values of $\varepsilon_1/kT_E \gtrsim 0.65$. An essentially singular surface occurs for values of $\varepsilon_1/kT_E > 1.25$. Of course, since we have neglected triplet, etc., formation, we should not consider data for $X^* \geqslant 0.1$ to be exact.

To evaluate the magnitude of $\varepsilon_1$ and relate it to some physically measurable quantity, we could assume that the measured surface free energy $\gamma'_{exp}$ minus

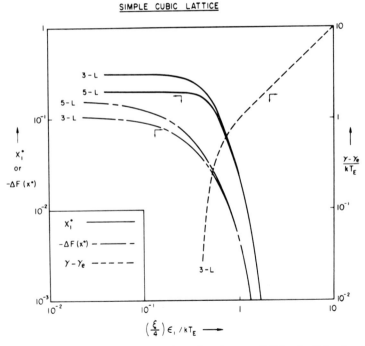

**Fig. 25.** Plot of (i) the equilibrium number of $+ 1$ states $X_1^*$ for the 3-level and 5-level model, (ii) the decrease in free energy $\Delta F(X^*)$ due to roughening for these models, and (iii) the interfacial energy $\gamma$ for the 3-level model all as a function of $\varepsilon_1/kT_E$, where $\varepsilon_1$ is the 1st nearest-neighbor bond strength.

the free electron contribution $\gamma_e$ is given approximately by the sum of $\gamma_c$ and $\gamma_t$ so that

$$\frac{\gamma_{\exp}}{kT_E} = \frac{\gamma'_{\exp}}{kT_E} - \gamma_e \approx \frac{v_1}{4} + \Delta F(X^*) \tag{55}$$

and $\varepsilon_1$ can be readily determined. In Fig. 25, we have also plotted $\gamma_t/kT_E$ and $\gamma_{\exp}/kT_E$ as a function of $\varepsilon_1/kT_E$. We can see that the value of $\gamma_t$ is negligible compared to $\varepsilon_1$ at small values of $X^*$ so that a linear relationship exists between $\varepsilon_1$ and $\gamma_{\exp}$. At a value of $X^* \sim 10^{-2}$, the magnitude of $\gamma_t$ begins to become significant with respect to $\varepsilon_1$ and $\gamma_{\exp}$ begins to decrease more rapidly. At a value of $X^* \sim 0.2$, the magnitude of $\gamma_t$ is almost equal to $\varepsilon_1$ and $\gamma_{\exp}$ is much less than $\varepsilon_1$. In fact, it soon goes negative, which signals a problem with the model.

From nucleation studies, we learned that $\gamma_{\exp} \approx 0.5\Delta H_f$ for nonmetallic systems. Thus, $\varepsilon_1/kT_E$ is related to the entropy of fusion per atom as might be expected. Assuming that Fig. 25 applies to the elements tin and germanium, for the sake of illustration, we find that $\gamma_{\exp} = 0.42\Delta H_f$ for Sn and $\gamma_{\exp} = 0.35\Delta H_f$ for Ge and that $\gamma_{\exp}/kT_E = 0.72$ for Sn and 1.06 for Ge. Thus, at these values of $\gamma_{\exp}/kT_E$, Fig. 25 gives $\varepsilon_1/kT_E \approx 0.8$ and 1 for Sn and Ge, respectively. It also gives $X^* \approx 0.05$ and 0.02 for Sn and Ge, respectively. This procedure illustrates the method of evaluating $\varepsilon_1/kT_E$ and $X^*$ for a particular system; however, the magnitudes are likely to be in error since Fig. 25 should not apply exactly to the close-packed faces of Sn and Ge.

## 2. Nonequilibrium Effect

When the system is departed from equilibrium an amount $\delta G = \Delta S_f \, \delta T$ per molecule, no changes in $\gamma_e$ occurs for the 1-level model. However, changes do occur in the 3-level model because the molecules changed to the $+1$ state reduce the free energy of the system by an amount $\delta G$, whereas molecules changed to the $-1$ state increase the free energy by the amount $\delta G$. This effect will lead to an increase in $X_1$ and a reduction in $X_{-1}$. In this case, neglecting $(1, 1)$ and $(-1, -1)$ neighbors, we have

$$\Delta F = X_1(4\varepsilon_1 + \delta G) + X_{-1}(4\varepsilon_1 - \delta G) - kT[X_1 \ln X_1 + X_{-1} \ln X_{-1}$$
$$+ (1 - X_1 - X_{-1})\ln(1 - X_1 - X_{-1})] \tag{56}$$

and

$$\frac{X_1}{1 - (X_1 + X_{-1})} = e^{-\delta G/kT} e^{-v_1(T_E/T)} \tag{57a}$$

$$\frac{X_{-1}}{1 - (X_1 + X_{-1})} = e^{+\delta G/kT} e^{-v_1(T_E/T)} \tag{57b}$$

Since $\Delta S_f/k \sim 1-3$ for most crystal structures, $-\delta G/kT$ is generally very small with respect to unity so that $X_1$ and $X_{-1}$ can be quite well approximated by Eq. (53).

## 3. Orientation and Structure Effects

As we expand our viewpoint to consider more than first nearest neighbors, different interface orientations, and different crystal structures, we can develop a general procedure for determining $\Delta u$ and thus $\Delta F$ and $X^*$ (since $\Delta S$ depends only on $X$). In general, we find that

$$\Delta u = 2(\eta_1 \varepsilon_1 + \eta_2 \varepsilon_2 + \eta_3 \varepsilon_3)[X(1-X)] = 8(\xi/4)\varepsilon_1 X(1-X) \qquad (58)$$

where $\eta_1$, $\eta_2$, and $\eta_3$ are the general number of 1st, 2nd, and 3rd N.N. bonds parallel to the interface so that $\xi$ is a constant depending upon the $\eta$'s. Thus, instead of using the $v_1$ of the earlier equations, we must use $(\xi/4)v_1$ in its place. We must also use the expression

$$\gamma_c = \bar{\omega}\varepsilon$$

Both $\xi$ and $\bar{\omega}$ are numbers that can be readily evaluated by bond-counting procedures. In Table VI, the values of $\xi$ and $\bar{\omega}$ are given for the (100), (110), and (111) planes of the S.C. lattice for both 1st N.N. and 123 N.N. interactions.

**TABLE VI**

BONDING DATA

|  | (100) | | | (110) | | | (111) | | |
|---|---|---|---|---|---|---|---|---|---|
|  | 1st | 2nd | 3rd | 1st | 2nd | 3rd | 1st | 2nd | 3rd |
| Bonds above plane | 1 | 4 | 4 | 2 | 5 | 2 | 3 | 3 | 4 |
| Bonds in plane | 4 | 4 | 0 | 2 | 2 | 4 | 0 | 6 | 0 |
| Bonds below plane | 1 | 4 | 4 | 2 | 5 | 2 | 3 | 3 | 4 |
| $\omega_1$ |  | 1 |  |  | 1 |  |  | 1 |  |
| $\xi_1$ |  | 4 |  |  | 2 |  |  | 0 |  |
| $\bar{\omega}_{123}$ |  | 1.64 |  |  | 2.7 |  |  | 3.52 |  |
| $\xi_{123}$ |  | 4.5 |  |  | 2.4 |  |  | 0.75 |  |

$^a$(S. C., $\varepsilon_2 = \varepsilon_1/8$; $\varepsilon_3 = \varepsilon_1/27$)

We can readily see that as $\xi$ decreases, both $X^*$ and $\Delta Y$ will increase. From Fig. 25, if $\varepsilon_1/kT_E = 1$, $X_{100}^* \approx 2 \times 10^{-2}$, $X_{110}^* \approx 12 \times 10^{-2}$, and $X_{111}^* \approx 27 \times 10^{-2}$. Thus, the greater is $\gamma_c$, the greater is $X^*$ and $|\Delta F(X^*)|$.

To see the effect of a different crystal structure, let us evaluate the diffuseness for the fcc and bcc structures. Here, there are 12 and 8 1st N.N., respectively, plus 6 and 6 2nd N.N. with $\varepsilon_2/\varepsilon_1 = 0.125$ and 0.263, respectively (neglect 3rd N.N. effects). We will compare only the most densely packed plane for these systems, i.e., the (111) and (110), respectively. We find $f_c = 3.4\varepsilon_1$ and $2.25\varepsilon_1$, respectively, with $\xi(\text{fcc}) = 6$ and $\xi(\text{bcc}) = 5$. In order to compare the $X^*$ for these systems, we must first evaluate $\varepsilon_1/kT_E$ from the experimental data on surface energies, i.e., $\gamma'_{\text{exp}}$, then we read $X^*$ directly from Fig. 25, at the appropriate value $(\xi/4)\varepsilon_1/kT_E$.

Let us choose Pb and Fe as our examples of the elements in the fcc and bcc systems, respectively. For Pb and Fe, $\gamma'_{\text{exp}} = 0.39\Delta H'$ and $0.45\Delta H'$, respectively with $\gamma'_{\text{exp}}/kT_E = 0.433$ and 0.432, respectively. If we presume that $\gamma_e \approx 0$ in these cases, then $(\xi/4)\varepsilon_1/kT_E \approx 0.38$ and 0.35 for Pb and Fe, respectively, so that $X^* = 0.22$ and 0.24, respectively. Of course, $\gamma_e < 0$ so that $\varepsilon_1/kT_E$ increases and $X^*$ decreases in both cases.

## 4. Other Important Aspects

The foregoing treatment can be readily extended to consider $5, 7, 9, \ldots N$ levels of diffuseness. The excess free energy $\Delta F'(X_1{}^*, X_2{}^*)$ over that of the 1-level model (S.C. system) for the 5-level model is also plotted in Fig. 25. The variation of $X_1{}^* = X^*$ with $\varepsilon_1/kT_E$ is also given in this figure. It is found that $X_2{}^* \approx (X_1{}^*)^2$ at small values of $X_1{}^*$ and that $X_2{}^*/X_1{}^* = 0.1, 0.25, 0.5,$ and 1.0 at $X_1{}^* = 0.062, 0.12, 0.18,$ and 0.2, respectively. We note from Fig. 25 that $X^*$ (5-level) departs from the results of the 3-level model at $\varepsilon_1/kT_E < 1$ and reaches its maximum value of 0.2 in the vicinity of $\varepsilon/kT_E \sim 0.2$. We note, also, that $-\Delta F(X_1{}^*, X_2{}^*)$ has increased over that of the 3-level model, indicating that $\gamma_t$ would be further lowered by roughening on other levels. Temkin (1966) has treated the $N$-level model (S.C. and 1st N.N. only with G.B.W. approximation) in a very elegant fashion. This work presents a classical illustration of the canonical ensemble approach of statistical mechanics and qualitatively supports the thermodynamic calculations of Cahn (1960) which will be mentioned below.

For the $1, 3, 5, \ldots, N$ layer model, the thickness of the diffuse layer $\Delta Y/\lambda$, where $\lambda$ is the lattice distance is given most generally by the equation

$$\frac{\Delta Y_N}{\lambda} = \left[ 1 - 2 \sum_{i=1}^{(N-1)/2} X_1{}^* \right]^{-1}; \qquad N = 3, 5, \ldots \tag{59}$$

This formula is obtained by considering the density of a layer as the sum of the fraction of "solid" molecules times the density of solid and the fraction of "liquid" molecules times the density of the liquid.

Using the 3-level, 5-level, etc., model, one can extend the above treatment to include the additional diffuseness generated at a layer edge in order to evaluate the lowering of the edge energy $\Delta\gamma_E$ due to this relaxation. One finds that the layer edge is considerably more diffuse in the lateral direction than in the direction normal to the interface. For the 3-level model, it is relatively easy to show that $X^*(\text{edge}) \sim [(X^*(\text{flat})]^{1/2}$.

In the absence of a driving force, i.e., $T_i = T_E$, an interface held parallel to a low index crystallographic plane will assume an equilibrium configuration. An identical configuration, displaced by an integral number of lattice planes, will also be an equilibrium configuration. We can conceive of the interface as being forced to move normal to itself and, at all intermediate stages of advancement between two equilibrium positions, must assume states of nonequilibrium diffuseness and thus higher interfacial energy. This oscillating excess free energy $G(x/\lambda)$, of the system as the interface moves through an integral number of atomic configurations, is illustrated in Fig. 26. Using the 3-level approach, one can readily evaluate the height of the barrier $\Delta G_{max}$ in Fig. 26; however, the result depends somewhat upon the specfic path one chooses to follow through the nonequilibrium states. For example, the interface can be caused to move from position $\frac{1}{2}$ to position 1 in Fig. 24b by adding $+2$ levels at the $-1$ positions and additional $+1$ states at the 0-state positions if needed. Many alternate devices for moving the centroid of the interface could be selected. For the example chosen, $\Delta G_{max}/kT_E$ is linear with $\varepsilon_1/kT_E$ at large values of $\varepsilon_1/kT_E$ and has a value larger than unity. As $\varepsilon_1/kT_E$ decreases, $\Delta G_{max}/kT_E$ decreases rapidly in a strongly nonlinear fashion towards zero. The work of both Cahn (1960) and Temkin (1966) support this conclusion. In real systems, the value of $\Delta G_{max}$ can never go to zero, and if one's calculation for a 3-level model suggests that it does, then one must consider a 5-level, 7-level, etc., model.

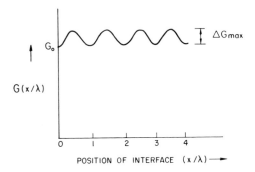

**Fig. 26.** Variation of system free energy, $G$ as a flat interface between solid and liquid (at the melting temperature) is moved through successive equilibrium positions.

## C. *Inadequacies of Present Bond Models*

From X-ray measurements and the fusion properties of simple liquids, it is found that the liquid has about 10% fewer 1st nearest neighbors with each located at a slightly greater distance than in the solid. These two factors determine the enthalpy and the small vibrational entropy difference between the liquid and the solid. Thus, the large entropy of fusion must reside in the fact that (1) there is no long range order but rather a chaotic atomic arrangement in the liquid, and (2) a greatly enhanced atomic mobility exists indicating that the random structure is loose packed rather than dense packed.

According to Cohen and Turnbull (1959), static atomic disorder cannot alone account for the entropy of fusion since a dense amorphous phase is only slightly less stable than a dense crystalline phase. The number of near neighbors in the "glassy state" is somewhat decreased from that of the crystalline state, the bond distances are about the same and the entropy change due to the randomized but immobile atomic configuration is only a small fraction of the entropy of fusion. The "Free Volume" model of Cohen and Turnbull shows that the glass–liquid transition is a change from dense amorphous to loose amorphous packing and indicates that the major contribution to the entropy of the loose random structure of the liquified state comes from the redistribution of free volume.

In the liquid state, each atom vibrates in a cage composed of its randomly arranged nearest neighbors. The average equilibrium distance between atoms forming a typical bond is $\bar{R}$ (see Fig. 27) and depends upon temperature via the thermal expansion coefficient. As $T$ increases, $\bar{R}$ moves to the right, whereas it is near $R_0$ at low temperatures. At low temperatures, if the cell fluctuates in size to $R > \bar{R}$, nearby cells will be contracted by a compensating volume. In this case, the increase in potential energy due to the expansion outweighs the decrease due to the contraction because the slope $R > \bar{R}$ is much greater than the slope at $R < \bar{R}$. However, as $T$ increases and $\bar{R}$ approaches the linear part of the curve $(R)$, this situation no longer holds. At this point, when the net potential energy increase associated with a fluctuation is $kT$ or less, spatial fluctuations of the size of a bond length are spontaneously thermally generated and a portion of each cell can be redistributed amongst other cells. The redistribution of this free volume gives rise to an extra configurational entropy for the liquid state; it allows greatly enhanced mobility of the atoms so it is denoted as communal entropy to distinguish it from the configurational entropy change associated with the crystalline–amorphous transition. The critical density range in which the transition is predicted to occur is in satisfactory agreement with information concerning the glass–liquid transition threshold.

Some people feel that the full communal entropy appears discontinuously

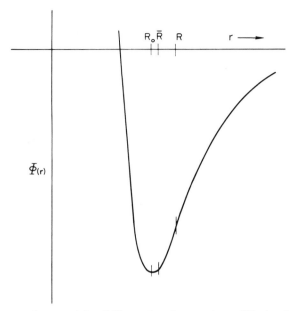

**Fig. 27.** Interatomic potential well illustrating the curvature of the local potential as the interatomic distance expands (by heating) from $R_o$ (greatest curvature) to $R$ (relatively flat portion of curve).

at the melting point and contributes an extra amount of entropy, $R$ per mole, to the liquid. Others feel that it appears more gradually since it should be a function of both volume and temperature. Experimentally, the average entropy of fusion for a large number of monatomic substances (fcc and bcc metals) is found to be about 2.2 e.u., which is very close to the total communal entropy contribution of 2.0 e.u. The internal energy of a fluid wherein the molecules are confined to move in a restricted volume cell is calculated to be only negligibly different from one wherein the communal entropy is fully developed. Thus, a proper model of a solid–liquid transitional interface must concern itself with the existence and development of communal entropy in the interface region in order to have a basis for being physically and numerically relevant.

The major objection to the bond model discussed in Section IV, B is that it treats a lattice liquid rather than a real liquid and thus does not include the prime factor that differentiates a liquid from a solid. It would be a suitable model for describing a crystal–glass interface, a magnetic domain interface, or even a solid–solid interface (with suitable adjustment for strain and dislocations, etc.); however, for the solid–liquid case, it can provide only qualitative insight at best.

If we begin to question the actual distinguishability of the $-1, -2, ...,$ liquid states at the interface, we find that they are not really distinguishable from the surrounding solid states because the $kT$ vibration fluctuations in the ordered atom positions produce local volume changes of about 5% quite readily. However, the $+1, +2, ...,$ solid states do indeed produce distinguishable states compared to the local liquid atoms since they are relatively fixed in space and have an ordered arrangement. Thus, the configurational entropy change associated with roughening via the present bond models should only be about one half of the presently calculated values.

It is not difficult to show that the bond-model calculations of Jackson (1958), Temkin (1966) and those of Section IV, B lead to negative values of $\gamma$ at the equilibrium interface roughness for certain materials and certain crystallographic orientations. This is undoubtedly due to neglecting the loss of some communal entropy for molecules in the transition region and to overestimating the configurational entropy of the roughened states (by insufficiently questioning distinguishability). Further, the use of 1st N.N. bonds only in the calculation of internal energy can be a significant numerical error as pointed out by Nicholas (1968). Likewise, the use of the G.B.W. approximation in the analysis can lead to significant numerical error as illustrated in Figs. 23e and 23f.

Finally, we must recognize that the latent heat of fusion $\Delta H_f$ and thus the entropy of fusion $\Delta S_f$ is comprised of two parts. One part is due to the change of free electron energy states and the other to the rearrangement of the atomic cores. Using free electron theory, one can show that

$$\Delta H_f(\text{electron}) \sim -\Delta H_f(\text{expt}) \tag{60a}$$

so that

$$\Delta H_f(\text{core}) \sim 2\Delta H_f(\text{expt}) \tag{60b}$$

In most cases, we expect the free electrons in both solid and liquid to be well correlated with the ion cores (clustered around the cores) so that $\Delta H_f$ is the appropriate value of enthalpy change to be considered in evaluating $\varepsilon_1$. For other cases, the free electrons may not be well correlated and one must evaluate $\varepsilon_1$ by considering just the $\Delta H_f(\text{core})$ contribution. Of course, for very diffuse interfaces, we must expect changes in $\gamma_e$ to be associated with the changes in diffuseness (especially for metals).

Although it is not wise to rely on the quantitative conclusions of the simple bond models presented in this section, they are able to provide satisfactory qualitative guidance. We can feel assured that $\varepsilon_1/kT_E$ or $\Delta H_f/kT_E$ is a major parameter in the interface relaxation situation and that $|\gamma_t|$ increases, $\Delta Y$ increases, $\Delta G_{max}$ decreases and $\gamma_{edge}$ decreases as $\Delta H_f/kT_E$ decreases.

## V. Kinetics

In this area, our qualitative understanding is good and is consistent with the view put forward some years ago by Cahn (1960) and reviewed by Cahn *et al.* (1964). However, our quantitative understanding, both theoretical and experimental, is in dire need of some new techniques and more careful investigations. This section builds on the foundation of the previous section, and since our quantitative understanding of the interface transition layer is slight, our quantitative understanding of interface attachment kinetics is still limited to a coarse-level parametric description. A reasonably, yet probably numerically inaccurate, description of the situation is that given by Cahn *et al.* (1964). Here, only a brief review will be presented.

For a nucleus of critical size to grow, or for a macroscopic crystal to increase in size, molecules of the liquid phase must become molecules of the solid phase via an attachment and an incorporation mechanism taking place at the interface between the two phases. Two major processes by which this occurs can be distinguished. First, the surface advances by the lateral motion of steps $n$ interplanar distances in height ($n \gtrless 1$). A molecule is not incorporated into the solid, and a surface undergoes no change, except during the passage of a step, i.e., a molecule can attach itself to the solid only at the edge of a step. Then, the surface advances normal to itself by $n$ interplanar spacings. The second process is one whereby the surface advances normal to itself without needing steps, i.e., a molecule can attach itself to the solid at any crystallographically identified lattice point of the entire surface. This means that in the presence of a driving force for solid formation, every element of surface is capable of continued change leading to advance of the interface.

The distinction between the processes is twofold. One deals with the geometrical nature of lateral step motion versus the motion of the whole surface normal to itself. The second deals with the time dependence of the motion of an element of surface, i.e., no motion except when a step passes versus continual motion. The first process is termed *nonuniform or lateral growth* and the second *uniform or normal growth* (Cahn, 1960).

Our working model is that represented by Fig. 26 wherein a free energy fluctuation of $\Delta G_{max}$ repeatedly occurs as the diffuse interface moves through successive equilibrium positions at an infinitesimally small driving force. During normal crystal growth, when the interface is departed from equilibrium by a small amount, $\Delta G_K \ll \Delta G_{max}$, a molecule attaching itself to the solid "sees" an activation barrier to interface motion. Thus, a layer generation process is necessary for molecular attachment to occur. On the other hand, at large departures from equilibrium, $\Delta G_K \gg \Delta G_{max}$, a molecule attaching itself to the solid leads to a decrease in free energy greater than the magnitude of the activation barrier $\Delta G_{max}$ and thus does not "see" any resistance to interface motion by this uniform attachment process.

To evaluate the molecular attachment kinetics in general, we must proceed to answer the following questions. (a) How diffuse is the interface between the crystal and the fluid from which it grows and how does it depend upon orientation, alloy content, etc.? (b) What is the critical driving force $\Delta G_K^*$ for the onset of the uniform attachment mechanism? (c) What are the kinetics of layer generation from the various kinds of sources potentially available to the crystal? The first question has, to some degree, been answered in the last section and more completely by Nason (1970). Here, we will devote ourselves to questions (b) and (c).

In the presence of sufficient driving force, the uniform attachment mechanism is operative and the variation in total free energy of the system per unit area of interface $\delta G$ due to interface motion through a distance $dx$ is given by

$$\delta G = [\Delta G_V + (d\gamma/dx)]\,dx \tag{61}$$

where $\Delta G_V = \Delta S_f \Delta T_K$ is the driving force per unit volume. The layer edge attachment mechanism is required only when there exists a value of $X$ for which $\delta G > 0$. The uniform attachment mechanism is operative when $(d\gamma/dx)_{max}$ is such that $\delta G \leqslant 0$ for all $X$. This critical value of $\Delta T_K = \Delta T_K^*$ depends sensitively upon the shape of the $G$ vs. $x/\lambda$ curve in Fig. 26. If, for the moment, we assume that this relationship is a harmonic one, i.e.,

$$G = G_0 + (\Delta G_{max}/2)[1 - \cos(2\pi x/\lambda)] \tag{62}$$

per atomic site of the interface, then $\Delta T_K^*$ is given by

$$|\Delta T_K^*| = \pi \Delta G_{max}/\Delta S' \tag{63a}$$

where $\Delta S'$ is the entropy of fusion per molecule. At $|\Delta T_K| > |\Delta T_K^*|$, the interface advance does not require layer motion. In general, one cannot expect Eq. (62) to hold and the free energy relationship may have one of the forms illustrated in Fig. 28 where curve 2 is the harmonic representation. Thus, in general, Eq. (63a) must be replaced by

$$|\Delta T_K^*| = m_1(\Delta G_{max}/\Delta S') \tag{63b}$$

where $m_1$ is a constant which is proportional to the steepest slope of the curves in Fig. 28.

When $\delta G > 0$ in Eq. (61), the surface must advance normal to itself by the passage of steps across the surface. In the absence of screw dislocations intersecting the surface, the steps are rapidly exhausted and the surface becomes exactly parallel to the crystallographic plane. Due to two-dimensional nucleation, layer sources will be generated as islands of solid of radius $r_c \gg -\lambda\gamma_E'/\Delta S'\Delta T_K$ where $\gamma_E'$ is the edge energy per edge atom. Because the step may have a considerable width (diffuseness) $W$ much larger than that normal to the interface $\Delta Y$, this expression must be modified when $r_c \gtrsim W$. Since $\gamma_E'$ will decrease for $r_c < W/2$, we should expect $r_c$ to decrease with increase of $|\Delta T_K|$ in a classical fashion until it reaches the value of $W/2$ and then to decrease more rapidly with further increase of $|\Delta T_K|$. This transition point occurs at $\Delta T_K = \Delta T_K^{**}$ given by

$$|\Delta T_K^{**}| = m_2 (\Delta G_{max}/\Delta S') \tag{64}$$

where $m_2$ is a constant. Using the Cahn model (1960), one finds $m_2 = 1$.

If we consider a screw dislocation intersecting an interface with a component of its Burger's vector normal to the interface, it gives rise to a step that must always terminate at the intersection. This self-perpetuating step is the source of spiral layers on the surface which have a separation distance of $2r_c$ in the vicinity of the dislocation intersection point. As $2r_c$ becomes less than $W$, these spiral steps begin to overlap producing a more or less continuous ramp on the surface and $\gamma_E'$ decreases.

From the foregoing, we note that as a function of $\Delta T_K$, there are three different regions of growth to be considered: (1) For $0 < |\Delta T_K| < |\Delta T_K^{**}|$, classical lateral growth mechanisms should be observed. In this range, the growth is characterized by a step energy $\gamma_E'$ that is independent of $\Delta T_K$. (2) For $|\Delta T_K^{**}| < |\Delta T_K| < |\Delta T_K^*|$, the lateral growth mechanism must be modified to allow for the change of $\gamma_E'$ with $\Delta T_K$ due to the overlap of the diffuse layer at step edges. In this region, a gradual transition is made from the classical lateral growth to uniform growth. (3) For $|\Delta T_K^*| < |\Delta T_K|$, the interface can advance normal to itself without the benefit of the lateral motion of steps. These three regions are illustrated in Fig. 29 for the two common methods of representing the kinetic data.

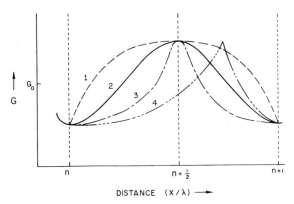

**Fig. 28.** Possible variations of system free energy $G$ between equilibrium interface positions.

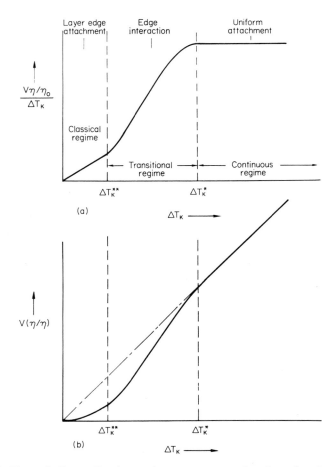

**Fig. 29.** Theoretically predicted growth rate curve as a function of undercooling $\Delta T_K$ for interfaces with emergent dislocations. The curve in the transitional regime is shown schematically. The ordinate is the growth rate adjusted for the temperature dependence of the melt viscosity and differs for curve (a) and (b) by the factor $\Delta T_K$.

The generalized growth laws for the different known mechanisms may be written as follows. In all cases, $V$ is the velocity of growth normal to the surface, $\Delta T_K$ is the interfacial undercooling required to drive the attachment process, and the $\mu_i$ are kinetic constants.

1. *Uniform interface attachment*

$$V = \mu_0 \Delta T_K \tag{65a}$$

2. *Layer edge motion*

$$V = \mu_\infty \Delta T_{\rm K} \tag{65b}$$

3. *Screw dislocation source*

$$V = \mu_1 \Delta T_{\rm K} |\Delta T_{\rm K}| \tag{65c}$$

4. *Two-dimensional nucleation (layer passage limited)*

$$V = \mu_4 \exp[\mu_3/3\Delta T_{\rm K}] \tag{65d}$$

5. *Two-dimensional nucleation (layer source limited)*

$$V = a_0 A\mu_2 \exp[\mu_3/\Delta T_{\rm K}] \tag{65e}$$

where $a_0$ = molecular step height normal to the interface and $A$ is the area of the crystal face.

6. *Twin-Plane Reentrant Corner Nucleation*

$$V = \mu_5 \exp[\mu_6/\Delta T_{\rm K}] \tag{65f}$$

It is of value to relate the various $\mu_i$ to one another so that when kinetic studies via different mechanisms are made on the same material, the correlation of data will provide a critical evaluation of our quantitative understanding in this area. The uniform attachment law is the simplest so all the $\mu_i$ will be related to $\mu_0$ (except for $\mu_5$ and $\mu_6$ which have not yet been critically evaluated). Cahn *et al.* (1964) give

$$\mu_0 = \frac{\beta^* D_{\rm L} \Delta H_{\rm f}}{a_0 R T_{\rm M}^2} \tag{66a}$$

where $a_0$ = advancement normal to interface for an atom attachment (usually $\lambda$), $D_{\rm L}$ = self-diffusion coefficient in the liquid, $T_{\rm M}$ = melting temperature, and $\beta^*$ is a constant introduced to relate the interface attachment transition frequency to the diffusional jump frequency in the liquid (and serves to correct for any other inadequacies of the model ). They also give

$$\mu_\infty = \mu_0[2+g_{\rm max}^{-1/2}] \tag{66b}$$

where $g_{\rm max}$, the diffuseness parameter, is related to the number of atomic layers $\Delta Y/\lambda$ in the transition region at $T_{\rm M}$. Using Turnbull's (1950) empirical relationship between $\gamma$ and $\Delta H_{\rm f}$, i.e.,

$$B = v_m \gamma/a_0 \Delta H_{\rm f} \tag{66c}$$

where $v_m$ is the molecular volume and $B$ is a constant ($0.2 \gtrsim B \gtrsim 0.5$), we find that

$$\mu_1 = \frac{\mu_0}{4\pi g_{max} BT_M} [1 + 2(g_{max})^{1/2}] \tag{66d}$$

$$\mu_2 = \frac{\mu_0 N}{v_m T_M} (\Delta H_f/R)^{1/2} |\Delta T_K|^{3/2} [2 + g_{max}^{-1/2}] \tag{66e}$$

$$\mu_3 = \mu_0 [\pi g_{max} B^2 a_0 T_M^2/\beta^* D_L] \tag{66f}$$

$$\mu_4 = \mu_0 (\Delta H_f/RT_M^2)^{1/6} |\Delta T_K|^{7/6} [2 + g_{max}^{-1/2}] \tag{66g}$$

where $N$ is Avogadro's number.

Using Eqs. (66) with approximately chosen parameters for the particular system of interest, kinetic coefficients for the different attachment mechanisms may be readily calculated. Using Hillig's corrected data (1958) for the $C$-axis growth of ice (where passage-limited, two-dimensional nucleation kinetics is considered to be the growth mechanism), Tarshis (1967) found $\mu^4 = 0.03$ cm/sec and $\mu_3 = -1.05°$ C and, using an average value of $\Delta T_K = -0.05°$ C, evaluated the following: $g_{max} = 3.8 \times 10^{-3}$, $\beta^* = 4.7 \times 10^{-2}$, $\mu_0 = 0.12$ cm/sec/°C, $\mu_\infty = 2.26$ cm/sec/°C, $\mu_1 = 3.0 \times 10^{-2}$ cm/sec/°C$^2$, and $\mu_2 = 7.5 \times 10^{19}$/sec/cm$^2$.

One can gather information concerning the molecular attachment kinetics via three discrete paths: (a) observations of the crystal morphology, (b) qualitative observations of the growth kinetics, and (c) quantitative observations of the growth kinetics.

(a) By observing the shape of crystals and the topology of their surfaces, one can detect dislocation spirals, twin planes with reentrant corners, polyhedral facets, and other artifacts that support a layer motion mechanism of growth. On the other hand, if a crystal is smooth and curved, one cannot deduce that continuous growth occurred since the molecular layers may still be present but on a scale below the limits of detection of your microscope. Such morphological evidence of layer motion is available for Zn, Ge, Si, Bi, Sb, $Bi_2Te$, $PbI_2$, $H_2O$, $\beta$-methylnaphthalene, and a very large number of other systems (Cahn et al., 1964). In fact, evidence abounds to support the layer growth mechanisms; however, it is not so readily available for the uniform advance mechanism.

(b) Certain qualitative kinetic observations can be used as evidence of a layer mechanism of crystal growth. At small undercoolings, $p$-toluidine platelets grow from the melt at nearly constant thickness to give lateral diameter:thickness ratios of 1000:1. The thickening rate is found to be a sensitive function of $\Delta T_K$. In addition, the introduction of defects when a platelet was rapidly growing into a wall caused an abrupt change in the growth behavior and an increase in the thickening rate. The following was observed for the growth of nearly perfect platelets of durene from the melt: at $\Delta T_K = -0.25°$ C, brightly interference-colored platelets existed at constant color for several minutes; at $\Delta T_K = -0.5°$ C, the colors faded in a few minutes; at $\Delta T_K = -3.0°$ C, the thickening was essentially instantaneous. When a platelet at $\Delta T_K = -0.5°$ C touched, and stuck to a second crystal, it thickened almost instantly.

From the foregoing we see that the lateral layer spreading mechanism of growth is supported by the "go/no-go" type of growth behavior where one sees that the growth rate increases astronomically with a small increase in $|\Delta T_K|$ and observes the marked effect of mechanical damage on perfect habit faces. The former characteristic indicates that the two-dimensional nucleation of layers is operative, whereas the latter indicates that when the dislocation layer source becomes operative, it is much more effective in generating layers.

(c) The quantitative evaluation is the most sensitive and the most powerful for illuminating our understanding of the various mechanisms; however, it also contains the most traps for the unwary. The experimental data is usually presented in the form

$$V = f_1(\Delta G) = f_2(\Delta T) = f_3(\Delta C) \tag{67}$$

where $f_i$ are the same functional dependence upon the total driving force $\Delta G$ or its representatives, the total undercooling $\Delta T$ and the total supersaturation $\Delta C$. In the past, many investigators followed the path of least effort and made the assumption that $\Delta G_K \equiv \Delta G$ or that $\Delta T_K \equiv \Delta T$ and $\Delta C_K \equiv \Delta C$. This can often lead to a great error in the analysis since the "coupling equations" (Tiller, 1964) in $\Delta T$ and $\Delta C$ are

$$\Delta T = \Delta T_S + \Delta T_H + \Delta T_E + \Delta T_K \tag{68a}$$

and

$$\Delta C = \Delta C_S + \Delta C_H + \Delta C_E + \Delta C_K \tag{68b}$$

The total driving force $\Delta G$ may be most generally divided into four segments, the one needed to drive matter transport $\Delta G_S$, that needed to drive heat transport $\Delta G_H$, that needed to provide the excess free energy of the crystal due to defects and surfaces $\Delta G_E$, plus that needed to drive the molecular attachment process at the interface $\Delta G_K$. In a great many crystal growth situations, $\Delta T_K \gtrsim 0.1\Delta T$ and the velocity dependence of $\Delta T_K$ is quite different from that of $\Delta T$. Thus, the equating of $\Delta G_K$ with $\Delta G$ can lead one up the proverbial "garden path" as far as detailed understanding is concerned. Many investigators do recognize the need to combine theoretical analyses with the experimental results in order to separate out a realtionship of the form given by Eqs. (65). However, to date, the numerical accuracy of these analyses provides only a marginal ability to discriminate differences between mechanisms or subtle variations in a particular mechanism.

Past work indicates that we have a reasonable qualitative understanding of this area and that we have some good theoretical guidelines for future work. We know that as $\Delta H_f/kT$ decreases, the interface diffuseness increases; both $\gamma$ and $\gamma_E$ decrease so that the rate of molecular attachment increases ($|\Delta T_K|$ decreases for fixed $V$) and the anisotropy of the attachment kinetics increases greatly, i.e., small changes in $\Delta Y/\lambda$ with orientation $\theta$, produce large changes in $|\Delta T_K(\theta)|$ for fixed $V(\theta)$. Although we are close to a satisfactory parametric description of these kinetic equations, we must realize that more effort is needed to reliably relate the $\mu_i$ to basic physical quantities. On the experimental front, it is to be realized that we do not yet have sufficiently exacting experimental data or techniques to transform these theoretical models into a more reliable description of the molecular attachment kinetics.

## VI. Topography

Almost 20 years ago, Herring (1951) showed that the excess free energy of a molecule on a curved surface over that of a plane surface was given by

$$\Delta G_E = \gamma \varkappa \tag{69}$$

where $\varkappa$ is the curvature of the surface. This result has been extensively used to evaluate the driving force for the migration of molecules from a surface region of one value of curvature to that of another. Since that time, it has become general practice to use Eq. (69) to account for the excess free energy of a molecule on a curved segment of surface during a phase transformation, i.e., during motion of the interface. One of the main purposes of this section is to show that the use of Eq. (69) for moving interfaces is an error which, when corrected, alters the quantitative features of perturbation and crystal morphology analyses. We shall proceed by first considering a continuum approach to the creation of surface and follow this with a layer and ledge approach (Tiller and Jindal, 1971).

### A. Continuum Model

#### 1. SMALL PARTICLE GROWTH

In contrast to phase transformation via the unidirectional translation of an infinitely flat interface, phase transformation via the growth of small particles involves the creation of substantial amounts of new surface area. In such cases, a portion of the total driving force for phase change must be expended in the creation of this new surface. If the excess surface free energy is isotropic, then, for certain shapes of particle, the driving force for particle growth can be calculated by averaging over the entire particle. Using the familiar example of a cylindrical particle, we recall that as the particle grows from $R$ to $R+dR$, the area of surface created is $\Delta A = 2\pi dR$ and the volume transformed is $\Delta V = 2\pi R dR$. If the volume free energy change per unit volume of transformed material at the new equilibrium transformation temperature is $\delta G_V$, the total free energy change $\Delta G$ is given by

$$\Delta G = \delta G_V + \Delta G_S$$
$$= 2\pi R dR \delta G_V + 2\pi dR\gamma \qquad (70)$$

The free energy change per unit volume of transformed material is

$$\frac{dG}{dV} = \frac{\Delta G}{\Delta V} = \delta G_V + \frac{\gamma}{R} \qquad (71)$$

and thus the new equilibrium temperature $T_E$ which derives from taking $dG/dV = 0$, is given by

$$T_E = T_M - (\gamma/\Delta S)\varkappa \qquad (72)$$

where $\varkappa = 1/R$ in this case and $T_M$ is the equilibrium transformation temperature of the bulk phase. This is the well-known Gibbs–Thomson equation.

If we next consider the growth of a cylinder of square cross section instead of circular cross section and compute the average free energy change per unit volume of transformed material, we again obtain Eqs. (71) and (72) provided $\gamma$ is isotropic and we replace $\varkappa$ with $1/R$ in Eq. (72). Thus, the average free energy change for a square particle is the same as for a circular particle. However, unlike the circular cylinder which exhibits a constant curvature at any point of its surface, the square cylinder exhibits a zero curvature along its faces and infinite curvature at its corners. The use of Eq. (72) on a point-to-point basis would obviously lead to a very different picture than that given by the averaging procedure.

The proper procedure for a surface of varying curvature like the square cylinder must be the point-to-point evaluation wherein one considers a surface increment *moving normal to itself*. In the present case, we note, as indicated in Fig. 30a, that the excess surface created at the faces is zero so that $T_E = T_M$ at the faces and that the excess surface created at each corner is $2dR$ yielding

$$dG/dV = \delta G_V + 2\gamma/R \qquad (73)$$

Thus, the equilibrium transformation temperature of the corner is lower, by a factor of two, than that given by the earlier averaging treatment and is very

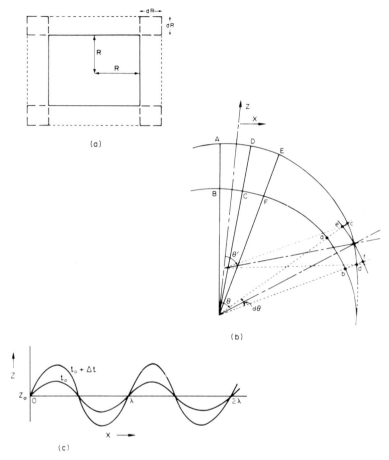

**Fig. 30.** (a) Representation of square crystal growth on a continuum model to illustrate that surface creation occurs only at the corners, (b) representation of cylindrical cap growth in $Z$ direction at constant $V$ with surface segment $ab$ becoming segment $cd$ and (c) change of harmonic surface shape between times $t_o$ and $t_o + \Delta t$.

different from that given by Eq. (72). We note also that such a sharp cornered particle requires a large temperature difference to exist along its surface in order to grow with a maintained shape (otherwise rounded corners will result). Careful attention must be paid to this difference between the nucleation-type procedure for evaluating $T_E$ and the growth procedure for evaluating $T_E$. They are both thermodynamic in nature and require the evaluation of $\Delta A/\Delta V$; however, averaging is proper for one, but point-to-point evaluation is necessary for the other.

## 2. STATIONARY-STATE SHAPES

In the past we have considered the unidirectional growth of a continuous surface of constant shape as a simple translation in the growth direction and have used Eq. (69) to derive the departure from equilibrium associated with the curvature. As the teachings of the previous section indicate, however, the proper procedure is to evaluate the point-to-point variation of $\Delta A/\Delta V$ which, for isotropic $\gamma$, is linearly related to the excess free energy per unit volume of the curved moving interface. The simplest illustrative example is that of the steady state growth of a cylindrical cap of radius $R$ as illustrated in Fig. 30b. In this case we know that the total rate of creation of surface over the entire cap is $2\gamma' V$, where $\gamma'$ is the specific excess free energy of the side faces of the slab and the total rate of formation of new phase volume is $2RV$. Thus, the average excess free energy per unit volume is given by $\gamma'/R$, i.e., it conforms to Eq. (69) and one would anticipate that the growing surface was isothermal. However, as we shall see from the point-to-point analysis, the surface is not isothermal, so that the rate of growth of the plate will be strongly affected (Bolling and Tiller, 1961).

Considering Fig. 30b, let us focus our attention on the segment of surface at the angle $\theta$ and time $t_0$, i.e., the segment $ab$. In the time $\Delta t$, this becomes segment $cd$. Thus $\Delta A = cd - ab$ and $\Delta V = abcd$. The segment $ef$ is what $ab$ would have become if the surface grew as a cylinder instead of translating in the $Z$ direction at constant velocity $V$. It is apparent that $cd \neq ef$ but that $abcd \approx abef$ so that $\Delta G_E$ will not be given by Eq. (69) which applies to the case where $ab$ grows as a cylinder with a fixed center. In this example, the proper expression for $\Delta G_E$ is found to be (Tiller and Jindal, 1971)

$$\Delta G_E = (\gamma/R)[1 - \tan^2 \theta][1 + \Delta^*] \tag{74}$$

where $\Delta^*$ is a correction factor that is small for all values of $\theta$ except those approaching $\pi/2$.

From Eq. (74), we note that Eq. (69) holds in the vicinity of $\theta = 0$, but that $\Delta G_E$ decreases with increasing $\theta$ until it is zero at $\theta = \pi/4$ and then becomes negative for $\theta > \pi/4$. This seems to be a rather startling result at first glance, but can be rationalized in physical terms by examining Fig. 30b. If we focus our attention on the segment $BC$ and treat it as growing independently of any segment at larger values of $\theta$ (since it has the lead position), we note that the net creation of area in time $\Delta t$ is $(AB + AD + CD - BC)$ and can readily evaluate the local $\Delta A/\Delta V$. Moving to the next segment $CF$, its growth leads to the annihilation of the surface $DC$ and a net creation of area of $(DE - CF + EF - DC)$. Thus, as we move to larger values of $\theta$, we find that the net creation of surface in any segment shrinks and that we have an effective "propagation" of surface from small $\theta$ to large $\theta$. This effect is even more obvious if we consider the surface as being composed of the edges of layers which translate in the $X$ direction (unit advance in the $Z$ direction requires unit creation of edge at $\theta = 0$).

The above procedure is found to be exact for the calculation of $\Delta A$ for any steady state two-dimensional interface shape. Further, the calculation of $\Delta V$ is expected to be correct, to first order, for any shape. Thus, Eq. (74) may be used for any steady state two-dimensional interface shape provided that $R$ is the radius of curvature at $\theta$, $R(\theta)$. This approach can be extended, by inspection, to the general three-dimensional steady state surface yielding, for isotropic $\gamma$,

$$\Delta G_E = \gamma\{(1/R_1)[1-\tan^2\theta][1+\Delta_\theta{}^*] + (1/R_2)[1-\tan^2\phi][1+\Delta_\phi{}^*]\}$$
(75)

where $\phi$ is the angular variation of the surface element in the plane perpendicular to the $(\theta, Z)$ plane.

## 3. Time–Varying Shapes

As an example of a time-varying surface shape to illustrate the error of using Eq. (69), consider the periodic interface of Fig. 30c (cylindrical in perpendicular direction). At time $t = 0$, let the shape be given as

$$Z(0) = Z_0 + \delta \sin \omega x$$

and to be moving with a velocity $\dot{Z}$ given by

$$\dot{Z} = V_0 + \dot{\delta} \sin \omega x$$

where $V_0$ is a constant so that at time $t$ the shape is given by

$$Z(t) = Z_0 + V_0 t + \left[\delta + \int \dot{\delta} dt\right] \sin \omega x$$

If we had used Eq. (69) to account for the excess free energy of surface creation, the total excess free energy in some length $b - a$ (see Fig. 30c) of the interface would for isotropic $\gamma$ be given by

$$\Delta G_T = \gamma \int_a^b \frac{Z''}{(1+Z'^2)^{1/2}} dl = \gamma [Z'(b) - Z'(a)]$$
(76)

where $l$ denotes length along the surface. If, in Eq. (76), we set $b - a = n\lambda$, then $\Delta G_T = 0$. However, we know that the system free energy has increased over that of a flat surface by $\gamma[l(b)-l(a)]$ and we see that the use of Eq. (69) does not properly account for the change in total area of surface as the interface changes shape. From the previous section, we saw that it did not account for the local creation of surface either. Tiller and Jindal (1971) have shown that for $\delta$ and $\omega$ such that $Z'$ is small ($\delta\omega \ll 1$),

$$\Delta G_E = \gamma\varkappa\left[1 + \left(\frac{\alpha\lambda\beta^2}{2\delta}\right)\frac{\cos^2\omega x}{\sin\omega x(1+\beta\sin\omega x)^2}\right] \qquad (77)$$

where $\beta = \delta/V_0$, $\lambda$ is the lattice spacing, and $\alpha$ is a constant greater than unity. The second term, being an even function in $x$ leads to the net creation of surface if one integrates over an integral number of wavelengths. Provided that $\beta$ is not negligible, the second term can be much larger than the first, especially for small $\delta$.

From the foregoing, it should be clear that the inclusion of this new factor into $\Delta G_E$ will significantly alter the time-response of the interface with respect to changes in interface shape compared to the calculations made in the past (Sekerka, 1968). Further, since the second term in Eq. (77) is proportional to $\delta^2$, it indicates that perturbation analyses should be carried out to second order in $\delta$ in order to include all the important physics of the problem and thus make a careful assessment of interface stability (the surface area is second order in $\delta$). Of course, it can be properly argued that $\overset{\circ}{\delta} = 0$ at the instability limit; however, to observe any physical manifestations of instability $\overset{\circ}{\delta} > 0$ and the second term in Eq. (77) becomes significant. Perhaps the most important consequence of this change in $\Delta G_E$ is that the perturbation problem becomes nonlinear and the analysis of an arbitrary distortion by considering the response of individual surface waves is not rigorously proper (surface area is not an algebraic quantity conforming to constructive and destructive interference). Thus, since surface area is an absolute quantity, it is necessary to treat the growth of a single bump (or distortion) on a flat surface to evaluate the stability of the bump. The bump must form initially via a nucleation type of process (Tarshis and Tiller, 1967). Since the probability of forming the initial bump via a nucleation event will probably be very small on a flat surface, the initial distortion may require some special surface site to aid its formation, i.e., perhaps the curved regions of reentrant surface segments bounding defects of one sort or another on the interface.

When one extends these considerations to include anisotropic surface energies, the situation is somewhat more complicated; however, the foregoing conclusions still hold. In addition, one finds terms depending upon the ratio $(\partial\gamma/\partial\theta)/\gamma$ which are larger in magnitude than the isotropic $\gamma$ terms provided the surface has a steep cusp in the $\gamma$-plot with $\theta$ (Tiller and Jindal, 1971). The foregoing conclusions apply most strongly to such surfaces.

## B. Layer Model

### 1. SURFACE CREATION ON GROWING BODIES

Using a continuum model for the interface, one finds that even for stationary state shapes such as are found during cellular or eutectic solidification, continuous surface creation occurs at the regions of positive curvature and continuous annihilation occurs at the regions of negative curvature. If we begin to consider a layer model of the interface, such as is illustrated in Fig. 31 a–d, the situation is more realistic and more interesting. In (a) to (c), the surfaces are plane curves, whereas in (d), the surface is considered to be roughly a body of revolution. In both case (a), where the layers are growing in the axial direction, and in case (b), where the layers are growing perpendicular to the axis of the body, the creation of new surface occurs entirely in

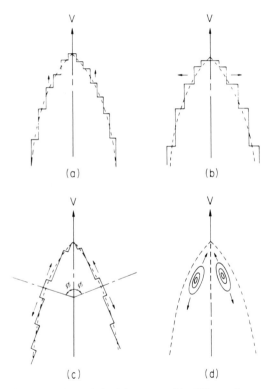

**Fig. 31.** Schematic illustration of dendrite caps with different layer configurations and growth directions, (a) plane curve, layers growing axially, (b) plane curve, layers growing perpendicular to axis, (c) plane curve, layers growing at angle $(\pi/2 - \phi)$ to axis, and (d) body of revolution, layers growing at angle to axis.

the axial region and is propagated to the outer edges of the body. In case (c), the layers are growing in a direction $(\pi/2 - \phi)$ to the axis and we have surface creation occurring at the poles in the $\phi$-directions and annihilation of half of the total created surface at the specimen axis. Of these three choices, case (c) is the most favorable for dendritic growth (everything else being equal) since it produces a heat dipole at the surface causing a temperature change in the opposite direction to that used in the continuum picture. In addition, the undercooling $\Delta T_K$ to drive the interface attachment at layer edges is minimum at the tip and maximum at the $\phi$-poles. Therefore, both effects tend to make the dendrite tip nonisothermal but in the opposite direction from that presently used in our continuum analyses, i.e., temperature is greatest at the tip with the present model. This configuration of layers would allow the body to transform at a greater rate than for other layer configurations.

The situation illustrated in case (d) is analogous to case (c) except that it is a body of revolution rather than a plate. Again, this is a highly favored layer configuration because of the advantageous temperature distribution it produces. In essence, if this is an fcc crystal lattice, there are four point sinks for free energy (sources of layer edges) at the $\langle 111 \rangle$ poles, one point source of comparable strength (sink for layer edges) at the axial $\langle 100 \rangle$ pole and four line sources of half strength (sink for layer edges) along the four 110 lines formed by the $\{111\}$ intersections. It is obvious that these considerations will make a significant alteration in the calculated dendrite velocity and tip radius (Bolling and Tiller, 1961) and will open the door to a more meaningful assessment of crystallographic features in cellular or eutectic solidification than has been possible heretofore.

## 2. EDGE INTERACTION

To properly evaluate the variation of $\gamma$ with orientation, it is necessary to describe both the density of layer edges as a function of orientation and the interaction energy between adjacent edges. The excess free energy of unit area of surface that makes an angle $\theta$ with its layer plane is given by

$$\gamma(\theta) = \gamma_F \cos \theta + (\gamma_E/h) \sin \theta + \Delta \gamma_E(\theta) \tag{78}$$

where $\gamma_F$ is the surface tension of the layer plane, $\gamma_E$ is the surface tension of an edge of height $h$ on that plane, and $\Delta \gamma_E(\theta)$ is the interaction energy associated with these multiple layers. Earlier in this article, we have discussed the various components of $\gamma_F$ and $\gamma_E$; now we must turn our attention to $\Delta \gamma_E$, the magnitude of which depends upon the dominant force-distance law operative in the system.

To illustrate the importance of the force-distance law in evaluating $\Delta \gamma_E$, let us suppose it is such that only 1st N.N. bonds are important and we consider

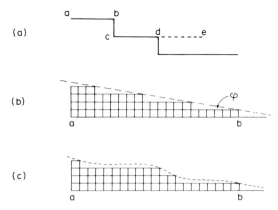

**Fig. 32.** Schematic representation of macroscopic interface (dashed line) created by microscopic layer structure and variation of macroscopic shape by variation of layer adjustment. Interaction of layer edges is developed from construction in (a).

the interface between a simple cubic crystal and its melt, as illustrated in Fig. 32. If the interface makes an angle $\phi$ with a layer plane on a macroscopic scale, then it can curve to any degree on a microscopic scale (changing its area significantly) without altering the bond energy count (perhaps this is why shape perturbations can develop easily at certain interface locations and not at others). Of course, these microscopic shape excursions lead to configurational entropy changes and thus provide a free energy factor that is important to the overall value of $\gamma(\theta)$. On such a 1st N.N. picture, one can show that a sphere of volume $v$ has a higher surface energy than a cube of the same volume. This is consistent with the fact that the equilibrium form of an S.C. crystal is predicted to be a cube rather than a sphere. As we allow the force-distance law to expand to more than first nearest neighbors, we note that normal to vicinal faces of a crystal, the formation of shape perturbations (probably of skewed shape) may not require the creation of surface in the conventional sense, but probably requires changes in $\gamma$ via the $\Delta\gamma_E(\theta)$ contribution in Eq. (78). However, for perturbation development normal to singular faces (layer planes), the creation of surface in the conventional sense is required and shape distortions are difficult to form. This important distinction can be more clearly understood by consideration of Fig. 33.

In Fig. 33, cylindrical half-wave distortions of the form

$$Z = \delta \sin \phi x, \qquad 0 \leqslant X \leqslant \lambda/2 \tag{79}$$

are given for the case where the layer plane is parallel to the solid–liquid interface (case a) and for the case where the layer plane makes an angle $\phi$ with the solid–liquid interface (case b). In the latter case, a physical limitation exists on the ratio of $\delta$ to $\lambda$ since the cap $Z$ may not penetrate the plane $A$-$A$ without significant change in energy. For case (a), the energy required to create the cap of dimensions given by Eq. (79) are

$$\Delta G = \frac{2\delta}{h}\gamma_E + \int_0^{\lambda/2} \Delta\gamma_E(\theta)\left[\frac{dl(\theta)}{dx}\right]dx$$

$$= \frac{2\delta}{h}\gamma_E + 2\int_0^{\lambda/4} \Delta\gamma_E(x)(1+Z'^2)^{\frac{1}{2}}\,dx \qquad (80a)$$

where

$$\Delta\gamma_E(x) \equiv \Delta\gamma_E(d) \qquad \text{with} \quad d = h/Z' \qquad (80b)$$

In Eq. (80b), $d$ is the spacing between edges. For the case (b), the energy required to create such a cap is given by

$$\Delta G = \int_0^{\lambda/2} \Delta\gamma_E(x)(1+Z'^2)^{\frac{1}{2}}\,dx \qquad (81a)$$

where

$$\Delta\gamma_E(x) \equiv \Delta\gamma_E(d) \qquad \text{with} \quad d = h\cot\psi$$

and $\qquad\qquad\qquad\qquad\qquad\qquad\qquad\qquad\qquad\qquad\qquad (81b)$

$$\psi = \theta + \phi \qquad \text{where} \quad \theta = \tan^{-1}Z'$$

Thus, we see that cap formation for case a requires considerable more energy than for case (b) and will thus be correspondingly more difficult to form by natural processes. This accounts, in part, for the stabilization of singular faces of crystals during growth from solution in comparison with the relatively easy breakdown of vicinal faces. If one considers caps of circular symmetry,

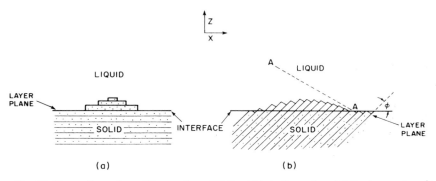

**Fig. 33.** Formation of a solid cap at a solid–liquid interface when (a) the layer plane is parallel to the interface and (b) the layer plane makes an angle $\phi$ with the interface.

then new layer edge is exposed for both cases and an additional increase in free energy of the order of Eq. (80) is required for both cases. If there are low energy lines in the $\gamma_E$ versus orientation plot, these caps are likely to be bounded by such directions. At this point, it should be clear to the reader how one proceeds to evaluate $\Delta G$ for a general cap of arbitrary shape. To complete the picture, we now need to gain an expression for $\Delta\gamma_E(d)$.

Considering Fig. 32b, the distance $d$ between steps is given by $na$, where $n = (h/a)\cot\phi$ and $a$ is the intermolecular distance. If one includes dipole effects and other Coulomb effects as well as the normal van der Waals attractive forces and diffuse interface effects, the force between layer edges could be either attractive or repulsive depending upon which factor dominates. However, if we follow Landau (1967) and restrict our attention to the interaction between two electrically neutral groups of atoms at a distance $d$ via van der Waals attractive forces, then the layers repel each other. To see this, consider the two steps in Fig. 32a. Their interaction is related to the finiteness of the distance $cd$, i.e., to the absence of interaction of atoms on $ab$ with the atoms which would be on $de$ if the second edge did not exist. It is not difficult to determine the dependence of this repulsion on $n$. The van der Waals interaction of two atoms at a distance $r$ is proportional to $r^{-6}$; summation over all pairs of atoms (which may be replaced by an integration over $ab$ and $de$) gives a quantity proportional to $n^{-3}$ provided we include the interaction with all molecules in the layer edge normal to the figure. Let the proportionality constant be $A$ $(A > 0)$, then the interaction energy $\Delta E$ of two steps is given by

$$\Delta E = An^{-3} \tag{82}$$

The additional surface tension $\Delta\gamma_E(\phi)$ due to the interaction of all the steps is thus

$$\Delta\gamma_E(\phi) = \frac{1}{an\cos\phi}\left(\frac{A}{n^3} + \frac{A}{(2n)^3} + \frac{A}{(3n)^3} + \cdots\right) \tag{83a}$$

$$= \frac{1.2A}{an^4\cos\phi} = \frac{1.2a^3 A\tan^4\phi}{h^4\cos\phi}; \qquad \phi < \pi/4 \tag{83b}$$

The evaluation of $A$ is straightforward and the expansion of this procedure to include other force-distance laws can be simply made. It is interesting to note that $\Delta\gamma_E(\phi)$ increases strongly with $\phi$ up to $\phi = \pi/4$ and then for $\phi > \pi/4$ the edges and faces reverse position and the same analysis follows, i.e., $\Delta\gamma_E(\phi)$ is symmetrical about $\phi = \pi/4$ for simple systems.

ACKNOWLEDGMENT

This work was supported by the Air Force Office of Scientific Research, Grant AF-AFOSR-68-1360.

# References

Abraham, F. F. (1969). *J. Chem. Phys.* **50**, 3977.
Bolling, G. F., and Tiller, W. A. (1960). *J. Appl. Phys.* **31**, 1345.
Bolling, G. F., and Tiller, W. A. (1961). *J. Appl. Phys.* **32**, 2587.
Bradshaw, F. J., Gasper, M. E., and Pearson, S. (1958). *J. Inst. Metals* **87**, 15.
Burton, W. K., Cabrera, N., and Frank, D. D. (1949). *Discuss Faraday Soc.* **5**, 33.
Cahn, J. W. (1960). *Acta Met.* **8**, 554.
Cahn, J. W., and Hilliard, J. E. (1958). *J. Chem. Phys.* **28**, 258.
Cahn, J. W., and Kikuchi, R. (1961). *J. Phys. Chem. Solids* **20**, 94.
Cahn, J. W., Hillig, W. B., and Sears, G. W. (1964). *Acta Met.* **12**, 1421.
Chadwick, G. A. (1970). Private communication.
Cohen, M. H., and Turnbull, D. (1959). *J. Chem. Phys.* **31**, 1164.
Dunning, W. J. (1966). *In* "The Solid Gas Interface" (E. A. Flood, ed.), p. 271. Dekker, New York.
Fowkes, F. (1967). "Treatise on Adhesion and Adhesives" (R. L. Patrick, ed.), Vol. 1, p. 325. Dekker, New York.
Fowler, R., and Guggenheim, E. A. (1936). "Statistical Thermodynamics," p. 445. Cambridge Univ. Press, London and New York.
Garbuny, M. (1965). "Optical Physics," p. 267. Academic Press, New York.
Glicksman, M. E., and Childs, W. J. (1962). *Acta Met.* **60**, 925.
Glicksman, M. E., and Vold, C. L. (1968). "The Solidification of Metals," p. 37. Iron and Steel Inst., London.
Gomer, R., and Schrieffer, R. (1969). Private communication.
Herring, C. (1951). *In* "The Physics of Powder Metallurgy" (W. E. Kingston, ed.), p.153. McGraw-Hill, New York.
Hildebrand, J. N., and Scott, R. L. (1962). "Regular Solutions," Prentice-Hall, Englewood Cliffs, New Jersey.
Hillig, W. B., (1958). *In* "Growth and Perfection of Crystals" (R. H. Doremus et al., eds.) p. 350 Wiley, New York.
Holloman, J. H., and Turnbull, D. (1950). "The Solidification of Metals and Alloys." Amer. Soc. Metals, Cleveland, Ohio.
Jackson, K. A. (1958). "Liquid Metals and Solidification." Amer. Soc. Metals, Cleveland, Ohio.
Landau, L. D. (1967). "Collected Papers of L. D. Landau." Gordon & Breach, New York.
Lifshitz, E. M. (1956). *Soviet Physics, JETP,* **2**, 73.
Miller, W. A., and Chadwick, G. A. (1967). *Acta Met.* **15**, 607.
Mullins, W. W. (1959). *Acta Met.* **7**, 746.
Mullins, W. W. (1963). *In* "Metal Surfaces," Chapter 2, p. 18. Amer. Soc. Metals, Metals Park, Ohio.
Mutaftschiev, B., and Kern, R. (1964). *C. R. Acad. Sci.* **259**, 572.
Mutaftschiev, B., and Kern, R. (1965). *C. R. Acad. Sci.* **260**, 533.
Nason, D. (1970). Ph.D. Thesis, Stanford University.
Nicholas, F. J. (1968). *Aust. J. Phys.* **21**, 21.
Onsager, L. (1944). *Phys. Rev.* **65**, 117.
Plenum Press. (1964). "Handbook of High Temperature Materials." Plenum Press, New York.
Pound, G. M., and LaMer, K. V. (1952). *J. Amer. Chem. Soc.* **74**, 2323.
Samsonov, G. P. (1964). "Handbook of Thermionic Properties." Plenum Press, New York.
Sekerka, R. F. (1968). *J. Cryst. Growth* **34**, 71.

Sundquist, B. E., and Mondolfo, L. F. (1961). *Trans. AIME* **221**, 157.

Tarshis, L. (1967). Ph.D Thesis, Stanford University.

Tarshis, L., and Tiller, W. A. (1967). *In* "Crystal Growth" (H. S. Peiser, ed.), p. 709. Pergamon, Oxford.

Temkin, D. E. (1966). *In* "Crystallization Processes" (D. D. Sirota *et al.*, ed.), p. 15. Consultants Bureau, New York,

Tiller W. A. (1964). *Science* **146**, 871.

Tiller, W. A. (1969). *Scr. Met.* **3**, 273.

Tiller, W. A., and Jindal, B. (1972). *Acta Met.* To be published.

Tiller, W. A., and Takahashi, T. (1969). *Acta Met.* **17**, 483.

Turnbull, D. (1950). *J. Appl. Phys.* **21**, 1022.

Turnbull, D., and Vonnegut, B. (1952). *Ind. Eng. Chem.* **44**, 1292.

Verwey, E. J. W., and Overbeek, J. T. G. (1948). "Theory of the Stability of Lyophobic Colloids." Elsevier, Amsterdam.

Wagner, R. S. (1962). "Metallurgy of Advanced Electronic Materials," AIME Conf. 19, p. 239. Wiley (Interscience), New York.

# Fracture of Composites

## A. S. ARGON

*Massachusetts Institute of Technology*
*Cambridge, Massachusetts*

## I. Introduction

The *fasces*, a bundle of birch rods tied together by a string, was a symbol of strength in ancient Rome, and the idea that the strength of a group of individuals is often greater than the sum of their individual strengths has been used not only for inanimate objects but also for societies.

Since the first use of twisted rope which is probably the earliest man made "composite" material, man has used composites widely and ingeniously throughout his civilized existence. It is only now, however, that some of the strength properties of composites are beginning to be understood, making rational predictions and new developments possible. In the following we will present some interesting recent developments in the understanding of the complex interactive phenomena in the strength of some filamentary and laminar composites. Since there have been a number of other recent assessments of the field by others, in this article we will concentrate particularly on the statistical aspects of the problem of strength of composites. For other general reviews, see, for instance, Corten (1967), Tetelman (1969), and Rosen and Dow (1972).

Most engineering materials, from reinforced concrete to age-hardened alloys, are composites of some kind in which the strength depends sensitively on the nature and dispersion of the second phase. A general rational introduction to this problem has been given by Kelly (1966). Here we will discuss the strength of only a very small class of composites made from a stiff, strong, and brittle reinforcing component and a matrix component having the capacity to undergo much larger elastic or plastic strains than can be carried by the reinforcement at its point of fracture. Even in this more restricted category the problems for which a reasonable understanding can be developed are limited either to the more regular laminates[†] and continuous element fibrous composites or to the quasi-isotropic quasi-homogeneous materials which can be treated by the well-developed methods of fracture mechanics.

We will first give a qualitative discussion of some of the more important modes of failure of composites in the restricted group described above. In the main body of the paper the following subjects will then be discussed in detail: the effect on its surroundings of an isolated failure in the initially continuous reinforcement, the statistical sequential process of subcritical crack growth leading to general fracture instability, the size effect on the strength of the composite, conditions for maximum strength, effect of element variability on the strength of the composite, the strength of a bundle, failure and fracture of composites in compression, and fatigue of composites.

## II. Types of Fracture in Composites

As discussed above, we will limit our attention to composites made from brittle fibrous reinforcement, in either continuous or discontinuous form, or alternatively to composites made from brittle sheet reinforcement—being bonded in both cases by a more extensible matrix than the reinforcement. We

---

[†] We will restrict the word "laminate" in this paper to ideal layered composites made by alternate planar layers (lamina) of reinforcement and matrix. Whenever the developments are also explicitly or implicitly extended to fibrous composites the word "composite" will be used.

will only consider in detail composites in sheet form where the thickness dimension will be much less than the next larger dimension of the body. This is not a severe limitation since much of the engineering interest in composites is in thin walled light shells and structures made from sheets. In such forms, in nearly all cases the planar properties of the sheets are of primary interest.

The simplest composite to consider is a two-dimensional laminate of alternating sheets of a reinforcing phase and a matrix phase, stressed by forces in the plane of the laminate.

We assume that in common with all brittle materials the strength of the reinforcing phase of the composite is governed by flaws distributed on the surface (or throughout the volume) and resulting either from manufacturing techniques or from handling. If the reinforcing phase were to be tested in tension individually, it would fail in a brittle manner from the worst of these flaws. It is well known that in this case not only is the strength of the individuals subject to scatter, but the average strength and scatter increases with decreasing size of the reinforcing element tested. If a collection of reinforcing elements of the same length were stressed in parallel as a bundle it is quite clear that the average fracture strength of the bundle will be neither as poor as the weakest element nor as good as the strongest element. Clearly the answer must also depend on the number of elements in the bundle since fracture in one element among a few would produce a larger increase of average stress in the remainder than for the case of a large number of elements in the bundle. In this case of an unbonded bundle the coupling between the elements is minimal. If the elements of the bundle are glued together or are embedded in an extensible matrix, it would appear intuitively obvious that the fractures in the weakest elements will be more effectively bridged by the adjoining elements and the composite strength should be improved. It will become clear later that this is not always true and that in some cases the strong mechanical coupling will actually result in a lower composite strength than is obtainable by a weaker coupling.

In either strong or weak coupling of elements, composite fracture under increasing stress will involve a certain amount of stable accumulation of damage in the reinforcing elements leading to an eventual fracture instability where the fracture of another element produces enough elevation of stress to give rise to an increasingly rapid succession of fractures of the remaining elements. In the weakly coupled bundle, the damage will accumulate in a nearly random manner with little correlation, while in a strongly coupled system of a composite with a matrix, the fractures in the elements will be correlated and will tend to spread out from a few weak primary origins across elements, in a crack-like manner. In the latter case the final fracture instability is of Griffith type in which the product of the number of adjacent fractured elements raised to some power and the fracture stress of the composite can be expected to be a constant. It is possible to view this fracture process in strongly coupled composites in another way as well. By a linear fracture mechanics approach, it can be expected that for any level of applied stress a fracture instability is possible if a certain critical number of adjacent fractures occur in neighboring elements. At low stress where the required critical number of adjacent fractures is large, this number is unlikely to develop by a sequential process of fractures emanating from a weak point in one element. As the stress is increased, requiring fewer adjacent fractured elements for instability, it becomes more and more likely that this number can be produced by a sequential process of coupled fractures in adjacent elements starting from a weak point in one element. Hence the problem of interest is the development of critical effective cracks in composites by sequential fractures in adjacent elements. In weakly coupled composites the final fracture instability would be preceded by a more random damage accumulation starting out from a large number of weak centers than the case in strongly coupled composites.

Fracture in composites having aligned, discontinuous elements of large aspect ratio is rather similar to the process discussed above, as was already pointed out by Zweben (1969). There the initially discontinuous elements behave as those in the continuous element composite subjected to a prestress which has produced some stable fragmentation.

Tensile fracture in fibrous composites involves spreading of the fracture in two dimensions, but is otherwise qualitatively similar to the one-dimensional spread of fracture in laminates. Quantitatively there are, however, important differences in the nature of the stress coupling of elements at isolated fractures, and in the number of modes of fracture propagation from element to element.

The process of accumulation of a critical amount of damage in composites with fibres aligned in more than one direction, leading to fracture instability, must be considerably more complex than the cases discussed above, since the coupling interactions also involve large rotations of fiber elements and interfacial couples between layers of fibers in addition to inhomogeneous extensions. In composites with random fiber mats, additional complexities of mechanical locking between fibers must also be present. The differential geometry of the interpenetrating deformation of the reinforcement during fracture would appear to be even more complicated than the corresponding problems in extension of elements in twisted yarn (see, e.g., McClintock and Argon, 1966, Chapter 22). In view of the great difficulties of the deterministic understanding of propagation of fracture processes in such materials, it is often more profitable to adopt a phenomenological fracture mechanics approach dealing with the scaling laws of only the conditions of fracture instability as discussed by Tetelman (1969).

Thin tensile panels of composite materials may fail in compression by overall buckling. If buckling is suppressed by appropriate stiffening, failure of the composite can occur by internal buckling or buckling of reinforcing elements by undergoing relative shear, bending, and rotation in a planar region. Once such a planar failure nucleus is formed it can spread by its own stress concentration much like a deformation twin or kink band in a metal crystal, producing delamination and element fracture in bending inside the deformation zone.

Composites subjected to repeated applications of stress can fatigue in a variety of mechanisms. In composites with uniformly strong reinforcing elements, failure may occur by fatigue in the matrix producing excessive delamination and loss of bending stiffness. In composites with considerable variability in element strength the initial maximum excursion of the tensile stress will produce a number of isolated fractures in the reinforcing elements. Repeated stress cycling then can produce delamination by matrix fatigue along the element–matrix interface between some of the isolated element fractures. Once enough of these isolated fractures have been joined by delamination to produce an effective critical transverse crack, the composite will fracture in a brittle manner.

These mechanisms of fracture of laminates will now be developed in detail below.

## III. Tensile Fracture in Laminates

### A. The Stress Singularity and the Statistically Independent Length

Consider a two-dimensional laminate of parallel reinforcing sheets and intervening extensible matrix material subjected to a tensile stress in the plane. Because of the strongly stress-dependent nature of fracture of the reinforcing phase, it can be expected that the mode of failure will consist of propagation of fracture from element to element as shown in Fig. 1. Clearly once such a crack has arisen by whatever process necessary, the deformation will be concentrated at the points A and A'. Whether the extensible matrix is elastic with a low modulus or plastic with a finite yield stress, a stress concentration will be produced on the two elements immediately facing the tip of the crack, making fracture in these elements more likely at points A and A' than would have been the case if the crack had not been there. We are therefore interested first in the distribution of the overstress along the crack tip elements. This problem has been considered by McClintock for both the case of the elastic and plastic matrix. Using the approach of Sih and Liebowitz (1968) and Paris and Sih (1965), for a planar, anisotropic elastic material, McClintock

(1969) has derived the distribution of tensile stress $\sigma_{22}$ ahead of a crack of length $2c$ to be

$$\sigma_{22}^e = \sigma + \frac{k_1}{2}\frac{\sqrt{x_1 + \sqrt{x_1{}^2 + \left(\dfrac{a_{22}}{a_{66}}\right)x_2{}^2}}}{\sqrt{x_1{}^2 + \left(\dfrac{a_{22}}{a_{66}}\right)x_2{}^2}} \tag{1}$$

with

$$a_{22} = \frac{1}{(1-V_r)E_m + V_r E_r} \simeq \frac{1}{V_r E_r}, \qquad a_{66} = \frac{(1-V_r)}{G_m} + \frac{V_r'}{G_r} \simeq \frac{(1-V_r)}{G_m}$$

where $V_r$ is the volume fraction of the reinforcing material, $G_m$ the shear modulus of the matrix material, $E_r$ the Young's modulus of the reinforcing material, and $k_1 = \sigma\sqrt{c}$ is the Mode I stress intensity factor. Clearly, there is a stress singularity at the crack tip in this quasi-homogeneous idealization of the laminate. In reality the singularity would be eliminated by blunting of the crack in the extensible elastic matrix adhering to the front faces of the two reinforcing elements framing the crack. The distribution of this stress along the first unfractured element at $x_1 = (t_r + t_m)/2$ is shown in Fig. 1.

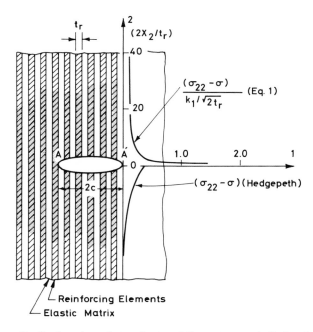

**Fig. 1.** Stress distribution along first unfractured element at crack tip in a laminate with an elastic matrix.

Adapting the analysis of Rice (1967) for planar yielding in front of a Mode III crack to the case of the laminate which can yield only in the matrix layers between the planar reinforcing elements, McClintock (1969) obtains a tensile stress distribution

$$\sigma_{22}^P = \sigma + \frac{2k}{\pi}\left(\frac{a_{66}}{a_{22}}\right)^{1/2} \ln\left\{\frac{\pi^2 (k_1/k)^2 (a_{22}/a_{66})}{4[x_1^2 + (a_{22}/a_{66}) x_2^2]^{1/2}}\right\} \tag{2}$$

ahead of the crack, where $k = $ is the yield stress in shear of the matrix material, and all other symbols have the same meaning as before. As shown in Fig. 2, in this case a planar plastic zone extends along the first reinforcing elements at the crack tip for a distance

$$d_p = (\pi^2 k_1^2/16k^2)\sqrt{\frac{a_{22}}{a_{66}}} \tag{3}$$

along which the shear stress $\sigma_{12} = k$ remains constant. The distribution of the tensile stress $\sigma_{22}$ along the first unfractured element at the crack tip is shown in Fig. 2.

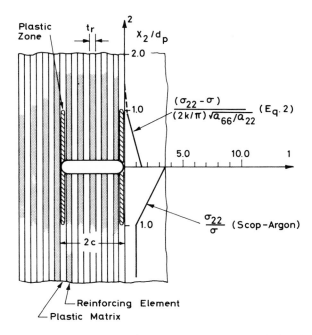

*Fig. 2.* Stress distribution along first unfractured element at crack tip in a laminate with a plastic matrix.

Other, less accurate analyses employing a shear lag approach were used in the past to characterize the stress distribution in front of cracks in laminates with elastic and plastic matrixes. For elastic matrixes Hedgepeth (1961) and Hedgepeth and Van Dyke (1967) have found that the tensile stress in the first element on either side of a crack spanning all adjacent fractured elements is given by a series of negative exponentials and can be readily obtained by numerical calculations. The resulting distribution of $\sigma_{22}$ along the first unfractured element at the crack tip is also shown in Fig. 1. Since in the shear lag analysis only overall equilibrium is observed there is no singularity in the tensile stress distribution as both $x_1$ and $x_2$ become very small. This, however, is not a serious limitation if the matrix phase does not separate from the reinforcement and can undergo large elastic strains to blunt the crack. For a plastic matrix a crude limit analysis has often been used in which all the intercepted force by the adjacent fractured elements along the length $2c$ has been assumed to be applied to only the two elements on either side of the crack leading to a tensile stress distribution in these elements of

$$\sigma_{22}(x_2) = \sigma\left[\left(1+\frac{2c}{t_r+t_m}\right)-\left(\frac{2c}{t_r+t_m}\right)\frac{x_2}{d_p}\right] \tag{4}$$

where $t_r$ and $t_m$ are the thicknesses of the reinforcing and matrix layers, respectively, and

$$d_p = c\sigma/k \tag{4a}$$

is the extent of the plastic zone parallel to the reinforcing elements at the crack tip. The distribution of stress given by Eq. 4 is also given in Fig. 2.

Whatever the level of the analysis described above, the extent of the stress singularity $d_e$ or the extent of the plastic zone $d_p$ in the laminates with elastic or plastic matrixes gives the influence range of a stress inhomogeneity caused by one or more adjacent fractures of reinforcing elements. Once a crack-like fracture nucleus is formed as shown in Fig. 1 or 2, and the stress in the two elements on either side of the nucleus has been elevated in a region $\delta = 2d$ above the nominal stress elsewhere, further fracture processes are much more likely to be concentrated in this laminate layer of length $\delta$ at the extension of the existing crack nucleus. Hence, as was pointed out first by Gücer and Gurland (1962), and extensively used by Rosen and co-workers (see Rosen, 1964), it is possible to consider the stressed laminate of total length $L$ as a series of $n = L/\delta$ statistically independent links connected in series as shown in Fig. 3, in each of which independent fracture nucleation and growth processes can occur.

In this idealization of the laminate in the form of $n$ series chain links, failure of the chain occurs when at least one of the links is broken. If we let $F(\sigma)$ be

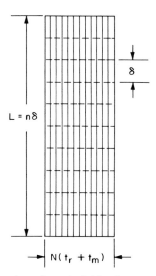

**Fig. 3.** Schematic representation of a uniaxial laminate as a chain of $n$ links in series.

the probability that a link or a strip of length $\delta$ breaks when the applied stress is equal to or less than $\sigma$, then it follows directly that the probability $G(\sigma)$ for the failure of the entire chain at a stress equal to or less than $\sigma$ is given by the well known chain-link formula

$$G(\sigma) = 1 - [1 - F(\sigma)]^n \tag{5}$$

(see, e.g., McClintock and Argon, 1966).

In the following sections we will consider in detail the processes affecting fracture in a statistically independent layer of length $\delta$.

## B. Crack Formation by Correlated Fracture of Elements

### 1. GENERAL CONSIDERATIONS

Consider now a laminate of width $w$ made up of $N$ alternating parallel sheets of reinforcing elements of thickness $t_r$ and $(N-1)$ sheets of extensible matrix of thickness $t_m$. Following Scop and Argon (1969) we will idealize, from this point on, the laminate by merely the $N$ parallel reinforcing elements without any matrix in between, save for the characteristic interfacial tractions which the matrix produces between the reinforcing elements which, of course, are retained. In this model any calculated laminate strength is for 100%

reinforcement, from which the actual strength of the laminate can be found by multiplication with the actual volume fraction of the reinforcing phase. We assume that the reinforcing elements have a surface density of strength limiting flaws characterizable by a simple distribution function $\xi(\sigma) = C\sigma^m$ (see, e.g., Weibull, 1939a, b), giving the number of flaws per unit surface area limiting the tensile strength to $\sigma$ or less. Both the exponent $m$ and the coefficient $C$ can be determined experimentally by testing a large number of individual reinforcing elements.

When such a two-dimensional laminate made of elements with a certain variability in strength is subjected to a tensile stress $\sigma$ parallel to the reinforcing elements, isolated fractures can occur in the elements where the worst flaws are located. The fractured element will then unload over a certain length $\delta/2$ on each side of the break. The portion of the load carried by the fractured element is then transferred to the neighboring unfractured elements where it produces a longitudinal tensile stress concentration which was discussed in the preceding section and is shown in Figs. 1 and 2 for elastic and plastic matrices in the two adjacent elements on either side of the fractured element. The overstress in the adjacent elements increases the probability of fracture in the overstressed region of extent $\delta$ considerably above the level of fracture in an element of the same length with uniform stress. The increased probability of fracture in the overstressed region can be calculated by integrating the probabilities of fracture in each element of area of the overstressed region of the reinforcing element. It is now convenient to consider this increased probability as if it had been the result of a $K$-fold uniform rise in stress $\sigma$ in the affected region. For the case of $r$ adjacent fractured reinforcing elements, the factors $K_r$ are

$$K_r^e = \left[ \lim_{y \to \infty} \frac{1}{y\sigma^m} \int_0^y (\sigma_{er}(\sigma, x_2))^m \, dx_2 \right]^{1/m} \tag{6}$$

$$K_r^p = \left[ \frac{1}{d_p \sigma^m} \int_0^{dp} (\sigma_{pr}(\sigma, x_2))^m \, dx_2 \right]^{1/m} \tag{7}$$

for elastic and plastic matrices, respectively, where $\sigma_{er}$ and $\sigma_{pr}$ are given by Eqs. (1) and (2) for $x_1 = t_r/2$ in which $c = rt_r/2$. Since the factors $K_r$ involve a surface averaging of the enhanced stress, with the flaw distribution as a weighting function, they have been called "stress enhancement factors" by Scop and Argon (1969) to distinguish them from stress concentration factors. If the cumulative strength distribution $\xi$ of the surface density of flaws can be characterized by the simple power form given above, it can be seen from Eqs. (6) and (7) that the stress enhancement factors for the two elements on either side of a fracture nucleus depend only on the exponent $m$ and the number $r$ of adjacent fractured elements, but do not depend on the scale factor $C$ of the flaw distribution. Stress enhancement factors $K_1$, $K_2$, and $K_3$ have been calculated by Scop and Argon (1969) for elastic and plastic matrices for several types of flaw distributions using the shear-lag approach of stress analysis.

The calculation of the probability $F(\sigma)$ of fracture in a strip of $N$ parallel reinforcing elements of length $\delta$ requires the enumeration of very many separate and differently weighted modes of fracture. To investigate which of these modes are dominant, Scop and Argon (1969) have considered the fracture process in a strip of length $\delta$ in a way in which the initial stages of nucleation of fracture are accounted for exactly but the final stages where the fracture nears the condition of instability are treated approximately. Scop and Argon assume in this fracture propagation model that under an applied stress $\sigma$ or below, in a strip of $N$ parallel reinforcing elements, a small number of isolated primary fractures can occur. Each fractured element raises the stress in its two immediate neighbors to a new enhanced stress value, $K_1 \sigma$. It is assumed that a certain number of elements with enhanced stress $K_1 \sigma$ fracture at a stress which lies between $\sigma$ and $K_1 \sigma$. The fracture of these sheets representing a secondary stage can occur by several different paths. As a first alternative, all secondary fractures could occur around different initial fractures, i.e., none of the secondary fractures occur around the same primary fractures. As a second alternative, all secondary fractures could occur in pairs around the primary fractures. Finally, as a third alternative, the secondary fractures could divide up in various ways between single and double fractures around the primary fractures. They assumed then that after the various combinations of single and double secondary fractures of elements around the isolated primary fractures occur, the remaining elements in the strip will fracture simultaneously (in one way) between a stress of $\sigma$ and either $K_2 \sigma$ or $K_3 \sigma$ (where $K_3 > K_2 > K_1$), depending upon whether all the secondary fractures have been single or at least one pair has been double.

Computer experimentation by Scop and Argon with a large range of flaw distributions established that of the three paths of failure discussed above, the one for double secondary fractures, is always the most probable; furthermore, the greatest contribution to this mode comes from the first term describing the mode in which a single primary fracture is followed by simultaneous secondary fracture of the two neighboring sheets on both sides.

If the simple strip fracture model of Scop and Argon were to be generalized to include many successive steps in the fracture chain, it can be expected that the dominant mode will still be the one in which the fracture nucleus propagates by a symmetrical pairwise fracture process until, for the given applied stress, the required $r$ adjacent elements have fractured, creating an effective crack of length $2c = rt_t$ for the tensile fracture instability to set in.

At an applied stress $\sigma$ or less the probability $Q(\sigma)$ of fracture of an element of length $\delta$, and width $w$, having a flaw density distribution $\xi(\sigma) = C\sigma^m$ per unit surface area is

$$Q(\sigma) = 1 - \exp(-w\delta\xi(\sigma)) = 1 - \exp(-s) \tag{8}$$

where $s$ has been introduced as a useful nondimensional stress for future calculations. Similarly the probability of fracture of the first unfractured element on either side of a series of $r$ adjacent fractured elements is

$$Q(K_r\sigma) = 1 - \exp(-K_r^m s). \tag{9}$$

The probability of fracture $F(\sigma)$ in a strip of $N$ elements of length $\delta$, at a stress $\sigma$ or less, by the expected dominant pairwise fracture mode described above would then be

$$F(\sigma) = NA_0 A_1^2 A_3^2 \ldots A_r^2 (1-A_0)^{N-r} \tag{10}$$

where

$$A_0 = Q(\sigma)$$
$$A_1 = Q(K_1\sigma) - Q(\sigma)$$
$$A_3 = Q(K_3\sigma) - Q(\sigma)$$
$$\vdots \qquad \vdots \qquad \vdots$$
$$A_r = Q(K_r\sigma) - Q(\sigma) \tag{11}$$

are the probabilities of (1) fracture of an element at a stress of $\sigma$ or less; (2) fracture of an element at a stress between $\sigma$ and $K_1\sigma$; (3) fracture of an element at a stress between $\sigma$ and $K_3\sigma$; $\cdots$; and (4) fracture of an element at a stress between $\sigma$ and $K_r\sigma$, where the probability of fracture is very nearly unity. The last term in Eq. (10) represents the probability of fracture of the remaining $(N-r)$ elements simultaneously or in succession between a stress $\sigma$ and stresses above $K_r\sigma$. Since the cascade of fracture can start from any of the $N$ parallel elements, $F(\sigma)$ is obtained by multiplying the product of the probabilities by $N$.

The probability of fracture of the entire laminate is then obtained from the chain-link formula of Eq. (5). As the stress is increased $G(\sigma)$ rises rapidly in a narrow range of $\sigma$, between near zero to near unity. The value of $\sigma$ where $G(\sigma)$ reaches, say, $G_0 = 0.5$, can be picked to represent the fracture stress of the laminate. Hence the chain-link formula

$$G_0 = 1 - [1 - (NA_0 A_1^2 A_3^2 \ldots A_r^2 (1-A_0)^{N-r})]^n = 0.5 \tag{12}$$

becomes an implicit equation for the laminate strength $\sigma_c$.

The evaluation of the laminate strength by this method is tedious and is worthwhile only when accurate information is desired and when reliable information exists for the strength of the individual elements. In the absence of these conditions the somewhat less accurate but convenient simple method below will be more attractive.

## 2. A SIMPLIFIED MODEL

As discussed in the previous section, in their study of the dominant modes of fracture propagation, Scop and Argon (1969) considered a contracted fracture cascade which considered only a detailed accounting of the steps of fracture through the secondary stage and lumped the fracture of all other $N-3$ elements into one step, giving for Eq. (3.12)

$$G_0 = 1 - (1 - NA_0 A_1{}^2 A_3^{N-3})^n \approx nNA_0 A_1{}^2 A_3^{N-3} \qquad (12a)$$

where the above approximate equality is acceptable for small values of $G_0$.

Substitution of the appropriate forms of Eqs. (8) and (9) into Eqs. (11) and these into (12) gives

$$G_0 = nN(w\delta C\sigma^m)^N (K_1{}^m - 1)^2 (K_3{}^m - 1)^{N-3}$$

which simplifies further for very large $N$ into an explicit expression for the laminate strength

$$\sigma_c = f(N)[w\delta C(K_3{}^m - 1)]^{-1/m} \qquad (13)$$

where $f(N)$ which is near and less than unity depends on $N$, $m$, $n$, and asymptotically approaches unity for very large $N$. The length $\delta$ in Eq. (13) should be taken to be $2d_p = 2t_r\sigma/k$ for a plastic matrix with shear strength $k$. Equation (13) can be given in other forms which make it more useful for immediate application when the average tensile strength $\bar{S}_T$ and coefficient of variation $\alpha$ is known for elements of a large population for a constant test area $A_T$ (or test length $L_T$ for fibers). Thus,

$$\sigma_c = \left[ \frac{\Gamma(1 + 1/m)}{m^2} \frac{w\delta}{A_T} \frac{(K_3{}^m - 1)}{\bar{S}_T{}^m} \right]^{-1/m} \qquad \text{for elastic matrix} \qquad (13a)$$

$$\sigma_c = \left[ \frac{\Gamma(1 + 1/m)}{m^2} \left( \frac{2wt_r}{A_T} \right) \frac{(K_3{}^m - 1)}{k\bar{S}_T{}^m} \right]^{-1/(m+1)} \qquad \text{for plastic matrix} \qquad (13b)$$

where the exponent $m$ of the flaw density distribution can be obtained from the coefficient of variation by means of Fig. 4 (McClintock and Argon, 1966).

As will become clear, the simple expression developed above applies best to elements of small strength variability in which critical crack lengths are short. Since $f(N)$ was taken as unity, the expression also fails to show the effect of change in strength with increasing numbers of reinforcing elements to be discussed below. It nevertheless forms a very convenient explicit upper bound expression on the composite strength and will be used later in the discussion on the effect of strength variability of the reinforcing elements on the composite strength.

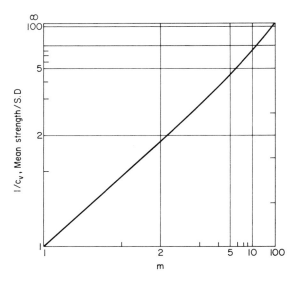

**Fig. 4.** Relation between the element coefficient of variation ($c_v$ = S.D./Mean) and the flaw distribution exponent $m$. (From McClintock and Argon, 1966, courtesy Addison-Wesley.)

The utility of the simple expression given above becomes clear from its good agreement with experimental results where enough information is available to afford accurate evaluation.

### 3. EXPERIMENTAL VERIFICATION

Proper verification of the theory with experiments requires not only accurate information on the density of surface flaws in the reinforcing elements, and the shear strength of the matrix, but also employment of lamination techniques which do not degrade the reinforcing elements or otherwise alter their individual properties.

Glass ribbons made by the Corning Co. lend themselves ideally for use in testing. Laminates of 5 and 10 elements of this material were made by Scop and Argon (1967) by gluing them together with an epoxy glue, the shear strength of which was measured by a lap shear experiment. The flaw distribution parameters $C$ and $m$ of the individual elements, and the composite strength $\sigma_c$, all measured in tensile experiments with a 1-inch gage length are shown in Table I together with the calculated strength values using the detailed theoretical model discussed in Section III, B, 1. As is clear, the agreement between theory and experiments is quite satisfactory.

Another set of experiments giving satisfactory agreement with our simple

**TABLE I**

EXPERIMENTAL AND THEORETICAL RESULTS FOR RIBBON GLASS LAMINATES[a, b]

| Sample type | No. of samples | Mean strength (ksi) | | Standard deviation (ksi) | |
|---|---|---|---|---|---|
| | | Exp. | Theory | Exp. | Theory |
| Single Sheets | 28 | 26.9 | 26.9 | 7.0 | 7.0 |
| 5 Layer Laminate | 7 | 27.8 | 28.1 | 2.6 | 2.2 |
| 10 Layer Laminate | 6 | 32.7 | 29.4 | 2.2 | 2.1 |

[a] From Scop and Argon (1967).
[b] Sheet dimensions: $w = 0.462$ in., $L = 1.0$ in., $t_r = 0.0017$ in., $\xi(\sigma) = 3 \times 10^{-9}\sigma^6$ in.$^{-2}$ ($\sigma$ in ksi), $k = 2.0$ ksi

theory is that of Armenakas *et al.* (1970) who have tested S-glass fibers (a) individually to obtain their flaw distribution parameters, (b) in unbonded bundles, and (c) in aligned composites with an Epon 828 Epoxy binder. Their experimental results and our theoretical strengths are shown in Table II. Examination of the results in Table II shows that the theoretical values lie consistently higher than the experimental results. One reason for this can be the uncertainty of the shear strength of the epoxy resin; another reason could be some inadvertant damage in lamination.

**TABLE II**

EXPERIMENTAL AND THEORETICAL RESULTS FOR THE STRENGTH OF S-GLASS FIBER LAMINATES[a]

$$\xi(\sigma) = \begin{cases} 2.9 \times 10^{-7}\sigma^{2.625} \text{ in.}^{-1} & \text{for } \dot{\varepsilon} = 0.0265 \text{ min}^{-1} \\ 1.24 \times 10^{-6}\sigma^{2.4} \text{ in.}^{-1} & \text{for } \dot{\varepsilon} = 0.660 \text{ min}^{-1} \\ 4.13 \times 10^{-6}\sigma^{2.165} \text{ in.}^{-1} & \text{for } \dot{\varepsilon} = 26.5 \text{ min}^{-1} \end{cases} \quad (\sigma \text{ in ksi})$$

| Strain rate min$^{-1}$ | Mean fiber strength (ksi) | | Composite strength (ksi) | | Composite strength S.D. (ksi) |
|---|---|---|---|---|---|
| | Exp. | Theory | Exp. | Theory | |
| 0.0265 | 237 | 253 | 279 | 309 | 38.9 |
| 0.660 | 226 | 242 | 296 | 322 | 52.0 |
| 26.50 | 215 | 254 | 256 | 346 | 21.2 |

[a] From Armenakas *et al.* (1970). Fiber dimensions: $d = 0.00485$ in., $L = 1.516$ in., [$k = 3.25$ ksi. From Broutman (1967)].

## C. Failure of an Unbonded Bundle

The failure of a bundle of parallel noncontacting elements of some strength variability has been of interest as an idealized model of tensile failure to textile technologists and early workers on composites. The problem has been investigated in great detail by a number of investigators such as, for instance, Daniels (1945) and Coleman (1957a, b). In this approach it is recognized that the elements of the bundle have a distribution of tensile strengths, and that as the bundle is loaded in tension the elements begin to break in order of their increasing strength. The failure of each element increases the load on the $r$ surviving elements by a factor $(r+1)/r$, so that the surviving elements feel an increase of load resulting both from the increase of the external load on the bundle and the increase of load resulting from a decrease in the number of elements which share this load. As can be readily visualized, in this loading process a stage will come when the failure of an additional element at constant external load will increase the load on the survivors enough to make the progressive failure of the remaining elements possible. This level of load divided by the area of the initial number of elements is the strength of the bundle.

Because of its utility in the discussion below we will reproduce the result of Daniels (1945, see also Rosen, 1964, for a compact treatment of the problem) for the bundle strength.

$$\sigma_b = [1/emAC]^{1/m} \tag{14}$$

where $m$ and $C$ are the parameters of the flaw density distribution, $e = 2.72...$ is the Naperian number, and $A = wL$ for lamina and $\pi dL$ for fibers.

Rosen (1964) has pointed out the utility of this expression for bonded element composites by using for the test area the product of the element width $w$ and the statistically independent length $\delta$. In this model the composite strength is identical to the average strength of elements with the very much shorter test length of $\delta$. Clearly, since the Rosen analysis does not consider the strong coupling effect of the matrix in producing local stress concentrations, it is useful only as a high upper bound expression—giving better results as the matrix coupling becomes weaker.

The expression in Eq. (14) for the strength of an unbonded bundle has been verified many times. One recent successful attempt is that by Armenakas *et al.* (1970) for glass fibers. Their data is reproduced in Table III.

## D. Fracture Instability

As discussed earlier in Section II, for each laminate under an applied stress $\sigma$, there is a critical number of adjacent fractured elements which acts as a

**TABLE III**

Experimental and Theoretical Results for the Strength of S-Glass Bundles[a,b]

| Strain rate min$^{-1}$ | Bundle strength (ksi) | | Bundle strength S.D. (ksi) |
|---|---|---|---|
| | Exp. | Theory | |
| 0.0265 | 122 | 127 | 35.7 |
| 0.660 | 157 | 113 | 41.4 |
| 26.5 | 121 | 113 | 35.9 |

[a] Armenakas *et al.* (1970).
[b] Dimensions and flaw distribution parameters are the same as in Table II.

crack that can be propagated to fracture the laminate, and that this number decreases with increasing stress. Our statistical considerations herein deal with the possible successive events which can produce such a fracture nucleus from weak sites in the reinforcing elements. Equation (12) for the strength of the composite allows us to state explicitly the critical number of adjacent element fractures for over-all fracture instability. The required number $r$ of fractured elements is that which makes Eq. (9), the probability of fracture of the $r$th element in a cascade of fracture, nearly unity (say, 0.99), where $r-1$ adjacent elements have already fractured, i.e.,

$$Q(s) = 1 - \exp(-K_r^m s) = Q_0 \approx 0.99 \tag{9}$$

From which

$$\sigma = \frac{1}{K_r(r)}\left[\frac{\ln(1/(1-Q_0))}{Cw\delta(r)}\right]^{1/m} \tag{15}$$

for an elastic matrix in which $\delta$ is independent of stress. For a plastic matrix in which $\delta$ is also a function of $\sigma$, a different expression would hold which, for the simple shear lag analysis of Eqs. (4) and (4a), gives

$$\sigma = \left[\frac{2k(m+1)\ln(1/(1-Q_0))}{wCt_r}\left\{\frac{((r/2)+1)-1}{((r/2)+1)^{m+1}-1}\right\}\right]^{1/(m+1)} \tag{16}$$

where the explicit form of

$$K_r = \left[\frac{1}{(m+1)}\frac{((r/2)+1)^{m+1}-1}{((r/2)+1)-1}\right]^{1/m} \tag{17}$$

was used for the stress enhancement factor (Scop and Argon, 1969) based on the shear-lag analysis of the stress distribution given by Eq. (4) in which $t_m$

was set equal to zero in keeping with the idealization outlined in Section III, B, 1.

For low stresses (large $r$) and reinforcing elements with little variability in strength (large $m$) Eq. (16) simplifies to

$$\sigma \approx \left(\frac{2k(m+1)\ln(1/(1-Q_0))}{wCt_r}\right)^{1/(m+1)} \left(\frac{1}{r/2+1}\right)^{m/(m+1)} \tag{18}$$

which indicates that the effective crack length $rt_r$ for large $m$ is inversely proportional to the stress. This is different from the Griffith-type fracture instability for homogeneous isotropic continua, and shows that in our case the instability condition depends on the surface strength distribution on the reinforcing elements.

There has yet been no experimental verification of this departure from the normal continuum behavior.

### E. Size Effect in Tensile Fracture of Laminates

The strength of the laminate of $N$ parallel layers of reinforcement coupled by interfacial tractions characteristic of the behavior of the matrix, and given implicitly by Eq. (12) will, in general, depend on the number of parallel reinforcing elements in the laminate. Note that as already discussed above, in this idealization the matrix phase has been removed but its behavior as it affects the interfacial tractions has been retained. Hence the above statement has nothing to do with the actual volume fraction of the reinforcing phase but indicates instead that the strength of the laminate depends on the number of segments the fracture cascade has to travel through and the number of sites from which it can initiate.

Consider Eq. (12) for the strength of the laminate. Substitution of Eqs. (9) and (11) into (12) gives

$$1 - (1-G_0)^{1/n} = N[1-\exp(-s)][\exp(-s)-\exp(-K_1{}^m s)]^2$$

$$\times [\exp(-s)-\exp(-K_3{}^m s)]^2 \dots [\exp(-s)-\exp(-K_r{}^m s)]^2$$

$$\times [\exp(-s)]^{N-r} \tag{19}$$

where

$$s = w\delta C\sigma^m \tag{19a}$$

Equation (19) indicates that for a given reinforcing element with constant $m$, the nondimensional stress $s$ is a function of $N$.

We show first that when the number of elements is very large the laminate strength decreases monotonically with increasing $N$. Differentiation of both sides of Eq. (19) gives

$$\frac{dN}{N}\left[\frac{1}{N}-s\right]+\left[\frac{1}{N}\left(\frac{\exp(-s)}{1-\exp(-s)}+2\frac{K_1{}^m\exp(-K_1{}^m s)-\exp(-s)}{\exp(-s)-\exp(-K_1{}^m s)}+\cdots\right.\right.$$

$$\left.\left.+2\frac{K_r{}^m\exp(-K_r{}^m s)-\exp(-s)}{\exp(-s)-\exp(-K_r{}^m s)}+r\right)-1\right]ds=0 \qquad (20)$$

which for $N\to\infty$ goes as

$$-s(dN/N)-ds=0.$$

Substitution for $s$ the equivalent form given by Eq. (19a) and $\delta=\sigma t_r/2k$ into the last equation gives

$$\frac{d\ln\sigma}{d\ln N}=-\frac{1}{m+1} \qquad \text{for a plastic matrix}$$

$$\frac{d\ln\sigma}{d\ln N}=-\frac{1}{m} \qquad \text{for an elastic matrix}$$

(21)

The monotonic decrease in the laminate strength given by Eq. (21) is primarily a result of the increasingly many sites from which the fracture cascade can start as $N$ increases. A similar result has also been obtained by more elementary arguments by Kies (1964) and by a similar development discussed above by McKee (1967) and McKee and Sines (1969).

The decrease of composite strength with increasing number of reinforcing elements at nearly constant volume fraction has been established in a series of painstaking experiments on glass fiber wound pressure vessels of increasing size by Kies (1964), whose results are shown in Fig. 5. Note that the number of elements increases from right to left in the figure. The effect was also observed experimentally by McKee (1967).

We now proceed to show that the strength of the laminate has a maximum at a certain $N$, and that therefore for small $N$ it increases with $N$. To find the condition for maximum strength we set $ds/dN=0$ in Eq. (20). This requires simply that

$$(1/N)-s=0 \qquad \text{or} \qquad sN=1 \qquad (22)$$

for the strength maximum. In the derivatives in Eq. (20) $dr/dN$ was considered small on the anticipation that $N$ will always be large and that the critical crack length $rt_r$ will always be small in comparison with the size $Nt_r$ of the laminate. To obtain the maximum strength, or alternatively the number of elements

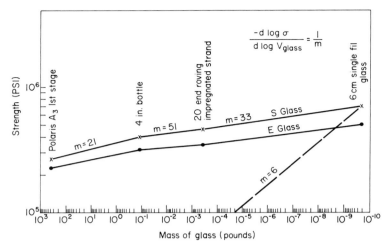

**Fig. 5.** Monotonic decrease of composite strength with increase in amount of reinforcement used. (From Kies, 1964, courtesy U.S. Naval Research Laboratory.)

resulting in maximum strength, we substitute the condition given in Eq. (22) into Eq. (19), which results in

$$[(1-(1-G_0)^{1/n})e]^{1/2}$$
$$= \{[1-\exp(-(K_1{}^m-1)s)][1-\exp(-(K_3{}^m-1)s)]\dots$$
$$\times [1-\exp(-(K_r{}^m-1)s)]\} \tag{23}$$

for the anticipated solution where $1/N \ll 1$. Equation (23) can now be solved simultaneously with the equation for the condition of fracture instability,

$$K_r{}^m s = \ln[1/(1-Q_0)] \tag{14}$$

which gives the relation of the critical crack size $rt_r$ to the applied stress $\sigma$ for the condition of fracture instability.

The results of a numerical solution of these equations for a plastic matrix utilizing the shear lag analysis for the local stress distribution [Eq. (4), for $t_m = 0$] are given in Table IV. Table IV also shows that the nondimensional maximum strength $s_m$ steadily decreases with increasing $m$. The actual maximum strength $\sigma_m$, however, increases with increasing $m$ in the range investigated, as the second to the last column of Table IV shows for a typical family of flaw density distributions based on the experimental data of Metcalfe and Schmitz (1964).

We showed that with increasing numbers of elements, the strength of the composite goes through a maximum, and for very large numbers of elements

## TABLE IV

NUMBERS OF REINFORCING ELEMENTS NECESSARY FOR A MAXIMUM IN
LAMINATE STRENGTH AS FUNCTION OF $m$

| $m$ | $K_r^m$ | $s_m$ | $r$ | $N_m$ | $\sigma_m/\sigma_0^a$ | $\sigma_m(r)^{1/2}/\sigma_0^a$ |
|---|---|---|---|---|---|---|
| 4 | 78 | $5.8 \times 10^{-2}$ | 6.3 | 17.2 | 0.203 | 0.510 |
| 6 | 200 | $2.3 \times 10^{-2}$ | 4.35 | 43.5 | 0.298 | 0.621 |
| 8 | 460 | $1.0 \times 10^{-2}$ | 3.4 | 100 | 0.365 | 0.673 |
| 10 | 1300 | $0.354 \times 10^{-2}$ | 2.95 | 282 | 0.409 | 0.700 |

[a] For a family of flaw distributions based on data of Metcalfe and Schmitz (1964), where $\sigma_0$ represents the extrapolated strength of a very short reference specimen.

monotonically decreases. For small numbers of elements, then, the strength of the composite must rise with increasing numbers of elements which is intuitively obvious for very small numbers.

We note further from Table IV that as the flaw density exponent $m$ increases, the critical crack length $rt_r$ at the maximum laminate strength decreases monotonically.

The increase of composite strength with increasing numbers of reinforcing

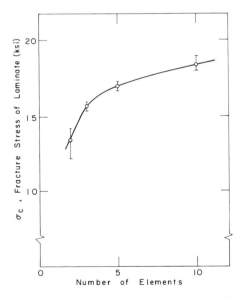

**Fig. 6.** Increase in laminate strength with increasing number of parallel elements for a boron–polyimide–epoxy film laminate. (From Scop and Argon, 1969, courtesy *J. Compos. Mater.*)

elements has been verified by Scop and Argon (1969) in experiments on laminates made from vacuum-deposited boron sheets on polyimide film, which were subsequently laminated together with epoxy glue. The increase in composite strength with increasing numbers of elements (constant volume fraction of reinforcement) is clearly seen in Fig. 6. Another, but less perfect, verification was obtained by testing different widths of glass fiber reinforced packaging tape. As shown in Fig. 7 here too a monotonic increase of fracture stress was observed with increasing numbers of reinforcing elements.

In their study on composite strength, Zweben and Rosen (1970) have found that the strengths of certain boron fiber composites could be bounded by stresses for which the probabilities of two and three adjacent element fractures

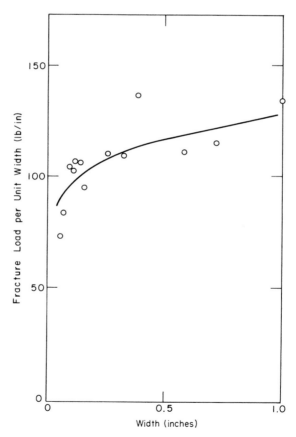

**Fig. 7.** Increase in composite strength with increasing width of glass fiber tape. (From Scop and Argon, 1969, courtesy *J. Compos. Mater.*)

became appreciable. Table IV shows that for elements with flaw density distributions of $m > 10$ this is probably a satisfactory criterion. For the large majority of glass fibers used in industry for which the exponent $m$ lies between 4 and 6 (Kies, 1964) the criterion of Zweben and Rosen would be overly pessimistic.

## F. Effect of Element Variability on Composite Strength

The statistical theory for laminate strength also lends itself to the investigation of the effect of element variability on the strength of the composite. To do this we consider an ideal continuous manufacturing process for reinforcing elements, e.g., glass ribbon or fiber in which the as-manufactured product has very high strength with little variability. In keeping with our previous discussion we represent the variability in strength by a flaw density distribution

$$g(\sigma) = d\xi(\sigma)/d\sigma = mC\sigma^{m-1} \tag{24}$$

which gives the density of flaws with strength-limiting properties between $\sigma$ and $\sigma + d\sigma$. If the "perfect" as-manufactured elements could be tested in varying lengths, their strength $\sigma_i$ would show a decrease with increasing length $L$, which should follow a relationship (Weibull, 1939a, b)

$$\sigma_i = [1/wLC]^{1/m} \tag{25}$$

where for fibers $w$ can be taken as $\pi d$. If the ultimate strength $\sigma_0$ of the elements in very short lengths comes from a flaw distribution $g_0(\sigma_0)$, then it is more convenient to relate the constant $C$ in Eq. (25) to $g_0(\sigma_0)$, $m$, and $\sigma_0$ by using Eq. (24), giving for the individual element strength

$$\frac{\sigma_i}{\sigma_0} = \left[\frac{m}{wLg_0(\sigma_0)\sigma_0}\right]^{1/m} \tag{26}$$

The values for $\sigma_0$, $m$, and $g_0(\sigma_0)$ can be obtained from the variation with length of the element strength (Kies, 1964; Rosen, 1964).

If the elements become damaged by handling (see Ernsberger, 1960), additional flaws with strength-limiting properties in the low stress ranges are introduced without altering the inherent flaws governing $g_0(\sigma_0)$, i.e., the flaw distribution is enriched in the low strength regions only. If the new strength distributions of the damaged elements can still be represented by power function forms of flaw distributions with different constants $C$ and exponents $m$, they would all have one point in common at the stress $\sigma_0$. The entire family of distributions could then be given by a form

$$g(\sigma) = g_0(\sigma_0)[\sigma/\sigma_0]^{m-1} \tag{27}$$

and since the product of $g_0(\sigma_0)\sigma_0$ would be constant, the changes in element strength with length, in damaged elements, would come entirely from the exponent $m$, as Eq. (26) indicates.

Now for the strength of the composite we take the result of the simple fracture model of Scop and Argon (1969) given by Eq. (13), which with the current symbols can be written as

$$\frac{\sigma_c}{\sigma_0} = \left[\frac{m}{\delta w g_0(\sigma_0)\sigma_0(K_3{}^m - 1)}\right]^{1/m} \qquad (28)$$

where the statistically independent length $\delta$ can be obtained from the range of the elastic stress singularity or from Eqs. (3) or (4a) for the plastic zone in a plastic matrix.

The ratio of the composite strength to the individual element strength can be obtained from Eqs. (26) and (28) by division as

$$\frac{\sigma_c}{\sigma_i} = \left[\frac{L}{\delta(K_3{}^m - 1)}\right]^{1/m} \qquad (29)$$

We find it instructive now to consider a specific case for the special case of a row of fibers taken as a laminate. In this case the product $w g_0(\sigma_0)\sigma_0 = \beta$ (where $w = \pi d$) in Eq. (26) remains constant for fibers of constant diameter. In a typical case of fibers investigated by Metcalfe and Schmitz (1964) $\beta = 4.8 \times 10^3$ cm$^{-1}$ for $\sigma_0 = 48$ kbars and $d = 10^{-3}$ cm. Using these values of $\beta$ and $\sigma_0$ the calculated values of $\sigma_i/\sigma_0$ based on Eq. (26) are plotted in Fig. 8 for a range of possible values of $m$, for two test lengths of 10 cm and 100 cm. The calculated values of $\sigma_c/\sigma_0$ based on Eq. (28) and using $\delta = 0.05$ cm for a plastic matrix are also plotted for the same range of $m$ in Fig. 8, together with the ratios of the composite strength to the average individual element strengths calculated from Eq. (29).

It is interesting to note that the average strength of the individual elements increases steadily with increasing $m$ or increasing perfection of fibers (decreasing variability). The strength of the composite, on the other hand, increases with increasing $m$ or fiber perfection only for small $m$ (fibers of large variability) and levels off sharply with increasing $m$ when $m$ is large (fibers of small variability). The effect shows up even more dramatically in the ratio of the composite strength to the individual element strength. For values of $m$ less than about 10, the composite strength is greater than the element strength. Here the coupling of the matrix is beneficial, the breaks at the bad flaws at small stress are effectively bridged by strong regions of the adjacent elements. When $m$ becomes significantly larger than 10, element variability is small and bad flaws are negligibly few. The stress is much higher when isolated fractures occur in individual elements. Now, however, the stress concentration more

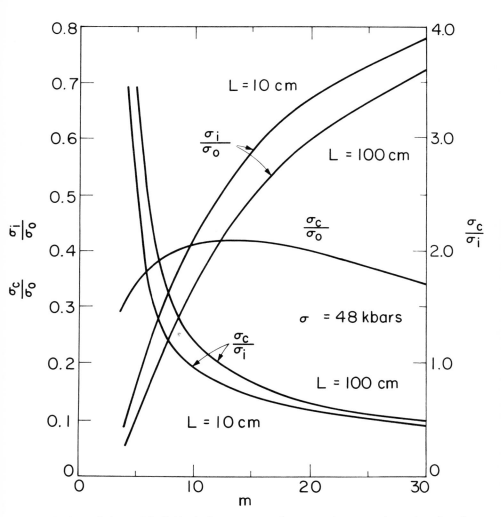

**Fig. 8.** Variation of individual element strength, composite strength, and ratio of composite strength to element strength as a function of flaw density exponent $m$.

than likely also overstresses the adjacent elements and fracture propagates. The composite now is weaker than the average of the individual elements since its strength is governed by the few elements with low strength. Stated in other words, composites with uniformly strong elements are more susceptible to stress concentrations than those with elements with more variable strength.

### G. Bonded Laminates vs. Unbonded Bundles

In a composite the matrix provides strong coupling between elements. As Fig. 8 shows, this strong coupling is an asset in composites with elements of large variability in strength, but a liability in composites with strong elements with low variability. It would seem, therefore, that apart from stiffness considerations in structural parts, strong elements could be used better in weakly coupled bundles with no matrix at all, because in the latter there can be no stress concentrations.

To demonstrate our assertion, let us compare the strength of unbonded bundles with the strength of matrix-bonded laminates as a function of element variability. To obtain the ratio of the composite strength to the bundle strength we divide Eq. (13) by Eq. (14) to obtain

$$\frac{\sigma_c}{\sigma_b} = \left[\frac{emL}{\delta(K_3{}^m - 1)}\right]^{1/m} \tag{30}$$

This ratio is plotted in Fig. 9 for a bundle of 10 cm length for the same data of Metcalfe and Schmitz (1964) used in Section III, F. As suspected, the bundle becomes more and more efficient with increasing $m$. Since the asymptotic strength of the composite is relatively insensitive to its length (Eq. 13), there

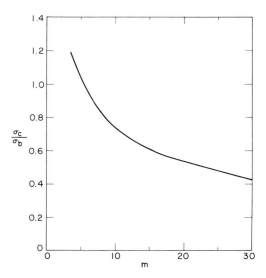

**Fig. 9.** Variation of ratio of composite strength with plastic matrix to strength of unbonded bundle for a 10 cm test length of the latter.

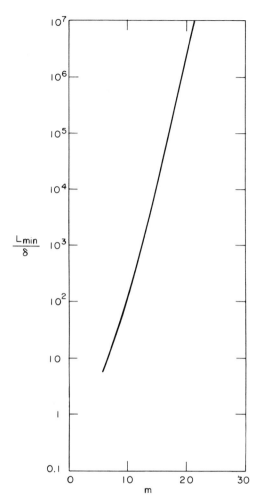

**Fig. 10.** Dependence of minimum length of bundle on flaw distribution exponent *m* for condition of equal strength for composite and bundle.

would be a certain length, however, where the strength of the bundle is equal to the strength of the composite. Above this length the composite would be stronger than the corresponding bundle. The change of this minimum length with the flaw distribution exponent *m* is given in Fig. 10. Evidently, in comparison to a composite with a plastic matrix, the bundle is stronger over a very large range of element lengths.

## IV.  Compressive Failure in Laminates

Rosen and co-workers (summarized by Rosen, 1970) have proposed that a unidirectionally reinforced composite will fail in compression by an internal buckling mode of the reinforcing elements embedded in the more extensible matrix material in a manner discussed in general by Biot (1965), and shown in Fig. 11. The in-phase and out-of-phase buckling modes, called the shear and extensional modes of compressive failure by Rosen, are characteristically of a magnitude about equal to the matrix shear modulus. The validity of this model has been verified by carefully designed experiments (Boeing Aircraft Co., 1968). More routine experimental observations, however, suggest that the internal buckling mode of failure is an upper bound applicable to ideal composites in which the reinforcing elements are perfectly parallel and aligned with the stress axis. In reality, specimens made by normal manufacturing techniques will always have regions in which the reinforcing elements are not fully aligned with the compression axis. Such regions will form a failure nucleus by undergoing a process of kinking similar to the well-known kinking of metal crystals (Orowan, 1942) and resembling in form the in-phase internal buckling mode of Rosen but operating at a stress much below the ideal buckling strength.

Consider a region of initial misalignment $\phi_0$ with the compression axis. In such a region the applied compressive stress produces an interlaminar shear component of $\tau = \sigma\phi_0$. When this shear stress becomes equal to the interlaminar shear strength $k$, the lamellae in the region will slide on each other and rotate to increase the resolved shear stress further, producing a local instability which tends to propagate outward as a shear collapse band by virtue of the shear stress concentrations which develop at the tips of the band. The compressive strength of the composite then should be simply

$$\sigma = k/\phi_0 \tag{31}$$

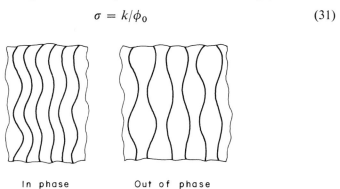

In phase                    Out of phase

*Fig. 11.* Schematic view of in-phase and out-of-phase internal buckling modes of continuous fiber composites in compression. (After Rosen, 1970.)

**Fig. 12.** A kinking-collapse mode of compressive failure in a boron–polyimide–epoxy film laminate. (Courtesy of Dr. Padawer, Norton Research Corp.)

which is independent of the volume fraction of the reinforcing element over the relatively wide range in which the model applies. As is shown in the detailed analysis in Appendix A, the elastic resistance of the surroundings of the band and the elastic bending of the lamellae do not materially change this simple result.

A region of kinking of this type observed in a boron sheet laminate is shown in Fig. 12. Evidently, once the collapse occurs and the shearing inside the band results in large rotations $\Delta\phi$, the normal tensile strain

$$\varepsilon_{xx} = \frac{1 - \cos \Delta\phi}{\cos \Delta\phi}$$

inside the band is relieved by delamination outside the band.

There has been no systematic study of this mechanism of compression failure as a function of local misalignment. The direct proportionality between compressive strength and matrix shear strength, suggesting the presence of the kinking mechanism, has, however, been amply demonstrated (see Broutman, 1967).

## V. Fatigue Fracture in Laminates

Fatigue of laminates is far less well understood than their tensile fracture. Generally it has been well established that any material which is incapable of undergoing plastic deformation does not fatigue. The apparent exceptions to this rule have been identified to be static fatigue in which no plastic deformation is involved; the crack grows by a slow crack tip corrosion during the tensile crack opening displacement of the fatigue cycle. For this reason, fatigue in composites made of plastically nondeformable reinforcing materials must by necessity result from fatigue in the matrix.

Several mechanisms of fatigue failure based on the above rationalization have been proposed.

First, if the reinforcing elements are continuous and strong and have a fairly narrow range of fracture strain which is considerably larger than the elastic strain at yield for the matrix, the latter can fatigue and disintegrate, rendering the unfractured but unbonded reinforcing elements useless. In this case, composite failure occurs by loss of bending stiffness much before any actual fracture has occurred. Such a process was discussed by Kelly and Bomford (1969) and has been widely encountered in use (Salkind, 1971).

Second, in initially continuous element composites, cycled in a tension-tension mode, the first extreme excursion of the stress will produce a number of isolated stable fractures in some reinforcing elements if the surface strength distribution function has a low exponent. Large interlaminar plastic strains occur at the tips of these fractures. Cyclic deformation of the composite will produce concentrated cyclic shear deformation at these sites which can then produce rapid delamination tending to link the primary cracks across the reinforcing elements by secondary delamination. Once enough of such primary crack linkage has occurred to result in a critical effective Mode I crack length, the laminate will fail by tensile fracture instability in a quarter cycle. Such a process has been studied by Argon and Bailey (1971) in specially prepared glass ribbon–polyethylene laminates.

Third, if the laminate is reinforced with initially discontinuous elements, all the processes described above will happen and the composite will fatigue by linkage of cracks starting from strain concentrations at element matrix interface.

A novel mode of fatigue crack propagation in compression cycling of composites in their stiff direction has recently been observed by Berg (1971), in which a small "process zone" ahead of the crack undergoes a localized kinking mode of failure (described in Section IV). The excessive local bending of lamellae fractures a few elements per cycle, resulting in fatigue crack growth.

## VI. Discussion and Conclusions

Rosen (1964) has pointed out that a first-order explanation for the success of composites of brittle reinforcing elements in an extensible matrix is that the strong coupling of the matrix phase limits the influence of any local fracture process in a reinforcing element to a small region. Thus, propagation of fracture from element to element requires the overstressing and fracture of only very short lengths of the reinforcing elements over which the latter are much stronger than over their entire length, by virtue of the well-known negative size effect for brittle substances. As we have demonstrated, however, the strong coupling of a matrix also results in concentration of stresses in the immediate vicinity of the fractured elements and can raise the probability of fracture there above the level of what it might have been had the increased stress been shared by all elements. These stress concentrations are a key to much of the interesting behavior of composites.

We have seen that at constant volume fraction, increasing the number of reinforcing elements first increases the composite strength, and then after a maximum strength is achieved, the strength of the composite monotonically decreases, which for large numbers of elements results in a fractional decrease in strength proportional to the fractional increase in size—with a proportionality constant equal to the reciprocal of the exponent of the surface strength distribution function of the elements.

In common with other materials of limited ductility, there is a critical number of adjacent fractured elements in a composite, which, acting as a crack, makes the fracture propagation under decreasing stress possible. We have examined the correlated processes of sequential element fracture in a composite which under constant applied stress can give rise to such an effective crack to make the fracture auto-catalytic. Not unexpectedly, our analysis shows that the critical crack length for fracture instability for a constant applied stress depends on the strength distribution of the reinforcing elements and on the mechanical properties of the matrix which governs the distribution of over-stress in the region ahead of any crack made up of one or more adjacent fractured elements. Examination of Table IV shows that, for the specific case of the strength maximum, the length of the critical crack for fracture instability decreases with increasing exponent $m$ (decreasing element variability). This increase is partly due to the rise in the composite strength with increasing $m$. As the last column in Table IV shows, however, in the conventional sense the "fracture toughness"[†] $\sigma_c(rt_r)^{1/2}$ is increasing in this range. The rate of increase in the fracture toughness, however, decreases sharply, and as our more approximate analysis on the effect of variability on composite strength shows $\sigma_c/\sigma_0$ actually levels off and decreases for $m > 10$.

Table IV and our statements above show that at fracture instability the critical fracture nucleus contains a decreasing number of fractured adjacent elements with increasing $m$. Hence, any fracture condition based on a high probability of encountering a given number of adjacent fractures, as has been advocated by Zweben (1969) and by Zweben and Rosen (1970), can be only of limited utility in a narrow range of $m$ values. The work of Armenakas et al. (1970), where fibers of very low $m$ values were used, has demonstrated this fact.

The strong coupling of the reinforcing elements by an extensible matrix producing local stress concentrations as discussed above affects the composites with elements of small strength variability more than those with large variability. In the former, the premature fracture in one reinforcing element below the mean strength of the elements has a high probability of setting up a chain of fracture processes leading to catastrophic fracture. As we have seen in such cases, characterized by flaw distribution exponents $m$ larger than 10, the strength of the composite is lower than the mean strength of the individual elements. The opposite is true for elements with large variability (small $m$ values, less than 10) where premature fracture well below the element mean strength has a high probability of being surrounded by elements which are stronger and therefore render the premature break temporarily harmless. For such composites our analysis shows that the composite strength is higher than the element mean strength. In such cases bonding the elements achieves a composite of improved strength.

The problems of fracture in unidirectionally reinforced composites we have dealt with form only a small class in the wide area of usage of composites. In most commercial applications, composites are used in mats and multidirectional layers. Fracture in these composites is far more complex in that it involves not only fracture and retraction of elements, but also large local rotations of fiber ends. In this range the more conventional approach of experimental

---

[†]Note that the term fracture toughness is really not appropriate here since at the start of the test the specimen is not provided intentionally with a crack. A crack of length $t_r r$ develops by statistical correlated fracture processes at the time the composite strength $\sigma_c$ is reached.

fracture mechanics is presently far more profitable—with the developments on the fracture of unidirectional composites serving only as a guide to provide working assumptions.

There is considerable evidence that failure in compression, even in uni-directionally reinforced composites, in practice is governed by local imperfections in alignment and volume fraction of reinforcement.

Fatigue damage production in composites, which in the long run is likely to be the most critical factor in the use of advanced composites, is hardly understood at all.

### Appendix: Compressive Strength of a Laminate

Consider a region of initial element misalignment as shown in Fig. A-1, where the reinforcing elements make an angle $\phi_0$ with the compression axis. Because of this misalignment, the applied compressive stress $\sigma$ will produce a shear stress between the elements in this region, which tends to shear and rotate the elements—in a sense to increase the shear stress acting between them. Hence a shear collapse deformation instability exists in the region. As the elements in the initial region of misalignment shear and rotate, shear stress concentrations appear on the extremities of the zone which tends to propagate the shear collapse region outward across the part. In crystals with well-defined slip planes, aligned parallel to a compression axis, this type of instability is well known and has been called a kink band. Below we develop a simple relation for this shear collapse mode of failure in composites by analogy with kink bands in metal crystals.

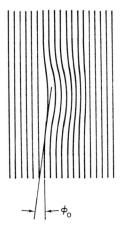

*Fig. A-1.* Schematic view of a local region of misalignment in a laminate.

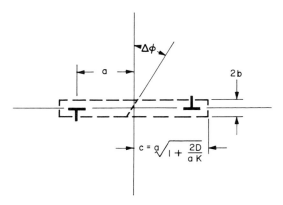

**Fig. A-2.** Schematic view of a kink-collapse nucleus of compressive failure.

When the region outlined in Fig. A-2 begins to shear and the angle of misalignment changes from $\phi_0$ to $\phi = \phi_0 + \Delta\phi$ a certain amount of work is done by the external compressive stress

$$\Delta W = (\cos\phi_0 - \cos\phi)(4cb/V)\sigma V$$

$$\Delta W = 4bc\,\phi_0\,\Delta\phi\sigma \tag{A-1}$$

where the first term in parenthesis is the local compressive strain in the band, $V$ is the total volume of the part, and $c$ is the region over which the interlaminar shear stress equals the shear strength $k$ of the matrix.

When the region shears, plastic work is being done inside the band; elastic energy due to bending the lamellae is stored at the top and bottom band boundaries; and elastic stresses outside the region develop, tending to resist the expansion of the band. Hence, both plastic work is done inside the region and additional elastic energy is stored both at the boundaries and outside the band. There is no change in elastic energy inside the band of comparable order. We estimate these quantities by the analog of a dipole of dislocations a distance $2a$ apart and each having a strength of $2b\Delta\phi$. The change in elastic energy outside the band and the plastic work done inside the band can be calculated by the well-known procedure of surface work along the plane of cut of the dislocations and is

$$\Delta E_1 = 2\frac{2\Delta\phi bD}{2}\int_c^\infty \left(\frac{1}{(x-a)} - \frac{1}{(x+a)}\right)dx + 4bc\Delta\phi k \tag{A-2}$$

where the first term is the elastic energy in the region external to the collapse band (note that the effect of the edge dislocation dipole arising from the change in thickness of the band will be of a magnitude $\phi_0$ times that for the shear field and therefore is negligible), and the second term is the plastic work done in the band. Furthermore,

$$D = (2b\Delta\phi)\, G_c/2\pi(1-v) \qquad \text{(A-3)}$$

is the scale factor of the stress field of an edge dislocation, where $G_c$ is the shear modulus of the composite; and

$$c = a[1+(2D/ak)]^{\frac{1}{2}} \qquad \text{(A-4)}$$

the region over which the shear stress equals $k$ is obtained directly by equating the dipole stress field to $k$.

The energy due to the double bending of the lamellae can be calculated readily by the well-known formula for bending slender beams and is

$$\Delta E_2 = cE_r\, I(\Delta\phi)^2/bt_r(1-v^2) \qquad \text{(A-5)}$$

where $E_r$ is the Young's modulus of the reinforcement, and $I = t_r^3/12$ is the bending area-moment of inertia per unit depth of the lamellae.

In an adiabatic collapse process we require

$$\Delta E_1 + \Delta E_2 - \Delta W = 0$$

which gives upon simplification the compressive strength of the laminate as

$$\sigma = \frac{k}{\phi_0}\left[1 + \frac{bG_c\,\Delta\phi}{2\pi ak(1-v)}\ln\frac{2\pi ak(1-v)}{bG_c\,\Delta\phi} + \frac{E_r\,\Delta\phi}{48k}\left(\frac{t_r}{b}\right)^2\right] \qquad \text{(A-6)}$$

At the point where the collapse starts, $\Delta\phi$ is very small so that the second and third terms in the equation produce a negligible contribution, the compressive strength of the laminate becomes simply

$$\sigma = k/\phi_0 \qquad \text{(A-7)}$$

which is dependent only on the plastic shear strength of the matrix and the angle of misalignment and is independent of the volume fraction of the reinforcement—to within the level of approximation in our analysis. Naturally when the volume fraction of the reinforcement becomes very small many of our assumptions will no longer hold.

Our discussion above applies to the condition of initiation of a collapse nucleus. Once such a nucleus is established, it will spread at 45° to the applied compressive stress where the shear stresses can do work proportional to $\Delta\phi$.

### ACKNOWLEDGMENT

Much of the material discussed in this paper is a result of a fruitful professional working arrangement with the Composites Development Group in the organization which has been known alternately as the National Research Corporation and the Norton Research Corporation. In the development of these ideas discussions with Drs. Scop, Feakes, Padawer, and Beecher of this organization have been most fruitful. I am also grateful for general discussions to my colleagues, Professors McClintock and Berg.

# References

Argon, A. S., and Bailey, D. (1971). *Symp. Adv. Composites, 5th. 1971.* Unpublished data.

Armenakas, A. E., Garg, S. K., Sciammarella, C. A., and Svalbonas, V. (1970). "Strength of Glass-Fiber Bundles and Composites Subjected to Slow Rates of Straining," AFML-TR-69-314. Wright-Patterson Air Force Base, Ohio.

Berg, C. A. (1971). Private communication.

Biot, M. (1965). "Mechanics of Incremental Deformations." Wiley, New York.

Boeing Aircraft Co. (1968). Advanced Composites Status Rep. No. D6-23317.

Broutman, L. J. (1967). *In* "Modern Composite Materials" (L. J. Broutman and R. H. Krock, eds.), pp. 337-411. Addison-Wesley, Reading, Massachusetts.

Coleman, B. D. (1957a). *J. Appl. Phys.* **28**, 1058.

Coleman, B. D. (1957b). *J. Appl. Phys.* **28**, 1065

Corten, H. T. (1967). *In* "Modern Composite Materials" (L. J. Broutman and R. H. Krock, eds.), pp. 27-92. Addison-Wesley, Reading, Massachusetts.

Daniels, H. E. (1945). *Proc. Roy. Soc., Ser. A* **183**, 405.

Ernsberger, F. M. (1960). *Proc. Roy. Soc., Ser. A* **257**, 213.

Gücer, D. E., and Gurland, J. (1962). *J. Mech. Phys. Solids* **10**, 365.

Hedgepeth, J. M. (1961). *NASA Tech. Note* **NASA-TND-882**.

Hedgepath, J. M., and Van Dyke, P. (1967). *J. Compos. Mater.* **1**, 294.

Kelly, A. (1966). "Strong Solids." Oxford Univ. Press (Clarendon), London and New York.

Kelly, A., and Bomford, M. J. (1969). *In* "Physics of Strength and Plasticity" (A. S. Argon, ed.), pp. 339–350. MIT Press, Cambridge, Massachusetts.

Kies, J. A. (1964). "The Strength of Glass Fibers and the Failure of Filament Wound Pressure Vessels," NRL Rep. No. 6034. U.S. Nav. Res. Lab., Washington, D. C.

McClintock, F. A. (1969). *High Performance Composites, 4th, 1969.* Unpublished data.

McClintock, F. A., and Argon, A. S. (1966). "Mechanical Behavior of Materials." Addison-Wesley, Reading, Massachusetts.

McKee, R. B., Jr. (1967). Ph.D. Thesis, U.C.L.A., Los Angeles.

McKee, R. B., Jr., and Sines, G. (1969). *J. Elastoplast.* **1**, 185.

Metcalfe, A. G., and Schmitz, G. K. (1964). *Amer. Soc. Test. Mater., Proc.* **64**, 1075.

Orowan, E. (1942). *Nature (London)* **149**, 643.

Paris, P. C., and Sih, G. C. (1965). *Amer. Soc. Test. Mater., Spec. Tech. Publ.* **381**, 30–81.

Rice, J. R. (1967). *Amer. Soc. Test. Mater., Spec. Tech. Publ.* **415**, 247–309.

Rosen, B. W. (1964). *AIAA J.* **2**, 1985.

Rosen. B. W. (1970). *In* "Mechanics of Composite Materials" (F. W. Wendt, H. Liebowitz, and N. Perone, eds.), pp. 621–652. Pergamon, Oxford.

Rosen, B. W., and Dow, N. F. (1972). *In* "Fracture" (H. Liebowitz, ed.), Vol. VII. Academic Press, New York.

Salkind, M. (1971). Private communication.

Scop, P. M., and Argon, A. S. (1967). *J. Compos. Mater.* **1**, 92.

Scop, P. M., and Argon, A. S. (1969). *J. Compos. Mater.* **3**, 30.

Sih, G. C., and Liebowitz, H. (1968). *In* "Fracture" (H. Liebowitz, ed.), Vol. II, pp. 68–190. Academic Press, New York.

Tetelman, A. S. (1969). *Amer. Soc. Test. Mater. Spec. Tech. Publ.*, **460**, 473–502.

Weibull, W. (1939a). *Ingenioers vetenskaps akad., Handl.* **151**.

Weibull, W. (1939b). *Ingenioers vetenskaps akad., Handl.* **153**.

Zweben, C. (1969). *Amer. Soc. Test. Mater. Spec. Tech. Publ.*, **460**, 528–539.

Zweben, C., and Rosen, B. W. (1970). *J. Mech. Phys. Solids* **18**, 189.

# Theory of Elastic Wave Propagation in Composite Materials

### V. K. TEWARY and R. BULLOUGH

*Theoretical Physics Division, A.E.R.E.*
*Harwell, United Kingdom*

## I. Introduction

The development of composite materials, or simply composites, as they are often called, is one of the most significant achievements in the field of materials science and technology in the past decade. A composite is a solid which is densely reinforced by fibers or precipitates of another solid. The host solid is usually referred to as the matrix. Although many combinations of various matrices and fibers have been investigated, the most important from a technological point of view is probably a matrix of epoxy resin reinforced by carbon fibers. Because of its low density and high tensile strength, this particular composite has a very high specific strength, which makes it potentially very useful for industrial applications. A review of various applications of this composite has been given by Langley (1970).

In spite of a considerable technological progress in the design and application of composites, very little work has been done on the theory of composites. Most of the work published so far deals with the various methods of calculation of the elastic moduli and other static properties of the composites (Heaton, 1968, 1970; Hashin and Rosen, 1964; Hill, 1964, 1965; Behrens, 1967a). Clearly, a lot more needs to be done for even a reasonably adequate understanding of the basic physical properties of the composites. Such a study would provide a great stimulus to further technological development in the field and its importance, therefore, cannot be overemphasised.

The present article deals with a theoretical investigation of the propagation of elastic waves (of wavelengths of the order of fiber spacings or larger) in a composite. The propagation of elastic waves is a very important tool for a study of the physical properties of a composite and we shall show how one can get useful information about the bonding strength and the alignment of the fibers by a study of various properties associated with the propagation of elastic waves.

The propagation of elastic waves of wavelengths much larger than the average spacing between the fibers has been studied by Behrens (1967a, b) and Behrens and Kremheller (1969) for a simple model of the composite—the lamellar periodic structure. Some experimental study using long ultrasonic waves has been made by Reynolds (1969, 1970). For the long waves, the composite behaves like an elastic continuum. In this case one can use the Green–Christoffel equations for studying the elastic wave propagation (Musgrave, 1954a, b; Fedorov, 1968).

However, the most characteristic features of the composite that can be attributed to the discrete arrangement of the fibers are not apparent in such studies. Such features can only be studied by using short elastic waves, i.e., of wavelength comparable with the fiber spacings. For example, the propagation of long waves cannot give any information about the bonding strength of the fibers. As we shall see below, a measurement of dispersion, when the wavelength is of the order of the spacing between the fibers, is most sensitive to bonding strength.

We shall present a model which is essentially an adaptation of the Born–von Karman model for a crystal lattice, and which has been extremely successful in the field of solid state physics (Born and Huang, 1954; Maradudin *et al.*, 1963). The main difference in the present model and the Born–von Karman model of a crystal lattice is that the present model is only semidiscrete to allow for the characteristic feature of the composite. On a plane perpendicular to the fibers the present model behaves like a discrete crystal lattice, whereas it behaves like a Debye elastic continuum in a direction parallel to the fibers. This is physically reasonable for a composite because the only non-continuum effects in a direction parallel to the fibers are those arising from the atomistic structure of the fibers for which a sound wave is too long. The present model for a composite is "dispersive" for short [†] waves in the same sense as the Born–von Karman model is for the electromagnetic waves (neutron beams, X rays). The limiting case of long waves is obtained by using Born's method of long waves (Born and Huang, 1954) when the formulas obtained here reduce to those obtained by using the Green–Christoffel equations for an elastic continuum (Musgrave, 1954b; Fedorov, 1968).

Two-dimensional translation symmetry has been assumed for the arrangement of the fibers in a plane perpendicular to the fibers—an assumption which does not seem to be too unrealistic for a densely reinforced composite. This assumption has also been made in most of the previously quoted papers. In our model the fibers act as the main "building blocks" of the composite. The matrix acts as a carrier of interactions between the fibers, i.e., the fibers "see" each other through the matrix. The interaction between the fibers is assumed to travel through the matrix instantaneously. This assumption is equivalent to the adiabatic approximation in the Born–von Karman model. Thus, the effect of the matrix is included into the model in terms of Born–von Karman-type force constants between the fibers and also in the definition of an effective mass for the fibers. It may be mentioned here that the assumption of "adiabatic approximation" may be of questionable validity for a composite. This would be a major problem if the force constants between the fibers were to be calculated from first principles. In the present paper, however, we shall consider the force constants only as parameters to be determined in a phenomenological fashion. It is therefore not considered worthwhile to bother about the finer details of the effective interaction between the fibers. Finally the amplitude of the wave is assumed to be small enough so that the harmonic approximation equivalent to the linear theory of elasticity is valid.

Composites with two different symmetries have been studied: tetragonal and hexagonal, which correspond to an arrangement of fibers on the vertices of a square and hexagon, respectively, on a plane perpendicular to the fibers. Numerical results have been obtained for a carbon fiber–epoxy resin composite. The static elastic constants for this composite have been calculated by Heaton (1970). No experimental measurements are available on any composite using short waves. It is, therefore, not possible at the moment to compare the

---

†Throughout this paper, the term short wave will refer to an elastic wave of wavelength of the order of the spacing between the fibers.

theoretical results obtained in this paper with any experimental results. The experimental results as obtained by Reynolds (1970) on the directional dependence of the velocity of long waves in a carbon fiber–epoxy resin composite are used in this paper to derive some information about the average state of alignment of fibers.

In the present model the force constants serve as a measure of the strength of the fiber–matrix bond. In the framework of this model it is possible to calculate the effect of bonding defects on the velocities of wave propagation. This and the thermodynamic properties of the composite will be dealt with in another paper.

The present theory predicts an interesting phenomenon—that of reflection of elastic waves of a particular wavelength from the composite which is analogous to the Bragg reflection of X rays from crystals. The theory also predicts some very interesting collimating and polarizing properties of the composite for elastic waves which may be important from an application point of view.

The formulas and the qualitative inferences derived should, of course, apply to all fibrous composites. However, all the numerical results reported here are for a carbon fiber–epoxy resin composite with 50% fiber loading (in volume) unless otherwise stated.

A brief resume of the Born–von Karman theory is given in Section II. The present model for a composite is described in Section III. Calculations of the phase velocities and the group velocities are described in Sections IV through IX for a composite with tetragonal symmetry. Results for a composite with hexagonal symmetry are given in Section X. A comparison with the experimental results on the phase velocity of long waves in a composite has been made in Section XI to obtain some information about the alignment of fibers. A short discussion of the results is presented in Section XII.


## II. Resume of the Born–von Karman Theory


Let us consider a bound assembly of $N$ objects arranged periodically in a lattice. The assumption of periodicity implies that a unit cell can be defined so that the whole lattice can be built up by joining all the unit cells together and also that all the unit cells are equivalent. We shall also assume the usual periodic boundary conditions so that the surface effects do not arise and the lattice has full translation symmetry. We shall consider here a monatomic lattice, which means that each unit cell contains only one of the $N$ objects. Thus we do not have to distinguish between the labels for the objects and the unit cells.

Let $\mathbf{r}(l)$ denote the position vector of the $l$th object at equilibrium with respect to some conveniently chosen system of Cartesian axes. In response to an external disturbance, such as an elastic wave, each object will be displaced from its position of equilibrium. Let $\mathbf{u}(l)$ denote the displacement vector of the $l$th object from its equilibrium position. The change in the potential energy of the system will obviously be a function of the displacements. For small displacements, the change $\Phi$ in the potential energy can be written in the form of a Taylor series.

$$\Phi = \sum_{al} \phi_\alpha(l) u_\alpha(l) + \tfrac{1}{2} \sum_{\substack{al \\ \beta l'}} \phi_{\alpha\beta}(l, l') u_\alpha(l) u_\beta(l') + \cdots \tag{1}$$

where $\alpha\beta$ refer to the Cartesian components $x$, $y$, and $z$ and

$$\phi_\alpha(l) = (\partial\Phi/\partial u_\alpha(l))_0 \tag{2}$$

$$\phi_{\alpha\beta}(l, l) = [\partial^2 \Phi/\partial u_\alpha(l)\, \partial u_\beta(l')]_0 \tag{3}$$

The subscript zero in Eq. (2) and (3) indicate that the derivatives have been obtained at the equilibrium position. We shall assume that the displacements are so small that the harmonic approximation is valid; thus we can neglect the cubic and higher order terms in Eq. (1).

There are various restrictions on $\phi_\alpha(l)$ and $\phi_{\alpha\beta}(l, l')$ imposed by the requirements of symmetry and stability of the lattice (Born and Huang, 1954; Maradudin et al., 1963).

Some of these are given below.

1. For a perfect lattice, free of internal stresses

$$\phi_\alpha(l) = 0 \qquad \text{for all } \alpha \text{ and } l \tag{4}$$

2. $\phi_{\alpha\beta}(l, l') = \phi_{\alpha\beta}(0, l - l') = \phi_{\alpha\beta}(0, l' - l)$   (5)

i.e., $\phi_{\alpha\beta}(l, l')$ depends only on the magnitude of the difference between $l$ and $l'$ and not separately on $l$ and $l'$. Thus it can be expressed in terms of a single index.

3. $\sum_{l'} \phi_{\alpha\beta}(l, l') = 0$   (6)

4. If $(S/O)$ denotes an operator in the point group of the lattice where $S$ is a matrix of proper or improper rotation, then we have

$$\phi_{\alpha\beta}(l) = \sum_{\gamma\delta} S_{\alpha\gamma} S_{\beta\delta} \phi_{\gamma\delta}(l') \tag{7}$$

where

$$r_\alpha(l) = \sum_\beta S_{\alpha\beta} r_\beta(l') \tag{8}$$

The constants $\phi_{\alpha\beta}(l-l')$ are known as the force constants. Physically $\phi_{\alpha\beta}(l-l')$ gives the force in the $\beta$ direction on the object at $l'$ when the object at $l$ is displaced by a unit amount in the $\alpha$ direction. For a particular value of $l-l'$, the force constant is a tensor of the second rank and can be written as a $3 \times 3$ matrix. Thus, Eq. (7) and (8) can be written in the matrix form as

$$\phi(l) = S^{-1}\phi(l')S \tag{9}$$

and

$$\mathbf{r}(l) = S\mathbf{r}(l') \tag{10}$$

The Fourier transform of $\phi(l)$ is defined as

$$\phi(\mathbf{k}) = \sum_l \phi(l)e^{+2\pi i\mathbf{k}\cdot\mathbf{r}(l)} \tag{11}$$

where $\mathbf{k}$ is a vector in the reciprocal space. The Fourier transform $\phi(\mathbf{k})$ is a $3 \times 3$ matrix.

The propagation of a wave in the lattice is determined by the $3 \times 3$ dynamical matrix $D(\mathbf{k})$ defined as

$$D(\mathbf{k}) = \mathbf{M}^{-\frac{1}{2}}\phi(\mathbf{k})\mathbf{M}^{-\frac{1}{2}} \tag{12}$$

where

$$\mathbf{M}_{\alpha\beta} = m\delta_{\alpha\beta}$$

$m$ being the mass of an object.

The eigenvalues of $D(\mathbf{k})$ are $\omega^2(\mathbf{k})$, where $\omega(\mathbf{k})$ is the frequency (angular) associated with the wave characterized by the vector $\mathbf{k}$ and the corresponding eigenvectors $\mathbf{u}(\mathbf{k})$ of the dynamical matrix define the polarization of the wave. The $\mathbf{u}(\mathbf{k})$ are also called the polarization vectors. In the long wavelength limit, i.e., for small values of $\mathbf{k}$, the matrix $D(\mathbf{k})$ must be identical with the Green–Christoffel matrix. This condition forms the basis of Born's method of long waves.

## III. Model for the Composite

In a typical composite, the fibers are arranged in a roughly aligned fashion along the length of the matrix. The fiber spacing is of the order of $10\mu$. All the fibers are not of equal length and therefore some of them may not extend through the length of the composite. The bonding of all the fibers to the matrix may not be perfect and the bonding strength of all the fibers may not be equal. Some of the fibers may be hollow and the alignment of all the fibers may not be perfect.

For the theoretical treatment given in this paper we shall first assume a perfect composite. A perfect composite is defined to be one in which all the fibers are of equal mass and length, extend throughout the length of the composite, have equally strong bonding with the matrix, are perfectly aligned, and intersect a plane perpendicular to their length in a regular periodic pattern. Note, that for the treatment given here it is not essential to assume that the bonding between the fibers and the matrix is perfect; the only assumption is that the bonding strength is equal for all fibers. However, if the bonding strength is equal for all fibers, it is likely that the bonding is perfect, or at least the imperfections in the bonding are not random. We shall also assume that such a composite behaves like a solid with well-defined effective bulk properties like the elastic constants, density, etc. The fibers are treated as structureless entities which do not absorb elastic energy. It should be emphasized that none of the assumptions made above for a perfect composite are particularly drastic. These are standard assumptions which have been made in most of the theoretical work published so far. Henceforth, unless otherwise stated, we shall consider only a perfect composite.

Since all the fibers have been assumed to be well aligned, the composite will behave like a discrete lattice in a plane perpendicular to the fibers and like an elastic continuum in the direction parallel to the fibers. If the $z$ axis is taken along the length of a fiber and $x$ and $y$ axes in a plane perpendicular to it, then $k_z$ will be continuously distributed from zero up to a certain maximum value whereas $k_x$ and $k_y$ will be allowed only certain discrete values which will be distributed in the appropriate Brillouin zone. Thus, the final picture of the composite we have proposed is that it behaves like a Born–von Karman solid in the $xy$ plane and a Debye solid in the $z$ direction. In a general direction it will have a mixed behavior. This implies that an elastic wave traveling through the composite will suffer dispersion in most of the directions except, of course, if the wavelength is much larger than the fiber spacing. Only the wave traveling in the $z$ direction will be free from dispersion. The dispersion effects will be most prominent for a wave traveling in the $xy$ plane.

In order to apply the Born–von Karman method to the present model of a composite, it is essential to define the force constants between the fibers or the change in the potential energy of the composite due to the fibers' displacements. The concept of potential energy itself at equilibrium may be a bit dubious because there is no direct interaction between the fibers. However, the force constants between the fibers, and therefore the change in the potential energy of the composite during an elastic disturbance, can be defined by noting that a fiber can feel the presence of other fibers through the matrix to which all the fibers are bound. If a fiber is displaced from its original position it would stretch or compress the matrix in its surrounding region, thus exerting a force on neighboring fibers. This is precisely the force constant, being the force on a fiber when another fiber is displaced by a unit amount. We shall assume that the matrix is stretched or compressed only locally and that the disturbance is transmitted through the composite in a series of local disturbances. Thus, we can restrict the force constants to few near neighbors. In practice, one may have to include the effect of several neighbors because the elastic disturbances in the matrix may extend quite far. However, the number of neighbors to be included can only be decided by comparing the results with the experiments.

We have thus been able to define an effective potential energy for the composite during an elastic disturbance. An effective potential between the fibers must exist simply because of the fact that the fibers form a bound assembly in a composite. We are treating the matrix as acting like the Born–von Karman springs connecting the fibers with an effective mass. The effect of the matrix in the present model of a composite is thus included in terms of the force constants and the effective mass of the fibers. The force constants are obviously a measure of the bonding strength or adhesion between the fibers and the matrix. The effective mass of the fibers will be defined later in terms of the density of the composite.

If the force constants between the fibers were known, we could construct the matrix $D(\mathbf{k})$ as given in Section II which would completely determine the propagation of an elastic wave through the composite. In the long wavelength limit, the matrix $D(\mathbf{k})$ reduces to the Green–Christoffel matrix for an elastic continuum. This correspondence enables us to express the effective elastic constants for the composite in terms of the force constants between the fibers. Alternatively, one could obtain the force constants with the help of these relations if the elastic constants were known. This is, generally not possible, however, because there are more force constants than elastic constants.

Since it is difficult to calculate the force constants from first principles, the usual procedure in the lattice dynamics of crystals is to treat the force constants as parameters and obtain their values in a phenomenological way by comparing the theoretical values with the observed dispersion data in symmetry directions. The same procedure could be used for the composites. Unfortunately, we do not have any experimental data on dispersion from composites for short waves; in most of the published experimental work only long waves have been used which do not exhibit dispersion, apart from that arising due to boundary and surface effects. At present, therefore, we have to determine the force constants in terms of the calculated values of the elastic constants. Since the elastic constants are fewer in number, we have to choose arbitrary values for some of the force constants.

In the present paper, two different symmetries for the configuration of fibers in a composite have been assumed, namely, tetragonal and hexagonal. This means that if a plane is cut perpendicular to the fibers, the points of intersection between the fibers and the plane will form a two-dimensional

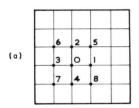 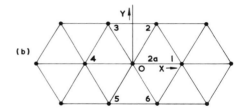

**Fig. 1.** Arrangement of the fibers, represented by dots in the $xy$ plane forming (a) a square lattice (b) a hexagonal lattice.

square and hexagonal lattice, respectively, for the two symmetries (Figs. 1 and 2). The wave motion along this plane will show characteristics of the respective geometry of the lattice. For a wave traveling in the direction of the fibers both the composites will behave similar to a Debye solid.

## IV. Force Constants for a Composite with Tetragonal Symmetry

Let us take the $z$ axis along the fibers and the $x$ and $y$ axes on a plane perpendicular to the fibers. The composite will behave like a square lattice for a

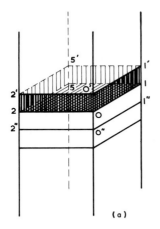 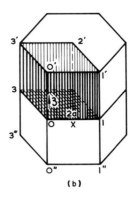

**Fig. 2.** The assumed three-dimensional lattice structure of a composite in (a) tetragonal symmetry (b) hexagonal symmetry. The vertical straight lines denote the fibers, and the numbers (with and without primes) label the "objects" into which the fibers have been divided. The shaded region denotes a unit cell. The vertical dimensions have been exaggerated for greater clarity.

## TABLE I

COORDINATES OF THE OBJECTS CONTRIBUTING TO THE DYNAMICAL MATRIX FOR A
COMPOSITE WITH TETRAGONAL SYMMETRY

| Object | Coordinates | | |
| --- | --- | --- | --- |
| | $x$ | $y$ | $z$ |
| 0 | 0 | 0 | 0 |
| 1 | $a$ | 0 | 0 |
| 2 | 0 | $a$ | 0 |
| 3 | $-a$ | 0 | 0 |
| 4 | 0 | $-a$ | 0 |
| 5 | $a$ | $a$ | 0 |
| 6 | $-a$ | $a$ | 0 |
| 7 | $-a$ | $-a$ | 0 |
| 8 | $a$ | $-a$ | 0 |
| 0' | 0 | 0 | $c$ |
| 1' | $a$ | 0 | $c$ |
| 2' | 0 | $a$ | $c$ |
| 3' | $-a$ | 0 | $c$ |
| 4' | 0 | $-a$ | $c$ |
| 5' | $a$ | $a$ | $c$ |
| 6' | $-a$ | $a$ | $c$ |
| 7' | $-a$ | $-a$ | $c$ |
| 8' | $a$ | $-a$ | $c$ |
| 0" | 0 | 0 | $-c$ |
| 1" | $a$ | 0 | $-c$ |
| 2" | 0 | $a$ | $-c$ |
| 3" | $-a$ | 0 | $-c$ |
| 4" | 0 | $-a$ | $-c$ |
| 5" | $a$ | $a$ | $-c$ |
| 6" | $-a$ | $a$ | $-c$ |
| 7" | $-a$ | $-a$ | $-c$ |
| 8" | $a$ | $-a$ | $-c$ |

two-dimensional wave traveling parallel to the $xy$ plane and like an elastic continuum for a wave traveling parallel to the $z$ axis. For such a structure it is mathematically convenient to divide the composite into $L$ equally spaced layers parallel to the $xy$ plane separated by a distance $c$ and then consider the limiting case as $c$ becomes very small (of the order of atomic spacing in a fiber). The composite problem is thus reduced to that of an ordinary crystal with tetragonal lattice structure. This structure is shown in Figs. 1a and 2a.

Effectively, we are dividing a fiber into $L$ portions equally distributed in $L$ planes. We shall refer to these portions as objects. Let the mass of each object be $m$. It should be emphasized that $Lm$ is not supposed to represent the real mass of the fiber. The quantity $m$ is defined in terms of $\rho$, the average density of the whole composite, viz.

$$m = \rho V \tag{13}$$

where $V$ is the volume per object.

The unit cell $01520'\,1'\,5'\,2'$ is denoted by the shaded region in Fig. 2a. This is a right prism with a square as its base and having one object at each corner. The number of objects per unit cell is one because the eight objects at each corner of a unit cell are shared by eight unit cells. If $a$ is the length of a side of the square and $c$ is the height of the prism, we have

$$V = a^2 c$$

and from Eq. (13)

$$m = \rho a^2 c \tag{14}$$

Later we shall be able to determine $c$ in terms of a Debye temperature. The quantity $m$ includes in its definition the properties of the composite as a whole and is therefore an effective mass.

The symmetry of the lattice and the operations in its point group are given in Appendix A. Let us take the origin at the point occupied by object O (Figs. 1a and 2a). For the calculation of $D(\mathbf{k})$ we shall include the object O, its first and second neighbors on the $xy$ plane (1–4 and 5–8, respectively), and the objects which are just above (O', 1'–4', and 5'–8', respectively) and just below (O'', 1''–4'', and 5''–8'', respectively) the objects O, 1–4, and 5–8, respectively. The coordinates of these objects with respect to the chosen system of axes are given in Table I.

By symmetry the force constants between O and the objects in the same invariant subspaces will be related as given by the Eq. (9) and (10). The general form of the force constant matrices, subject to symmetry restrictions imposed by Eqs. (9) and (10) using the group operations given in the Appendix A, are given below (the overbars denote minus sign).

$$\phi(O,O) = \begin{bmatrix} \mu_0 & 0 & 0 \\ 0 & \mu_0 & 0 \\ 0 & 0 & \lambda_0 \end{bmatrix}$$

$$\phi(O,1) = \phi(O,3) = - \begin{bmatrix} \mu_1 & 0 & 0 \\ 0 & \nu_1 & 0 \\ 0 & 0 & \lambda_1 \end{bmatrix}$$

$$\phi(O,2) = \phi(O,4) = - \begin{bmatrix} v_1 & O & O \\ O & \mu_1 & O \\ O & O & \lambda_1 \end{bmatrix}$$

$$\phi(O,5) = \phi(O,7) = - \begin{bmatrix} \mu_2 & \eta_2 & O \\ \eta_2 & \mu_2 & O \\ O & O & \lambda_2 \end{bmatrix}$$

$$\phi(O,6) = \phi(O,8) = - \begin{bmatrix} \mu_2 & \bar{\eta}_2 & O \\ \bar{\eta}_2 & \mu_2 & O \\ O & O & \lambda_2 \end{bmatrix}$$

$$\phi(O,O') = \phi(O,O'') = - \begin{bmatrix} \mu_0' & O & O \\ O & \mu_0' & O \\ O & O & \lambda_0' \end{bmatrix}$$

$$\phi(O,1') = \phi(O,3'') = - \begin{bmatrix} \mu_1' & O & \varepsilon_1' \\ O & v_1' & O \\ \varepsilon_1' & O & \lambda_1' \end{bmatrix}$$

$$\phi(O,1'') = \phi(O,3') = - \begin{bmatrix} \mu_1' & O & \bar{\varepsilon}_1' \\ O & v_1' & O \\ \bar{\varepsilon}_1' & O & \lambda_1' \end{bmatrix}$$

$$\phi(O,2') = \phi(O,4'') = - \begin{bmatrix} v_1' & O & O \\ O & \mu_1' & \varepsilon_1' \\ O & \varepsilon_1' & \lambda_1' \end{bmatrix}$$

$$\phi(O,2'') = \phi(O,4') = - \begin{bmatrix} v_1' & O & O \\ O & \mu_1' & \bar{\varepsilon}_1' \\ O & \bar{\varepsilon}_1' & \lambda_1' \end{bmatrix}$$

$$\phi(O,5') = \phi(O,7'') = - \begin{bmatrix} \mu_2' & \eta_2' & \varepsilon_2' \\ \eta_2' & \mu_2' & \varepsilon_2' \\ \varepsilon_2' & \varepsilon_2' & \lambda_2' \end{bmatrix}$$

$$\phi(O, 5'') = \phi(O, 7') = - \begin{bmatrix} \mu_2' & \eta_2' & \bar{\varepsilon}_2' \\ \eta_2' & \mu_2' & \bar{\varepsilon}_2' \\ \bar{\varepsilon}_2' & \bar{\varepsilon}_2' & \lambda_2' \end{bmatrix}$$

$$\phi(O, 6') = \phi(O, 8'') = - \begin{bmatrix} \mu_2' & \bar{\eta}_2' & \bar{\varepsilon}_2' \\ \bar{\eta}_2' & \mu_2' & \varepsilon_2' \\ \bar{\varepsilon}_2' & \varepsilon_2' & \lambda_2' \end{bmatrix}$$

$$\phi(O, 6'') = \phi(O, 8') = - \begin{bmatrix} \mu_2' & \bar{\eta}_2' & \varepsilon_2' \\ \bar{\eta}_2' & \mu_2' & \bar{\varepsilon}_2' \\ \varepsilon_2' & \bar{\varepsilon}_2' & \lambda_2' \end{bmatrix} \tag{15}$$

From Eq. (6) we get

$$\mu_0 = 2(\mu_1 + 2\mu_2 + 2\mu_1' + 4\mu_2' + \mu_0' + \nu_1 + 2\nu_1')$$
$$\lambda_0 = 2(2\lambda_1 + 2\lambda_2 + \lambda_0' + 4\lambda_1' + 4\lambda_2') \tag{16}$$

In terms of these force constants we can now construct the dynamical matrix. We notice from the Eqs. (15) and (16) that we have 16 independent force constants. However, as we shall see in the next section, in the limit of small $c$, all of them do not contribute independently to the dynamical matrix. Only ten independent combinations of the above force constants contribute to the dynamical matrix. We have therefore ten independent parameters in the present model.

## V. Dynamical Matrix for a Composite with Tetragonal Symmetry

The elements of the dynamical matrix as defined in the Eq. (11) and (12) are given by ($D_{ij} = D_{ji}$)

$$D_{ij}(\mathbf{k}) = \frac{1}{m} \sum_1 \phi_{ij}(1) \exp[2\pi i \mathbf{k} \cdot \mathbf{r}(1)] \tag{17}$$

Using Table I for the components of $\mathbf{r}(1)$ and Eqs. (15) and (16) for the force constant matrices $\phi(1)$, we get the following expressions for the elements of the dynamical matrix (the matrix elements are in units of $1/m$, $c$ is in units of $a$, and the subscripts 1, 2, and 3 refer to $x$, $y$, and $z$, respectively)

$$\begin{aligned} D_{11}(\mathbf{k}) = {} & 2\mu_1(1 - \cos k_1) + 4\mu_2(1 - \cos k_1 \cos k_2) \\ & + 2\nu_1(1 - \cos k_2) + 2\mu_0'(1 - \cos ck_3) \\ & + 4\mu_1'(1 - \cos k_1 \cos ck_3) + 4\nu_1'(1 - \cos k_2 \cos ck_3) \\ & + 8\mu_2'(1 - \cos k_1 \cos k_2 \cos ck_3) \end{aligned}$$

$$D_{22}(\mathbf{k}) = 2\mu_1(1-\cos k_2) + 4\mu_2(1-\cos k_1 \cos k_2)$$
$$+ 2v_1(1-\cos k_1) + 2\mu_0'(1-\cos ck_3)$$
$$+ 4\mu_1'(1-\cos k_2 \cos ck_3) + 4v_1'(1-\cos k_1 \cos ck_3)$$
$$+ 8\mu_2'(1-\cos k_1 \cos k_2 \cos ck_3)$$

$$D_{33}(\mathbf{k}) = 2\lambda_1(2-\cos k_1-\cos k_2) + 4\lambda_2(1-\cos k_1 \cos k_2)$$
$$+ 2\lambda_0'(1-\cos ck_3) + 4\lambda_1'(2-\cos k_1 \cos ck_3 - \cos k_2 \cos ck_3)$$
$$+ 8\lambda_2'(1-\cos k_1 \cos k_2 \cos ck_3)$$

$$D_{12}(\mathbf{k}) = 4\eta_2 \sin k_1 \sin k_2 + 8\eta_2' \sin k_1 \sin k_2 \cos ck_3$$
$$D_{13}(\mathbf{k}) = 4\xi_1' \sin k_1 \sin ck_3 + 8\xi_2' \sin k_1 \sin ck_3 \cos k_2$$
$$D_{23}(\mathbf{k}) = 4\xi_1' \sin k_2 \sin ck_3 + 8\xi_2' \sin k_2 \sin ck_3 \cos k_1 \tag{18}$$

The Brillouin zone for the present model is defined by

$$-\pi/c \leqslant k_3 \leqslant \pi/c \tag{19}$$

$$-\pi \leqslant k_1, \qquad k_2 \leqslant \pi \tag{20}$$

For a discrete lattice $k_i$ are allowed only discrete values. In the present case, to allow for the transition from a discrete lattice to a continuum in $z$ direction, we allow $k_3$ to assume continuous values in the above range.

For finite nonzero values of $c$, the composite will behave like a Debye solid rather than an ordinary elastic continuum. In a Debye continuum the components $k_i$ of the wave vector are confined in a finite zone and the maximum value of $k_i$, or the dimensions of the zone, are related to the sound velocities in different directions and can be used to define a characteristic temperature known as the Debye temperature. The usual approximation is to replace the zone of $\mathbf{k}$ values by a sphere and thus define a single Debye temperature. For an anisotropic solid it is more realistic to define different Debye temperatures in the $x$, $y$, and $z$ directions. Such an approach has been used for selenium and tellurium by Kothari and Tewary (1962), bismuth by Kothari and Tewary (1963), and for graphite by Kothari and Singwi (1957). Using the same principle we shall define $\theta_3$, the Debye temperature in the $z$ direction, in terms of the maximum value of $k_3$ as given in Eq. (19). Such a distinction between the $z$ direction and the $xy$ plane is consistent with the basic assumptions of the present model. For the $xy$ plane we do not have to introduce any characteristic temperature since we are explicitly taking into account the structure of the composite. It is of course possible to define a Debye temperature for the $xy$ plane as well with a parametric dependence on temperature as is usually done for the interpretation of experimental data on specific heat in the Born–von Karman theory.

We define $\theta_3$ with the help of following two relations

$$k_0\theta_3 = \hbar\omega_3^{\text{max}} \tag{21}$$

$$\omega_3(k_3) = (c_{33}/\rho)^{1/2}k_3 \tag{22}$$

where $k_0$ is the Boltzmann constant, $\hbar$ is the Planck constant, and $c_{33}$ is the elastic constant in appropriate units. From Eqs. (19), (21), and (22) we get

$$ca = \frac{\hbar(c_{33}/\rho)^{1/2}}{2k_0\theta_3} \approx 10^{-8}\,\text{cm} \tag{23}$$

which, as expected, is of the order of the atomic spacing. Substituting this estimate in Eq. (14) we get an estimate of the effective mass $m$ of $10^{-16}$ gm.

To make the transition from a Born–von Karman solid to a Debye solid in the $z$ direction we shall use Born's method of long waves and expand $\sin ck_3$ and $\cos ck_3$ as follows:

$$\sin ck_3 = ck_3 \tag{24}$$

$$\cos ck_3 = 1 - (c^2 k_3^2/2) \tag{25}$$

where cubic and higher powers of $k_3$ have been neglected. Substituting these expressions in the Eqs. (18) we get

$$
\begin{aligned}
D_{11}(\mathbf{k}) = {}& 2(\mu_1 + 2\mu_1')(1-\cos k_1) + 4(\mu_2 + 2\mu_2')(1-\cos k_1\cos k_2) \\
& + 2(\nu_1 + 2\nu_1')(1-\cos k_2) \\
& + c^2 k_3^2(\mu_0' + 2\mu_1'\cos k_1 + 2\nu_1'\cos k_2 + 4\mu_2'\cos k_1\cos k_2)
\end{aligned} \tag{26}
$$

$$
\begin{aligned}
D_{22}(\mathbf{k}) = {}& 2(\mu_1 + 2\mu_1')(1-\cos k_2) + 4(\mu_2 + 2\mu_2')(1-\cos k_1\cos k_2) \\
& + 2(\nu_1 + 2\nu_1')(1-\cos k_1) \\
& + c^2 k_3^2(\mu_0' + 2\mu_1'\cos k_2 + 2\nu_1'\cos k_1 + 4\mu_2'\cos k_1\cos k_2)
\end{aligned} \tag{27}
$$

$$
\begin{aligned}
D_{33}(\mathbf{k}) = {}& 2(\lambda_1 + 2\lambda_1')(2-\cos k_1 - \cos k_2) + 4(\lambda_2 + 2\lambda_2')(1-\cos k_1\cos k_2) \\
& + c^2 k_3^2(\lambda_0' + 2\lambda_1'\cos k_1 + 2\lambda_1'\cos k_2 + 4\lambda_2'\cos k_1\cos k_2)
\end{aligned} \tag{28}
$$

$$D_{12}(\mathbf{k}) = 4(\eta_2 + 2\eta_2' - \eta_2' c^2 k_3^2)\sin k_1\sin k_2 \tag{29}$$

$$D_{13}(\mathbf{k}) = 4(\xi_1' + 2\xi_2'\cos k_2)ck_3\sin k_1 \tag{30}$$

$$D_{23}(\mathbf{k}) = 4(\xi_1' + 2\xi_2'\cos k_1)ck_3\sin k_2 \tag{31}$$

To simplify the above equations further, let us recall that the distance between two neighboring objects on the same fiber is of the order of an angstrom, whereas on different fibers it is of the order of a micron. Hence $\mu_0'$ and $\lambda_0'$ will be several orders of magnitude larger than the other force constants. We shall, therefore, neglect all the terms in the coefficient of $k_3^2$ except $\mu_0'$ in

Eqs. (26) and (27) and $\lambda_0'$ in Eq. (28). Further, we notice in Eq. (29) that the last term in the parenthesis will make a significant contribution only when $k_3 \approx 1/c$ which corresponds to a wave length of the order of an angstrom. Since in the present paper we are not interested in such short waves, we can neglect the term containing $c^2 k_3^2$ in Eq. (29)

With these approximations, we finally get the following expressions for the elements of the dynamical matrix

$$D_{11}(\mathbf{k}) = 2A(1 - \cos k_1) + 2B(1 - \cos k_2)$$
$$+ 4D(1 - \cos k_1 \cos k_2) + E k_3^2 \qquad (32)$$

$$D_{22}(\mathbf{k}) = 2A(1 - \cos k_2) + 2B(1 - \cos k_1)$$
$$+ 4D(1 - \cos k_1 \cos k_2) + E k_3^2 \qquad (33)$$

$$D_{33}(\mathbf{k}) = 2F(2 - \cos k_1 - \cos k_2)$$
$$+ 4G(1 - \cos k_1 \cos k_2) + H k_3^2 \qquad (34)$$

$$D_{12}(\mathbf{k}) = 4K \sin k_1 \sin k_2 \qquad (35)$$

$$D_{13}(\mathbf{k}) = 4L(1 + p \cos k_2) k_3 \sin k_1 \qquad (36)$$

$$D_{23}(\mathbf{k}) = 4L(1 + p \cos k_1) k_3 \sin k_2 \qquad (37)$$

where

$$A = \mu_1 + 2\mu_1'$$

$$B = v_1 + 2v_1'$$

$$D = \mu_2 + 2\mu_2'$$

$$E = \mu_0' c^2$$

$$F = \lambda_1 + 2\lambda_1'$$

$$G = \lambda_2 + 2\lambda_2'$$

$$H = \lambda_0' c^2$$

$$K = \eta_2 + 2\eta_2'$$

$$L = \xi_1' c$$

$$p = 2\xi_2'/\xi_1'$$

Thus we see that $D(k)$, in its final form contains ten independent parameters. As we shall see in the next section, seven of these can be determined in terms of the elastic constants.

## VI. Dynamical Matrix in the Long Wavelength Limit

For a wave much longer than the fiber spacing, the composite will behave like an elastic continuum. In this limit ($|k| \to 0$) the matrix $D(\mathbf{k})$ must be identical with the Green–Christoffel matrix. Thus, using the method of long waves, we shall expand the sine and cosine functions in Eqs. (32)–(37), keeping only up to second-order terms and compare the resulting matrix with the Green–Christoffel matrix for a solid with tetragonal symmetry.

Substituting for

$$\sin k_i = k_i \quad \text{and} \quad \cos k_i = 1 - k_i^2/2$$

we obtain the following expressions for the elements of $D(\mathbf{k})$

$$D_{11}(\mathbf{k}) = (A+2D)k_1{}^2 + (B+2D)k_2{}^2 + Ek_3{}^2 \tag{38}$$

$$D_{22}(\mathbf{k}) = (B+2D)k_1{}^2 + (A+2D)k_2{}^2 + Ek_3{}^2 \tag{39}$$

$$D_{33}(\mathbf{k}) = (F+2G)(k_1{}^2+k_2{}^2) + Hk_3{}^2 \tag{40}$$

$$D_{12}(\mathbf{k}) = 4Kk_1 k_2 \tag{41}$$

$$D_{13}(\mathbf{k}) = 4L(1+p)k_1 k_3 \tag{42}$$

$$D_{23}(\mathbf{k}) = 4L(1+p)k_2 k_3 \tag{43}$$

The elements of the Green–Christoffel matrix $\Lambda(\mathbf{k})$ for a solid with tetragonal symmetry have the form (Fedorov, 1968)

$$\Lambda_{11}(\mathbf{k}) = c_{11}k_1{}^2 + c_{66}k_2{}^2 + c_{44}k_3{}^2 \tag{44}$$

$$\Lambda_{22}(\mathbf{k}) = c_{66}k_2{}^2 + c_{11}k_1{}^2 + c_{44}k_3{}^2 \tag{45}$$

$$\Lambda_{33}(\mathbf{k}) = c_{44}(k_1{}^2+k_2{}^2) + c_{33}k_3{}^2 \tag{46}$$

$$\Lambda_{12}(\mathbf{k}) = (c_{12}+c_{66})k_1 k_2 \tag{47}$$

$$\Lambda_{13}(\mathbf{k}) = (c_{13}+c_{44})k_1 k_3 \tag{48}$$

$$\Lambda_{23}(\mathbf{k}) = (c_{13}+c_{44})k_2 k_3 \tag{49}$$

Equating the coefficients of $k_i^2$ in the corresponding elements of the matrices $D(k)$ and $\Lambda(\mathbf{k})$, we get the following relations between the force constants and the elastic constants

$$A + 2D = c_{11}$$

$$B + 2D = c_{66}$$

$$E = c_{44}$$

$$F + 2G = c_{44} \tag{50}$$

$$H = c_{33}$$

$$4K = c_{12} + c_{66}$$

$$4L(1+p) = c_{13} + c_{44}$$

where the elastic constants are in units of $(4\pi a^2 \rho)^{-1}$, and the force constants, as may be recalled, are in units of $m^{-1}$.

To determine all the ten force constants it is necessary to have further information, such as measurements on dispersion in composites, i.e., sound velocities as a function of wavelength in the short wavelength region. In the absence of such measurements and to provide a basis for the numerical calculations in the present paper, we shall assume reasonable but arbitrary values for three of the force constants, namely, $D$, $G$, and $p$. The remaining force constants will be determined in terms of the elastic constants using the calculated values of Heaton (1970).

Heaton (1970) has calculated the elastic constants of composites with various fiber matrix combinations, by using a point matching technique, in terms of the elastic constants of the constituents of the composite. It may be remarked here that such a procedure involves the solution of a rather complicated boundary value problem and may depend sensitively upon the choice of matching points. As a result, the calculated values should be looked upon with some caution. An alternative and preferable procedure would be to determine all the force constants in terms of experimentally measured quantities.

The elastic constants and therefore the force constants for a composite depend upon the elastic constants of the constituents and the loading of the composite, i.e., the relative volume occupied by the fibers, in addition to, of course, several other factors. Heaton's values for the elastic constants of a tetragonal composite, are given in Table II for various percentages of the fiber loading; the constants are in units of $G^m$, where $G^m$ is the shear modulus of the matrix. Our final results for the frequencies will therefore be given in units of $\omega_t$, where $\omega_t$ is given by

$$\omega_t^2 = G^m / 4\pi a^2 \rho \tag{51}$$

**TABLE II**

ELASTIC CONSTANTS FOR A TETRAGONAL COMPOSITE[a, b]

| Elastic constant | Percentage of fiber loading | | |
|---|---|---|---|
| | 30 | 50 | 70 |
| $c_{11}$ | 6.118 | 8.164 | 11.576 |
| $c_{12}$ | 2.852 | 3.204 | 3.892 |
| $c_{13}$ | 3.599 | 4.917 | 7.170 |
| $c_{33}$ | 86.359 | 141.560 | 197.800 |
| $c_{44}$ | 1.712 | 2.593 | 4.738 |
| $c_{66}$ | 1.423 | 1.856 | 2.817 |

[a] In units of $G^m$.
[b] From Heaton (1970).

and the velocities are expressed in units of $c_t$, where

$$c_t = \omega_t 2\pi a \tag{52}$$

## VII. Wave Propagation in a General Direction

A knowledge of the dynamical matrix is sufficient to define all the dynamical properties of the composite. The three eigenvalues of the dynamical matrix for a value of **k** give the frequencies of the three types of waves, and the corresponding eigenvectors describe the polarization of those three waves. If $e_i$ denote the three Cartesian components ($i = 1$, 2, and 3) of an eigenvector corresponding to an eigenvalue $\omega^2$ we may write

$$D_{11}e_1 + D_{12}e_2 + D_{13}e_3 = \omega^2 e_1 \tag{53a}$$

$$D_{21}e_1 + D_{22}e_2 + D_{23}e_3 = \omega^2 e_2 \tag{53b}$$

$$D_{31}e_1 + D_{32}e_2 + D_{33}e_3 = \omega^2 e_3 \tag{53c}$$

with the condition

$$\begin{vmatrix} D_{11} - \omega^2 & D_{12} & D_{13} \\ D_{21} & D_{22} - \omega^2 & D_{23} \\ D_{31} & D_{32} & D_{33} - \omega^2 \end{vmatrix} = 0 \tag{54}$$

for a nontrivial solution. The frequencies of the three waves are given as the solution of the secular equation (54). The solution of the secular equation is considerably simplified for a symmetry direction in **k** space and fortunately, most of the information of interest about the composite can be obtained from the waves traveling in a symmetry direction or a symmetry plane. We shall discuss these special directions in the later sections. In this section, for the sake of completeness, we shall give the formal solution of the secular equation which is valid for any value of **k**.

The three, all real solutions of the secular equation (54), are

$$\omega_1{}^2 = S_1 + S_2 - P_2/3 \tag{55a}$$

$$\omega_2{}^2 = -\tfrac{1}{2}(S_1 + S_2) - P_2/3 + i\sqrt{3}(S_1 - S_2)/2 \tag{55b}$$

$$\omega_3{}^2 = -\tfrac{1}{2}(S_1 + S_2) - P_2/3 - i\sqrt{3}(S_1 - S_2)/2 \tag{55c}$$

where

$$S_1 = [R + (Q^3 + R^2)^{1/2}]^{1/3}$$

$$S_2 = [R - (Q^3 + R^2)^{1/2}]^{1/3}$$

$$Q = P_1/3 - P_2{}^2/9$$

$$R = [(P_1 P_2 - 3P_0)/6] - P_2{}^3/27$$

$$P_0 = -D_{11} D_{22} D_{33} + D_{11} D_{23}^2 + D_{22} D_{13}^2$$
$$\qquad + D_{33} D_{12}^2 - 2D_{12} D_{23} D_{31}$$

$$P_1 = D_{11} D_{22} + D_{22} D_{33} + D_{33} D_{11} - D_{12}^2 - D_{23}^2 - D_{31}^2$$

$$P_2 = D_{11} + D_{22} + D_{33}$$

The phase velocity $C_p$ and the group velocity $C_g$ can be obtained from the dispersion relations by using

$$C_p = \omega/|\mathbf{k}| \tag{56}$$

and

$$C_g = d\omega/dk \tag{57}$$

## VIII. Propagation in the $xy$ Plane: Bonding Strength of the Fibers

We shall see in this section that the propagation of an elastic wave in the $xy$ plane can yield complete information about the bonding strength of the fibers. It is obvious that only a wave traveling across the fibers will feel the discrete nature of the composite and will therefore be most sensitive to the bonding strength of the fibers. For a wave traveling in the $z$ direction, the discrete nature of the composite will manifest itself only through the distortions in the wavefront. This distortion is taken into account in the present model only in an average way by defining an effective value of the elastic constants for the whole composite. In any case, such a distortion will depend more on the fiber misalignments than on the bonding strength.

For a wave traveling in the $xy$ plane $k_3 = 0$, the matrix $D(\mathbf{k})$ assumes a block diagonal form with the following structure

$$D(\mathbf{k}) = \begin{pmatrix} D^p & 0 \\ 0 & D^z \end{pmatrix} \tag{58}$$

where the block $D^p$ is a $2 \times 2$ matrix and $D^z$ is a $1 \times 1$ matrix, i.e., a pure number. The elements of these matrices, obtained from Eqs. (32)–(37) are

$$D^p_{11}(\mathbf{k}) = 2A(1 - \cos k_1) + 2B(1 - \cos k_2) + 4D(1 - \cos k_1 \cos k_2)$$

$$D^p_{12}(\mathbf{k}) = D^p_{21}(\mathbf{k}) = 4K \sin k_1 \sin k_2 \tag{59}$$

$$D^p_{22}(\mathbf{k}) = 2A(1 - \cos k_2) + 2B(1 - \cos k_1) + 4D(1 - \cos k_1 \cos k_2)$$

and

$$D^z(\mathbf{k}) = 2F(2 - \cos k_1 - \cos k_2) + 4G(1 - \cos k_1 \cos k_2) \tag{60}$$

The eigenvalue corresponding to the element $D^z(\mathbf{k})$ is immediately known and is given by

$$\omega_z{}^2(\mathbf{k}) = 2F(2 - \cos k_1 - \cos k_2) + 4G(1 - \cos k_1 \cos k_2) \tag{61}$$

The eigenvector corresponding to this eigenvalue is $[0, 0, 1]$. This wave is always polarized perpendicular to the $xy$ plane and is therefore a true shear wave for all values of $k_1$ and $k_2$.

The frequencies associated with the other two waves, polarized in the $xy$ plane, can be obtained by diagonalizing the matrix $D^p(\mathbf{k})$. If $\omega_{p1}$ and $\omega_{p2}$ represent the frequencies of the two waves polarized at angles $\theta_p$ and $\theta_p + \pi/2$, respectively, from the $x$ axis in the $xy$ plane, then we have

$$\omega_{p1}^2 = (A + B)(2 - \cos k_1 - \cos k_2) + 4D(1 - \cos k_1 \cos k_2)$$
$$+ [(A - B)^2 (\cos k_2 - \cos k_1)^2 + 16K^2 \sin^2 k_1 \sin^2 k_2]^{\frac{1}{2}} \tag{62}$$

$$\mathbf{e}_{p1} = [\cos \theta_p, \sin \theta_p, 0]$$

$$\omega_{p2}^2 = (A + B)(2 - \cos k_1 - \cos k_2) + 4D(1 - \cos k_1 \cos k_2)$$
$$- [(A - B)^2 (\cos k_2 - \cos k_1)^2 + 16K^2 \sin^2 k_1 \sin^2 k_2]^{\frac{1}{2}} \tag{63}$$

and

$$\mathbf{e}_{p2} = [-\sin \theta_p, \cos \theta_p, 0]$$

where $\mathbf{e}_{p1}$ denote the polarization vectors and

$$\tan 2\theta_p = \frac{4K}{A - B} \left[ \frac{\sin k_1 \sin k_2}{\cos k_2 - \cos k_1} \right] \tag{64}$$

We notice from Eq. (64) that for a general value of $\mathbf{k}$ the waves $p_1$ and $p_2$ are not truly longitudinal or transverse waves, since the displacements, given by the polarization vectors are in a direction different than that of the incident wave, i.e.,

$$\theta_p \neq \theta_k$$

where

$$\tan \theta_k = k_2/k_1$$

However, it is customary to refer to the waves $p_1$ and $p_2$ as the quasilongitudinal and quasitransverse waves, respectively, because the angle between $\mathbf{e}_{p1}$ and $\mathbf{k}$ is usually small. This is true in the present case but not, as we shall see later, for the wave traveling in a plane perpendicular to the $xy$ plane.

The phase velocities for the three waves can be obtained very simply from the above equations by using Eq. (56). The group velocities, using Eq. (57) are given below.

(1) Quasilongitudinal wave ($p_1$-wave)

$$C_{gx} = \frac{\sin k_1}{2\omega_{p1}} \left[ A + B + 4D \cos k_2 + \frac{1}{x} \right.$$
$$\left. \times \{(A-B)^2 (\cos k_2 - \cos k_1) + 16K^2 \sin^2 k_2 \cos k_1\} \right]$$

$$C_{gy} = \frac{\sin k_2}{2\omega_{p1}} \left[ A + B + 4D \cos k_1 + \frac{1}{x} \right. \tag{65}$$
$$\left. \times \{(A-B)^2 (\cos k_1 - \cos k_2) + 16K^2 \sin^2 k_1 \cos k_2\} \right]$$

$$C_{gz} = 0$$

(2) Quasitransverse wave ($p_2$-wave)

$$C_{gx} = \frac{\sin k_1}{2\omega_{p2}} \left[ A + B + 4D \cos k_2 - \frac{1}{x} \right.$$
$$\left. \times \{(A-B)^2 (\cos k_2 - \cos k_1) + 16K^2 \sin^2 k_2 \cos k_1\} \right]$$

$$C_{gy} = \frac{\sin k_2}{2\omega_{p2}} \left[ A + B + 4D \cos k_1 - \frac{1}{x} \right. \tag{66}$$
$$\left. \times \{(A-B)^2 (\cos k_1 - \cos k_2) + 16K^2 \sin^2 k_1 \cos k_2\} \right]$$

$$C_{gz} = 0$$

where

$$x^2 = (A-B)^2 (\cos k_2 - \cos k_1)^2 + 16K^2 \sin^2 k_1 \sin^2 k_2$$

(3) Transverse wave ($z$-wave)

$$C_{gx} = \frac{\sin k_1}{\omega_z}[F+2G\cos k_2]$$

$$C_{gy} = \frac{\sin k_2}{\omega_z}[F+2G\cos k_1]$$

$$C_{gz} = 0 \qquad\qquad (67)$$

In the above expressions, the subscripts $x$ and $y$ represent the $x$ and $y$ components of the group velocity. The $z$ component in each case is zero. The magnitude $C_g$ and the angle $\theta_g$ from the $x$ axis in each case are given by

$$C_g^{\,2} = C_{gx}^2 + C_{gy}^2 \qquad\qquad (68)$$

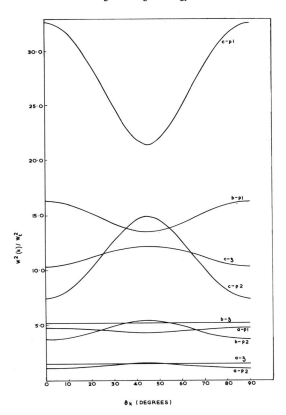

**Fig. 3.** Angular dependence of the dispersion relations for a tetragonal composite in the $xy$ plane. The values of $|k|/\pi$ are for (a) 0.25, (b) 0.5, and (c) 1.0. The angles are measured from the $x$ axis.

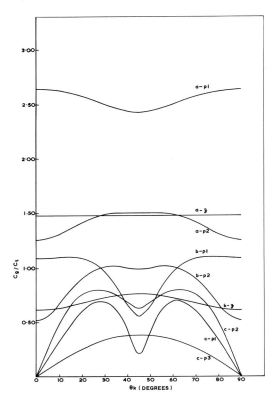

**Fig. 4.** Angular dependence of the group velocities of the elastic waves in the $xy$ plane in a tetragonal composite. The values of $|\mathbf{k}|/\pi$ are for (a) 0.25, (b) 0.5, and (c) 1.0. The angles are measured from the $x$ axis.

and

$$\tan \theta_g = \frac{C_{gy}}{C_{gx}} \tag{69}$$

The group velocity $C_g$ is the velocity with which the energy flux actually travels through the composite and the angle $\theta_g$ defines the direction of the flux. We notice that in general $\theta_g$ is different than $\theta_k$, the direction of the incident wave.

Experimentally, it may be often more convenient to measure the velocities as a function of $\theta_k$ for a fixed wavelength rather than as a function of wavelength itself. Such measurements will also yield the values of the force constants. For this purpose, we have plotted in Figs. 3 and 4 $\omega^2(\mathbf{k})$ and $C_g$ as a function of

$\theta_k$ for $|\mathbf{k}|/\pi = 0.25,\ 0.5$, and $1.0$. The angles $\theta_p$ and $\theta_g$ are also plotted in Fig. 5 as a function of $\theta_k$ for $|\mathbf{k}|/\pi = 0.25$. The force constants have been obtained with the help of Eq. (50) and Table II. The constant $p$ does not contribute independently to the results given here. For $D$ and $G$ we have used the following arbitrary values: $D = A/4,\ G = F/4$.

The above equations have particularly simple forms in the symmetry directions and provide convenient directions for the dispersion measurements. We shall give here explicit expressions for the frequency and the group velocity as a function of $\mathbf{k}$ in $\langle 100 \rangle$ and $\langle 110 \rangle$ directions. In these directions $p_1$ and $p_2$ waves become true longitudinal and transverse waves, respectively. The $p_2$ wave will be referred to as the $p$-transverse wave and the wave polarized perpendicular to the $xy$ plane will be referred to as the $z$-transverse wave.

### A. $\langle 100 \rangle$ direction

In this direction $k_1 = k$, $k_2 = 0$, $k_3 = 0$, and $\theta_k = \theta_p = \theta_g = 0$. The direction $\langle 010 \rangle$ is equivalent by symmetry. The frequencies and the group velocities of different waves are

(1) Longitudinal wave

$$\omega^2 = 2(A+2D)(1-\cos k)$$
$$C_g = (A+2D)^{\frac{1}{2}}\cos(k/2) \tag{70}$$

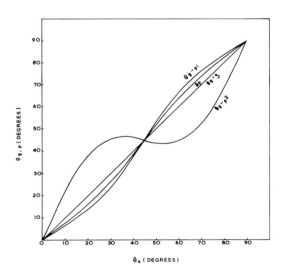

*Fig. 5.* Angular dependence of the directions of polarizsation ($\theta_p$) and the flux ($\theta_g$) for elastic waves in the $xy$ plane in a tetragonal composite for $|\mathbf{k}|/\pi = 0.25$. All angles are measured from the $x$ axis.

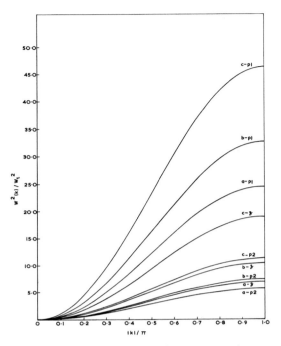

**Fig. 6.** Dispersion relations in $\langle 100 \rangle$ direction for a tetragonal composite with (a) 30% fiber loading, (b) 50% fiber loading, and (c) 70% fiber loading.

(2) $p$-transverse wave

$$\omega^2 = 2(B+2D)(1-\cos k)$$
$$C_g = (B+2D)^{\frac{1}{2}}\cos(k/2) \tag{71}$$

(3) $z$-transverse wave

$$\omega^2 = 2(F+2G)(1-\cos k)$$
$$C_g = (F+2G)^{\frac{1}{2}}\cos(k/2) \tag{72}$$

We note that in this direction the relevant combinations of the force constants are completely determined by the elastic constants. This is a consequence of taking only up to a second neighbor interactions. The calculated values of $\omega^2(\mathbf{k})$ and $C_g$ are plotted as a function of $k$ in Fig. 6 and Fig. 7, respectively.

## B. $\langle 110 \rangle$ direction

In this direction $k_1 = k_2 = k$, $k_3 = 0$, and and $\theta_k = \theta_p = \pi/4$, and the following expressions are obtained for the frequencies and group velocities of the three waves:

(1) Longitudinal wave

$$\omega^2 = 2(A+B)(1-\cos k) + 4(D+K)\sin^2 k$$

$$C_{\mathbf{g}} = V_1 \cos(k/2) \tag{73}$$

where

$$V_1 = \frac{A+B+4(D+K)\cos k}{[2(A+B)+4(D+K)(1+\cos k)]^{1/2}}$$

(2) $p$-transverse wave

$$\omega^2 = 2(A+B)(1-\cos k) + 4(D-K)\sin^2 k$$

$$C_{\mathbf{g}} = V_2 \cos(k/2) \tag{74}$$

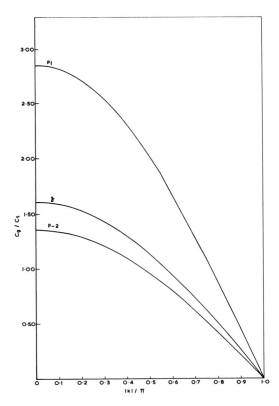

**Fig. 7.** Group velocities of the three elastic waves traveling in $\langle 100 \rangle$ direction in a tetragonal composite.

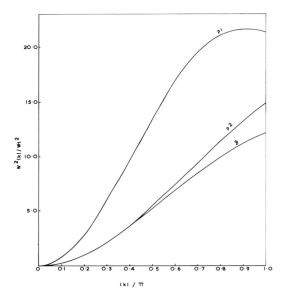

**Fig. 8.** Dispersion relations in $\langle 110 \rangle$ direction for a tetragonal composite. $|\mathbf{k}| = \pi$ is not the maximum allowed value of $|\mathbf{k}|$ in this direction.

where

$$V_2 = \frac{A+B+4(D-K)\cos k}{[2(A+B)+4(D-K)(1+\cos k)]^{\frac{1}{2}}}$$

(3) $z$-transverse wave

$$\omega^2 = 4F(1-\cos k) + 4G\sin^2 k$$

$$C_g = V_z \cos(k/2) \tag{75}$$

where

$$V_z = \frac{F+2G\cos k}{[F+G(1+\cos k)]^{\frac{1}{2}}}$$

$\omega^2(\mathbf{k})$ and $C_g$ are plotted as a function of $k$ for this direction in Fig. 8 and Fig. 9, respectively.

It is interesting to note in the above expressions that the group velocities $C_g$ all vanish when $k = \pi$. This means that a wave of this wavenumber will not be able to travel through the composite and will be reflected. This phenomenon is analogous to the Bragg reflection of X rays from crystals. This reflection is of course a consequence of the assumed translational symmetry and may not occur in a real composite. However, for a densely reinforced composite the assumption of translational symmetry is reasonable and, therefore, the group velocity should become quite small if not zero for $k = \pi$.

## C. Bonding Strength of the Fibers

As has been remarked earlier, we can get useful information about the bonding strength of the fibers from an analysis of the dispersion data for the waves traveling in the $xy$ plane. In the present model, the bonding strength will be defined in terms of the force constants between the fibers. The force constants can be obtained by a measurement of the dispersion in the symmetry directions (or the directional dependence of the velocity in the $xy$ plane), as is usually done in the lattice dynamics of crystals by using neutron scattering techniques.

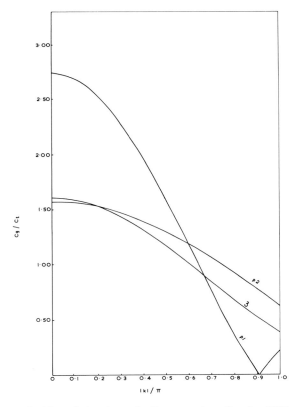

**Fig. 9.** Group velocities of the three elastic waves traveling in $\langle 110 \rangle$ direction in a tetragonal composite.

First, let us note that the matrix $D^p$ is exactly similar to the dynamical matrix of a two-dimensional square lattice of lattice constant equal to $a$ with nearest and next nearest neighbor interactions given by the following force constant matrices:

$$\phi(a,0) = \phi(\bar{a},O) = -\begin{pmatrix} A & O \\ O & B \end{pmatrix}$$

$$\phi(0,a) = \phi(O,\bar{a}) = -\begin{pmatrix} B & O \\ O & A \end{pmatrix}$$

$$\phi(a,a) = \phi(\bar{a},\bar{a}) = -\begin{pmatrix} D & K \\ K & D \end{pmatrix}$$

$$\phi(a,\bar{a}) = \phi(\bar{a},a) = -\begin{pmatrix} D & \bar{K} \\ \bar{K} & D \end{pmatrix}$$

and

$$\phi(O,O) = \begin{pmatrix} A_0 & O \\ O & A_0 \end{pmatrix}$$

where

$$A_0 = 2(A+B+2D)$$

As defined earlier, the force constant $\phi_{xy}(l_1,l_2)$ gives the force on $l_2$ in the $y$ direction when $l_1$ suffers a unit displacement in the $x$ direction. Thus $A_0$ is the force on the fiber when that fiber itself is moved by a unit amount. This force will act in the same direction as the displacement since the force constant matrix is diagonal. $A_0$, therefore, provides a convenient definition of the bonding strength.

The results obtained in this section should assist the development of non-destructive techniques for the evaluation of the strength of the fiber matrix bond. In this connection we note from, for example, Eq. (70) that

$$d\omega^2/dA = 1 - \cos k \tag{76}$$

Equation (76) shows the change in the wave frequency with respect to a change in the bonding strength as a function of the wavelength. We note that $d\omega^2/dA$ is maximum when $k = \pi$. Recalling the units of $k$, we may infer that a measurement of the frequency will be most sensitive to the bonding strength when

$$2\pi a/\lambda = \pi \qquad \text{or} \qquad \lambda \approx 2a \tag{77}$$

i.e., when the wavelength of the incident wave is of the order of the interfiber spacing. This result, of course, could have been expected intuitively.

## IX. Propagation in the $yz$ Plane:

In this section we shall study the propagation of a wave in the $yz$ plane. By symmetry, the $xz$ plane is equivalent to the $yz$ plane.

The elements of the dynamical matrix for the present case are obtained from Eqs. (32)–(37) on putting $k_1 = 0$. We get the block diagonal form for $D(k)$:

$$D(k) = \begin{pmatrix} D^x & O \\ O & D^p \end{pmatrix}$$

where

$$D^x = 2c_{66}(1 - \cos k_2) + c_{44} k_3{}^2 \tag{78}$$

is a number and the elements of the $2 \times 2$ matrix $D^p$ are

$$D^p_{11} = 2c_{11}(1 - \cos k_2) + c_{44} k_3{}^2 \tag{79}$$

$$D^p_{12} = D^p_{21} = (c_{13} + c_{44}) k_3 \sin k_2 \tag{80}$$

$$D^p_{22} = 2c_{44}(1 - \cos k_2) + c_{33} k_3{}^2 \tag{81}$$

As in Section VIII, one of the eigenvalues, the one corresponding to $D^x$ is immediately known. It is given by

$$\omega_x{}^2 = 2c_{66}(1 - \cos k_2) + c_{44} k_3{}^2 \tag{82}$$

The eigenvector corresponding to this eigenvalue is $[1, 0, 0]$. This wave is always polarized perpendicular to $yz$ plane for all values of $k$ and is a true transverse wave.

If $\omega_{p1}$ and $\omega_{p2}$ represent the frequencies of the two waves polarized at angles $\theta_p$ and $\theta_p + \pi/2$, respectively, from the $z$ axis in the $yz$ plane, then we have

$$\omega_{p1}^2 = (c_{11} + c_{44})(1 - \cos k_2) + (c_{44} + c_{33})(k_3{}^2/2)$$
$$+ [\{(c_{11} - c_{44})(1 - \cos k_2) + (c_{44} - c_{33})(k_3{}^2/2)\}^2$$
$$+ (c_{13} + c_{44})^2 k_3{}^2 \sin^2 k_2]^{1/2}$$

$$e_{p1} = [\sin \theta_p, \cos \theta_p, 0] \tag{83}$$

$$\omega_{p2}^2 = (c_{11} + c_{44})(1 - \cos k_2) + (c_{44} + c_{33})(k_3{}^2/2)$$
$$- [\{(c_{11} - c_{44})(1 - \cos k_2) + (c_{44} - c_{33})(k_3{}^2/2)\}^2$$
$$+ (c_{13} + c_{44})^2 k_3{}^2 \sin^2 k_2]^{1/2} \tag{84}$$

and

$$e_{p2} = [-\cos \theta_p, \sin \theta_p, 0]$$

where

$$\theta_p = \frac{\pi}{2} - \tfrac{1}{2} \tan^{-1} \frac{(c_{13} + c_{44}) k_3 \sin k_2}{(c_{11} - c_{44})(1 - \cos k_2) + (c_{44} - c_{33}) k_3{}^2/2} \tag{85}$$

We notice that due to a relatively large value of $c_{33}$, the propagation of waves in the $yz$ plane in the present model is almost completely nondispersive. A measurement of dispersion for waves in this plane is, therefore, not likely to yield any information about the bonding strength of the fibers. However, such a measurement can yield some useful information about the alignment of the fibers, which we shall discuss later in this section.

Due to a relatively large value of $c_{33}$, $\theta_p$ is nearly zero unless $k_3 \approx 0$. This means that for most of the angles of incidence the waves through the composite will be polarized in a direction close to the $z$ axis. Thus, in the present case we cannot label the waves as even quasilongitudinal and quasitransverse. We shall therefore, label them simply as $p_1$ and $p_2$. The wave given by Eq. (82) will be labeled as the $x$-transverse wave and is always a true transverse wave.

The group velocities for these waves are

(1) $p_1$ Wave

$$C_{gy} = \frac{\sin k_2}{2\omega_{p1}}\left[c_{11} + c_{44} + \frac{P_y}{X}\right]$$

$$C_{gz} = \frac{k_3}{2\omega_{p1}}\left[c_{44} + c_{33} + \frac{P_z}{X}\right]$$

$$C_{gx} = 0 \qquad\qquad (86)$$

(2) $p_2$ Wave

$$C_{gy} = \frac{\sin k_2}{2\omega_{p1}}\left[c_{11} + c_{44} - \frac{P_y}{X}\right]$$

$$C_{gz} = \frac{k_3}{2\omega_{p2}}\left[c_{44} + c_{33} - \frac{P_z}{X}\right]$$

$$C_{gx} = 0 \qquad\qquad (87)$$

where

$$X^2 = [(c_{11} - c_{44})(1 - \cos k_2) + (c_{44} - c_{33})(k_3{}^2/2)]^2$$
$$+ (c_{13} + c_{44})^2 k_3{}^2 \sin^2 k_2$$

$$P_y = [(c_{11} - c_{44})(1 - \cos k_2) + (c_{44} - c_{33})(k_3{}^2/2)](c_{11} - c_{44})$$
$$+ (c_{13} + c_{44})^2 k_3{}^2 \cos k_2$$

$$P_z = [(c_{11} - c_{44})(1 - \cos k_2) + (c_{44} - c_{33})(k_3{}^2/2)](c_{44} - c_{33})$$
$$+ (c_{13} + c_{44})^2 \sin^2 k_2$$

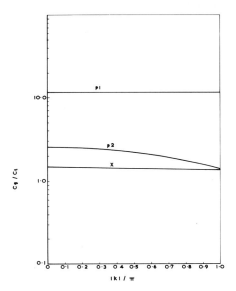

**Fig. 10.** Group velocities of the three elastic waves traveling in the $yz$ plane in a tetragonal composite for $\theta_k = 45°$.

(3) $x$-Transverse wave

$$C_{gy} = c_{66} \sin k_2/\omega_x$$

$$C_{gz} = c_{44} k_3/\omega_x$$

$$C_{gx} = 0 \tag{88}$$

As before, the magnitude and direction relative to the $z$ axis of the group velocity are given by

$$C_g^2 = C_{gy}^2 + C_{gz}^2$$

and

$$\tan \theta_g = C_{gy}/C_{gz} \tag{89}$$

Writing $k_3 = k\cos \theta_k$ and $k_2 = k \sin \theta_k$, we have plotted $C_g$, $\theta_p$, and $\theta_g$ as a function of $k$ for the above three waves in Fig. 10 and Fig. 11 for $\theta_k = 45°$. The dependence of the above quantities on $\theta_k$ is shown in Fig. 12 and Fig. 13 for $|\mathbf{k}|/\pi = 0.25$. Figure 13, showing variations of $\theta_p$ and $\theta_g$ with $\theta_k$ are very interesting from an application point of view. First, we notice from the curve for $\theta_p$ that for most of the directions of the incident wave, one of the outgoing waves is polarized in a direction close to the $z$ axis. Thus a composite can be

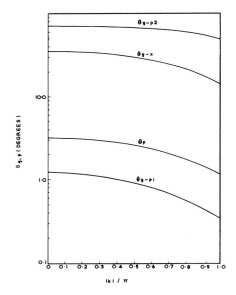

**Fig. 11.** k Dependence of the directions of polarization $(\theta_p)$ and flux travel $(\theta_g)$ for the elastic waves traveling in the $yz$ plane in a tetragonal composite for $\theta_k = 45°$. The angles are measured from the $z$ axis.

used as a polarizer for the stress waves. This effect arises because of a relatively large value of $c_{33}$ as discussed earlier.

Similarly, we notice from the curves for $\theta_g$, that for most of the directions of the incident wave, the outgoing wave travels close to the $z$ direction. If we impose a disturbance at one end of a composite, characterized by a set of waves incident at all angles with respect to the $z$ axis, the outgoing disturbance will be largely confined in a narrow cone around the $z$ axis. Thus, the intensity of the disturbance or the energy will increase by a large amount in this narrow region and will be practically zero across the fibers. Thus, let us assume that we have a set of waves incident at all angles ranging from 0 to $\pi/2$ with respect to the $z$ axis at one end of the composite and with a uniform angular density. For the outgoing waves, the angular distribution function, i.e., the number of waves emerging at an angle $\theta$ from the $z$ axis is given by

$$g(\theta) = \sum_{\theta_k} \delta(\theta - \theta_g(\theta_k)) \tag{90}$$

where $\theta_g(\theta_k)$ shows the functional dependence of $\theta_g$ on $\theta_k$. In this expression we shall consider only the $p_1$ wave, which, being mostly polarized along the $z$ axis, is of maximum interest.

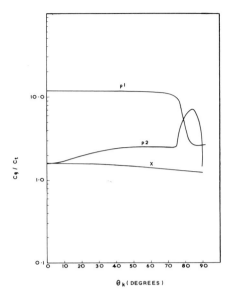

**Fig. 12.** Angular dependence of the group velocities of the elastic waves traveling in the $yz$ plane in a tetragonal composite for $|\mathbf{k}|/\pi = 0.25$. The angles are measured from the $z$ axis.

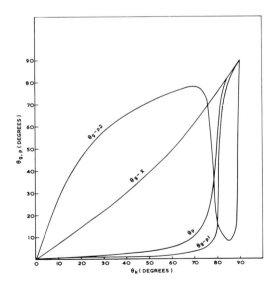

**Fig. 13.** Angular dependence of the directions of polarization $(\theta_p)$ and flux $(\theta_g)$ for elastic waves in the $yz$ plane in a tetragonal composite for $|\mathbf{k}|/\pi = 0.25$. All the angles are measured from the $z$ axis.

The function $g(\theta)$, obtained numerically from the Eqs. (90), (89), and (86), is plotted against $\theta$ in Fig. 14 for $|\mathbf{k}|/\pi = 0.0001, 0.5,$ and $1.0$. To see qualitatively the behavior of $g(\theta)$, let us consider a case when $c_{33}$ is very large and when the off-diagonal terms in $D_p$ are negligible. In this case the $p_1$ wave is always polarized along the $z$ axis. Its frequency is given by

$$\omega^2 = 2c_{11}(1-\cos k_2) + c_{33}k_3{}^2$$

and the group velocity by

$$C_{gy} = c_{11}(\sin k_2/\omega) \tag{91}$$

$$C_{gz} = c_{33}k_3/\omega \tag{92}$$

and

$$\tan\theta_g(\theta_k) = (c_{11}/c_{33})[\sin(\pi \sin \theta_k)/\pi \cos \theta_k] \tag{93}$$

where

$$|k| = \pi$$

From Eq. (93) we see that since $c_{33} \gg c_{11}, \tan\theta_g \approx \theta_g$ is nearly zero unless $\theta_k = \pi/2$ when $\theta_g = \pi/2$. Replacing the sum in Eq. (90) by an integral we have

$$g(\theta) \approx \int_0^{(\pi/2)-\eta} \delta[\theta-\theta_g(\theta_k)]\,d\theta_k + \int_{(\pi/2)-\eta}^{\pi/2} \delta[\theta-\theta_g(\theta_k)]\,d\theta_k$$

$$\approx [(\pi/2)-\eta]\delta(\theta) + \eta\delta[\theta-(\pi/2)] \tag{94}$$

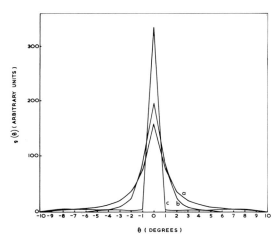

**Fig. 14.** Angular distribution function for a set of elastic waves traveling through a tetragonal composite in the $yz$ plane assuming a uniform distribution function for the angles of incidence. The values of $|\mathbf{k}|/\pi$ are for (a) 0.0001, (b) 0.5 and (c) 1.0. For $\theta > 10°$, $g(\theta) \approx 0$. The angle $\theta$ is measured from the $z$ axis. Only $p_1$ waves have been included.

where $\eta$ is an arbitrary small number. Since $\eta$ can be made as small as we like (depending upon the magnitude of $c_{33}/c_{11}$), it follows from Eq. (94) that $g(\theta)$ has a strong peak at $\theta = 0$.

These calculations show that the composite can be used as a collimator of stress waves to get stress waves of very high intensity and polarized in a direction parallel to the fibers. It can also be used as a filter of stress waves since the amount of energy passing across the fibers is small.

Finally, the large accumulation of strain energy within a small angular region in the composite may have some important implications. For example stress waves, after entering a closed continuous structure of carbon fibers (such as cylindrical, spherical, or conical shaped structures) cannot come out of it. Thus a large amount of strain energy may build up inside the structure causing an early failure. It is important, therefore, to study this property of a composite in detail before using it for such structural applications.

### Effect of Fiber Misalignment

As mentioned before, the propagation of elastic waves in the $yz$ plane can be used to derive useful information about the alignment of fibers. For this purpose it would be easier from an experimental point of view to use elastic waves of long wavelength compared to the fiber spacing. In the limit of long waves the dynamical matrix for the present model reduces to the Green–Christoffel matrix for an elastic continuum. A detailed account of the solutions of the Green–Christoffel matrix has been given by Fedorov (1968) for various symmetries.

If $C_p$ is the phase velocity, we can define an effective modulus of elasticity as follows

$$E = \rho C_p{}^2$$

Considering for example the $p_1$ wave [Eq. (83)]. We get, in the limit of long waves, for a wave incident at an angle $\theta_k$ from the $z$ axis

$$E(\theta_k) = \tfrac{1}{2}(c_{11}+c_{44})k_2{}^2 + \tfrac{1}{2}(c_{44}+c_{33})k_3{}^2$$
$$- [\{(c_{11}-c_{44})(k_2{}^2/2) + (c_{44}-c_{33})(k_3{}^2/2)\}^2 + (c_{13}+c_{44})^2 k_2{}^2 k_3{}^2]^{1/2}$$
$$(95)$$

where $k_2 = \sin\theta_k$, $k_3 = \cos\theta_k$ and by symmetry

$$E(\theta_k) = E(-\theta_k)$$

To represent the alignment of fibers, we introduce a normalized distribution function $F(\phi)$ so that the relative number of fibers aligned between the angles

$\phi$ and $\phi + d\phi$ from the $z$ axis is given by $F(\phi)\, d\phi$. The angle $\phi$ can have values in the range $-\pi/2$ to $\pi/2$.

For a composite in which all the fibers are perfectly aligned along the $z$ axis, $F(\phi)$ will have the form $\delta(\phi)$, whereas for a completely misaligned composite, $F(\phi)$ will be independent of $\phi$. We shall show here how one can get information about $F(\phi)$ by a comparison of experimental and theoretical values of $E(\theta_k)$ for a composite.

The contribution from a fiber at an angle $\phi$ from the $z$ axis to the modulus will be $E(\theta_k - \phi)$ so that the total effective modulus $E_e(\theta_k)$ for a wave incident at an angle $\theta_k$ from the $z$ axis in the $yz$ plane is given by

$$E_e(\theta_k) = \int_{-\pi/2}^{\pi/2} E_t(\theta_k - \phi)\, F(\phi)\, d\phi \qquad (96)$$

where the subscripts e and t denote the experimental and theoretical values of $E$, respectively, since the left-hand side is the one which can be measured experimentally.

Let us express the various functions in Eq. (96) in the form of a Fourier sine series in $\theta$ as follows:

$$E_e(\theta) = \sum_n E_{en} \sin 2n\theta \qquad (97)$$

$$E_t(\theta) = \sum_n E_{tn} \sin 2n\theta \qquad (98)$$

and

$$F(\theta) = \sum_n F_n \sin 2n\theta \qquad (99)$$

where the sum is over all integral values of $n$.

In practice only a few Fourier coefficients will be required since the Fourier series converges rapidly for functions which are not wildly varying. The coefficients $E_{en}$ and $E_{tn}$ can be obtained from $E_e(\theta)$ (observed values) and $E_{t\theta}$ [Eq. (95)], respectively, as follows

$$E_n = \frac{2}{\pi} \int_{-\pi/2}^{\pi/2} E(\theta) \sin 2n\theta \qquad (100)$$

where $E(\theta)$ stands for either $E_e(\theta)$ or $E_t(\theta)$. The integration in Eq. (100) has to be performed numerically at least for $E_{en}$.

The coefficients $F_n$ can be obtained in terms of $E_{en}$ and $E_{tn}$ with the help of Eq. (96). Since on the right of Eq. (96) we have a convolution of the functions $E_t(\theta)$ and $F(\theta)$, we get the following simple relation between the Fourier coefficients

$$F_n = E_{en}/E_{tn} \qquad (101)$$

Once $F_n$ are known, the distribution function $F(\theta)$ can be constructed from Eq. (99). Thus we see that a measurement of $E(\theta)$ can be used to derive the exact distribution function of the fibers in the composite.

As the Fourier coefficients are very sensitive to the details of the function, it is very important for this technique that $E_e(\theta)$ is accurately measured for many values of $\theta$, otherwise this approach can give results which are wildly in error. When the measurements of $E_e(\theta)$ are not sufficiently accurate the following simple approach may be more useful for a rough determination of $F(\theta)$.

We write $F(\theta)$ as a sum over delta functions

$$F(\theta) = \sum_n t_n \delta(\theta - \theta_n) \tag{102}$$

and substitute this expression into Eq. (96) to yield

$$E_e(\theta) = \sum_n t_n E_t(\theta - \theta_n) \tag{103}$$

Physically $t_n$ gives the number of fibers at an angle $\theta_n$ from the $z$ axis. For a perfectly aligned sample $t_n = 1$ for $\theta_n = 0$ and zero for all other values of $\theta_n$. For a completely misaligned sample $\theta_n$ is more or less continuously distributed in the range $-\pi/2$ to $\pi/2$ and $t_n$ is the same for all $n$.

For a real case, Eq. (103) can be used for a rough determination of $t_n$ and $\theta_n$ by using a least square fitting procedure. In actual practice, one can start with 2 or 3 terms on the right in Eq. (103), obtain the values of $t_n$ and $\theta_n$ that give the best fit with the observed values for $E(\theta)$. The number of terms can then be gradually increased to get a satisfactory agreement between the observed and calculated values of $E(\theta)$. The final values of $t_n$ and $\theta_n$ will give an idea of the alignment of the fibers. We shall use this technique for an actual case in Section XI.

## X. Composite with Hexagonal Symmetry

In this section we shall study the wave propagation in a composite with hexagonal symmetry. In this symmetry, the fibers are assumed to be arranged on the vertices of a regular hexagon in a plane perpendicular to the fibers. The basic assumptions and the technique for obtaining the dynamical matrix in this section will be the same as in the preceding sections on tetragonal symmetry. Most of the discussion given in the preceding sections also applies to this section, and we shall, therefore, usually give the formulas without any discussion. The relevant numerical results are given in Figs. 15–24.

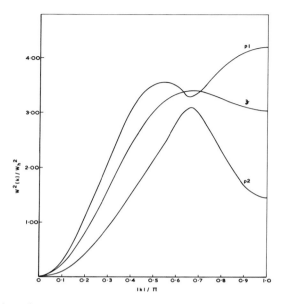

**Fig. 15.** Dispersion relations in the ⟨100⟩ direction for a hexagonal composite.

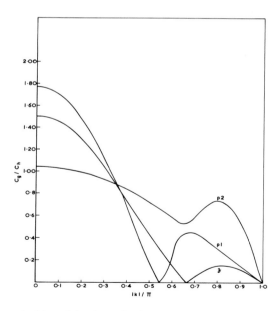

**Fig. 16.** Group velocities of the three elastic waves traveling in the ⟨100⟩ direction in a hexagonal composite.

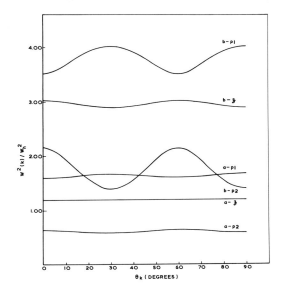

**Fig. 17.** Angular dependence of the dispersion relations in the $xy$ plane for a hexagonal composite. The values of $|\mathbf{k}|/\pi$ are for (a) 0.25 and (b) 0.50. The angles are measured from the $x$ axis.

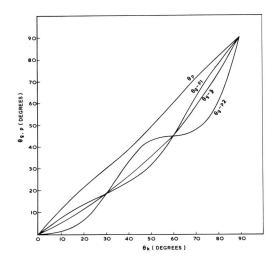

**Fig. 18.** Angular dependence of the directions of polarization ($\theta_p$) and flux ($\theta_g$) for elastic waves in the $xy$ plane in a hexagonal composite for $|\mathbf{k}|/\pi = 0.25$. All angles are measured from the $x$ axis.

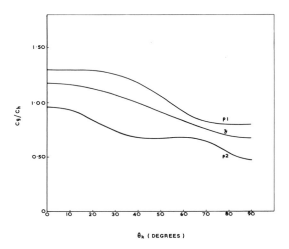

**Fig. 19.** Angular dependence of the group velocities of the elastic waves traveling in the $xy$ plane in a hexagonal composite for $|\mathbf{k}|/\pi = 0.25$. The angles are measured from the $x$ axis.

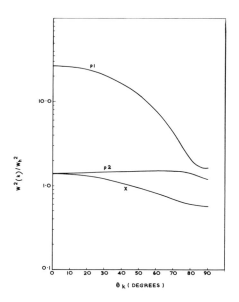

**Fig. 20.** Angular dependence of the dispersion relations in the $yz$ plane for a hexagonal composite for $|\mathbf{k}|/\pi = 0.25$. The angles are measured from the $z$ axis.

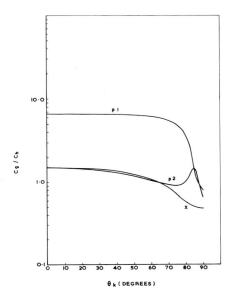

**Fig. 21.** Angular dependence of the group velocities of the elastic waves traveling in the $yz$ plane in a hexagonal composite for $|\mathbf{k}|/\pi = 0.25$. The angles are measured from the $z$ axis.

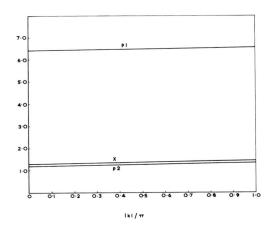

**Fig. 22.** $\mathbf{k}$ dependence of the group velocities for the elastic waves traveling in the $yz$ plane in a hexagonal composite for $\theta_k = 45°$.

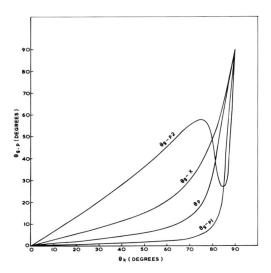

**Fig. 23.** Angular dependence of the directions of polarization ($\theta_p$) and flux ($\theta_g$) for elastic waves in $yz$ plane in a hexagonal composite for $|\mathbf{k}|/\pi = 0.25$. All the angles are measured from the $z$ axis.

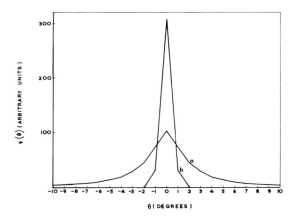

**Fig. 24.** Angular distribution function for a set of elastic waves traveling through a hexagonal composite in the $yz$ plane assuming a uniform distribution function for the angles of incidence. Only the $p_1$ waves have been included. The values of $|\mathbf{k}|/\pi$ are for (a) 0.0001 and (b) 0.5. For $\theta > 10°$, $g(\theta) \approx 0$. The angle $\theta$ is measured from the $z$ axis.

The operations of the point group for hexagonal symmetry are given in Appendix B. As before, the $z$ axis is taken in a direction parallel to the fibers and the $x$ and $y$ axes in a plane perpendicular to the fibers. Figure 1b shows the arrangement of the fibers in the $xy$ plane and Fig. 2b shows the three-dimensional lattice structure. The unit cell for the present model is a rhombohedral prism, shown as the shaded region in Fig. 2b, and each unit cell contains one object. If the length of each side of the rhombus base in $2a$ and $c$ is the height of the prism (distance between two adjacent layers) then the volume of a unit cell is given by

$$V = 2\sqrt{3}\,a^2\,c \qquad (104)$$

The effective mass of each object is defined in terms of $V$ and $\rho$, the average density of the composite, and $c$ is defined in terms of the Debye temperature for the composite in the $z$ direction.

For the calculation of the dynamical matrix we have assumed only nearest neighbor interactions between the fibers. The coordinates of the objects contributing to the dynamical matrix are given in Table III. In the present

**TABLE III**

COORDINATES OF THE "OBJECTS" CONTRIBUTING TO THE DYNAMICAL MATRIX FOR A
COMPOSITE WITH HEXAGONAL SYMMETRY

| | | Coordinates | | |
|---|---|---|---|---|
| Number | Object | $x$ | $y$ | $z$ |
| 1 | 0 | 0 | 0 | 0 |
| 2 | 1 | $2a$ | 0 | 0 |
| 3 | 2 | $a$ | $a\sqrt{3}$ | 0 |
| 4 | 3 | $-a$ | $a\sqrt{3}$ | 0 |
| 5 | 4 | $-2a$ | 0 | 0 |
| 6 | 5 | $-a$ | $-a\sqrt{3}$ | 0 |
| 7 | 6 | $a$ | $-a\sqrt{3}$ | 0 |
| 8 | 0′ | 0 | 0 | $c$ |
| 9 | 1′ | $2a$ | 0 | $c$ |
| 10 | 2′ | $a$ | $a\sqrt{3}$ | $c$ |
| 11 | 3′ | $-a$ | $a\sqrt{3}$ | $c$ |
| 12 | 4′ | $-2a$ | 0 | $c$ |
| 13 | 5′ | $-a$ | $-a\sqrt{3}$ | $c$ |
| 14 | 6′ | $a$ | $-a\sqrt{3}$ | $c$ |
| 15 | 0″ | 0 | 0 | $-c$ |
| 16 | 1″ | $2a$ | 0 | $-c$ |
| 17 | 2″ | $a$ | $a\sqrt{3}$ | $-c$ |
| 18 | 3″ | $-a$ | $a\sqrt{3}$ | $-c$ |
| 19 | 4″ | $-2a$ | 0 | $-c$ |
| 20 | 5″ | $-a$ | $-a\sqrt{3}$ | $-c$ |
| 21 | 6″ | $a$ | $-a\sqrt{3}$ | $-c$ |

case the dynamical matrix depends on only six independent parameters, which can be determined in terms of the elastic constants. For a more realistic calculation, one should include the effect of further neighbors. However, in the absence of any short wavelength dispersion measurements it does not seem worthwhile to introduce further unknown parameters. In this section numerical results have been obtained for a carbon fiber–epoxy resin composite using the elastic constants as obtained by Heaton (1970).

*Force Constants*

$$\phi(0,1) = \phi(0,4) = -\begin{bmatrix} \mu & 0 & 0 \\ 0 & \lambda & 0 \\ 0 & 0 & \delta \end{bmatrix}$$

$$\phi(0,2) = \phi(0,5) = -\frac{1}{4}\begin{bmatrix} \mu+3\lambda & \sqrt{3}(\mu-\lambda) & 0 \\ \sqrt{3}(\mu-\lambda) & 3\mu+\lambda & 0 \\ 0 & 0 & 4\delta \end{bmatrix}$$

$$\phi(0,3) = \phi(0,6) = -\frac{1}{4}\begin{bmatrix} \mu+3\lambda & -\sqrt{3}(\mu-\lambda) & 0 \\ -\sqrt{3}(\mu-\lambda) & 3\mu+\lambda & 0 \\ 0 & 0 & 4\delta \end{bmatrix}$$

$$\phi(0,0') = \phi(0,0'') = -\begin{bmatrix} \mu_0' & 0 & 0 \\ 0 & \mu_0' & 0 \\ 0 & 0 & \delta_0' \end{bmatrix}$$

$$\phi(0,1') = \phi(0,4'') = -\begin{bmatrix} \mu' & 0 & v' \\ 0 & \lambda' & 0' \\ v' & 0 & \delta' \end{bmatrix}$$

$$\phi(0,1'') = \phi(0,4') = -\begin{bmatrix} \mu' & 0 & -v' \\ 0 & \lambda' & 0 \\ -v' & 0 & \delta' \end{bmatrix}$$

$$\phi(0,2') = \phi(0,5'') = -\frac{1}{4}\begin{bmatrix} \mu'+3\lambda' & \sqrt{3}(\mu'-\lambda') & 2v' \\ \sqrt{3}(\mu'-\lambda') & 3\mu'+\lambda' & 2v'\sqrt{3} \\ 2v' & 2v'\sqrt{3} & 4\delta' \end{bmatrix}$$

$$\phi(0,2'') = \phi(0,5') = -\frac{1}{4}\begin{bmatrix} \mu'+3\lambda' & \sqrt{3}(\mu'-\lambda') & -2v' \\ \sqrt{3}(\mu'-\lambda') & 3\mu'+\lambda' & -2v'\sqrt{3} \\ -2v' & -2v'\sqrt{3} & 4\delta' \end{bmatrix}$$

$$\phi(0,3') = \phi(0,6'') = -\frac{1}{4}\begin{bmatrix} \mu'+3\lambda' & -\sqrt{3}(\mu'-\lambda') & -2v' \\ -\sqrt{3}(\mu'-\lambda') & 3\mu'+\lambda' & 2v'\sqrt{3} \\ -2v' & 2v'\sqrt{3} & 4\delta' \end{bmatrix}$$

$$\phi(0,3'') = \phi(0,6') = -\frac{1}{4}\begin{bmatrix} \mu'+3\lambda' & -\sqrt{3}(\mu'-\lambda') & 2v' \\ -\sqrt{3}(\mu'-\lambda') & 3\mu'+\lambda' & -2v'\sqrt{3} \\ 2v' & -2v'\sqrt{3} & 4\delta' \end{bmatrix}$$

and

$$\phi(0,0) = \begin{bmatrix} \mu_0 & 0 & 0 \\ 0 & \mu_0 & 0 \\ 0 & 0 & \delta_0 \end{bmatrix}$$

where

$$\mu_0 = 2\mu_0' + 3\mu + 3\lambda + 6\mu' + 6\lambda'$$

and

$$\delta_0 = 2\delta_0' + 6\delta + 12\delta'$$

## Dynamical Matrix

In what follows, $c$ is in units of $a$, $k_1$ and $k_3$ replace $2\pi a k_1$ and $2\pi a k_3$ respectively, and $k_2$ replaces $2\pi a \sqrt{3}k_2$.

$$D_{11}(\mathbf{k}) = 2A(1-\cos 2k_1) + (A+3B)(1-\cos k_1 \cos k_2) + Ek_3^2$$

$$D_{22}(\mathbf{k}) = 2B(1-\cos 2k_1) + (3A+B)(1-\cos k_1 \cos k_2) + Ek_3^2$$

$$D_{33}(\mathbf{k}) = 2D(1-\cos 2k_1) + 4D(1-\cos k_1 \cos k_2) + Hk_3^2$$

$$D_{12}(\mathbf{k}) = D_{21}(k) = \sqrt{3}(A-B)\sin k_1 \sin k_2$$

$$D_{13}(\mathbf{k}) = D_{31}(k) = Fk_3 \sin k_1(2\cos k_1 + \cos k_2)$$

$$D_{23}(\mathbf{k}) = D_{32}(k) = F\sqrt{3}k_3 \sin k_2 \cos k_1 \tag{105}$$

where $A = \mu+2\mu'$, $B = \lambda+2\lambda'$, $E = \mu_0'c^2$, $H = \delta_0'c^2$, $D = \delta+2\delta'$, and $F = 4v'$.

*Dynamical Matrix in the Long Wave length Limit*

$$D_{11}^0(\mathbf{k}) = \tfrac{3}{2}(3A+B)k_1{}^2 + \tfrac{1}{2}(A+3B)k_2{}^2 + Ek_3{}^2$$

$$D_{22}^0(\mathbf{k}) = \tfrac{3}{2}(A+3B)k_1{}^2 + \tfrac{1}{2}(3A+B)k_2{}^2 + Ek_3{}^2$$

$$D_{33}^0(\mathbf{k}) = 6Dk_1{}^2 + 2Dk_2{}^2 + Hk_3{}^2$$

$$D_{12}^0(\mathbf{k}) = D_{21}^0(k) = \sqrt{3}(A-B)k_1 k_2 \qquad\qquad (106)$$

$$D_{13}^0(\mathbf{k}) = D_{31}^0(k) = 3Fk_1 k_3$$

$$D_{23}^0(\mathbf{k}) = D_{32}^0(k) = \sqrt{3}Fk_2 k_3$$

*Green–Christoffel Matrix* (Fedorov, (1968)

$$\Lambda_{11}(\mathbf{k}) = c_{11}k_1{}^2 + (c_{66}/3)k_2{}^2 + c_{44}k_3{}^2$$

$$\Lambda_{22}(\mathbf{k}) = c_{66}k_1{}^2 + (c_{11}/3)k_2{}^2 + c_{44}k_3{}^2$$

$$\Lambda_{33}(\mathbf{k}) = c_{44}k_1{}^2 + (c_{44}/3)k_2{}^2 + c_{33}k_3{}^2$$

$$\Lambda_{12}(\mathbf{k}) = \Lambda_{21}(k) = [(c_{11}-c_{66})/\sqrt{3}]k_1 k_2$$

$$\Lambda_{13}(\mathbf{k}) = \Lambda_{31}(k) = (c_{13}+c_{44})k_1 k_3$$

$$\Lambda_{23}(\mathbf{k}) = \Lambda_{32}(k) = [(c_{13}+c_{44})/\sqrt{3}]k_2 k_3 \qquad\qquad (107)$$

where the elastic constants $c_{ij}$ are in units of $(4\pi a^2 \rho)^{-1}$.

*Relations between the Force Constants and the Elastic Constants*

$$A = (3c_{11}-c_{66})/12$$

$$B = (3c_{66}-c_{11})/12$$

$$E = c_{44}$$

$$D = c_{44}/6$$

$$H = c_{33}$$

$$F = (c_{13}+c_{44})/3 \qquad\qquad (108)$$

*Wave Propagation in the xy Plane* $(k_3 = 0)$

$$D(\mathbf{k}) = \begin{pmatrix} D^p & 0 \\ 0 & D^z \end{pmatrix} \qquad\qquad (109)$$

where

$$D_{11}^p = 2A(1-\cos 2k_1) + (A+3B)(1-\cos k_1 \cos k_2)$$

$$D_{12}^p = D_{21}^p = \sqrt{3}(A-B)\sin k_1 \sin k_2$$

$$D_{22}^p = 2B(1-\cos 2k_1) + (B+3A)(1-\cos k_1 \cos k_2)$$

and

$$D^z = 2D(1-\cos 2k_1) + 4D(1-\cos k_1 \cos k_2)$$

(1) *Quasi-longitudinal wave*

$$\omega_{p1}^2 = (A+B)(3-2\cos k_1 \cos k_2 - \cos 2k_1) + (A-B)X \qquad (110)$$

$$\mathbf{e}_{p1} = [\cos \theta_p, \sin \theta_p, 0] \qquad (111)$$

$$C_{gx} = \frac{\sin k_1}{2\omega_{p1}}\left[2(A+B)(1+\cos k_2) + \frac{(A-B)}{X}P_x\right] \qquad (112)$$

$$C_{gy} = \frac{\sin k_2}{2\omega_{p1}}\left[2(A+B)\cos k_1 + \frac{A-B}{X}P_y\right] \qquad (113)$$

$$C_{gz} = 0$$

(2) *Quasi-transverse wave*

$$\omega_{p2}^2 = (A+B)(3-2\cos k_1 \cos k_2 - \cos 2k_1) - (A-B)X \qquad (114)$$

$$\mathbf{e}_{p2} = [-\sin \theta_p, \cos \theta_p, 0] \qquad (115)$$

$$C_{gx} = \frac{\sin k_1}{2\omega_{p2}}\left[2(A+B)(2\cos k_1 + \cos k_2) - \frac{(A-B)}{X}P_x\right] \qquad (116)$$

$$C_{gy} = \frac{\sin k_2}{2\omega_{p2}}\left[2(A+B)\cos k_1 - \frac{(A-B)}{X}P_y\right] \qquad (117)$$

$$C_{gz} = 0$$

where

$$X^2 = (\cos k_1 \cos k_2 - \cos 2k_1)^2 + 3\sin^2 k_1 \sin^2 k_2$$

$$P_x = (4\cos k_1 - \cos k_2)(\cos k_1 \cos k_2 - \cos 2k_1) + 3\sin^2 k_2 \cos k_1$$

$$P_y = -\cos k_1(\cos k_1 \cos k_2 - \cos 2k_1) + 3\sin^2 k_1 \cos k_2$$

and $\theta_p$, the angle of polarization from the $x$ axis is given by

$$\tan 2\theta_p = \frac{\sqrt{3}\sin k_1 \sin k_2}{\cos k_1 \cos k_2 - \cos 2k_1} \qquad (118)$$

(3) *z-transverse wave*

$$\omega_z^2 = 2D(1 - \cos 2k_1) + 4D(1 - \cos k_1 \cos k_2) \tag{119}$$

$$\mathbf{e}_z = [0, 0, 1] \tag{120}$$

$$C_{gx} = \frac{2D \sin k_1}{\omega_z} [2 \cos k_1 + \cos k_2] \tag{121}$$

$$C_{gy} = \frac{2D \sin k_2 \cos k_1}{\omega_z} \tag{122}$$

$$C_{gz} = 0$$

We notice from Eq. (118) that the angle of polarization is independent of the force constants. In the continuum limit, using the method of long waves, we find that $\theta_p = \theta_k$. In this limit the waves $p_1$ and $p_2$ are true longitudinal and transverse waves, respectively, for all values of $\mathbf{k}$. This fact is a consequence of the transverse elastic isotropy of a hexagonal crystal.

$\langle 100 \rangle$ *Direction* $(k_1 = k, k_2 = k_3 = 0)$

(1) *Longitudinal wave*

$$\omega^2 = (1 - \cos k)[A(4 \cos k + 5) + 3B]$$

$$C_g = \frac{A(1 + 8 \cos k) + 3B}{[2A(4 \cos k + 5) + 6B]^{1/2}} \cos \frac{k}{2}$$

(2) *p-Transverse wave*

$$\omega^2 = (1 - \cos k)[B(4 \cos k + 5) + 3A]$$

$$C_g = \frac{B(1 + 8 \cos k) + 3A}{[2B(4 \cos k + 5) + 6A]^{1/2}} \cos \frac{k}{2}$$

(3) *z-Transverse wave*

$$\omega^2 = 4D(1 - \cos k)(\cos k + 2)$$

$$C_g = \frac{2D(2 \cos k + 1)}{[2D(\cos k + 2)]^{1/2}} \cos \frac{k}{2}$$

$\langle 010 \rangle$ *Direction* $(k_1 = 0, k_2 = k, k_3 = 0)$

(1) *Longitudinal wave*

$$\omega^2 = (A + 3B)(1 - \cos k)$$

$$C_g = \left(\frac{A + 3B}{2}\right)^{1/2} \cos \frac{k}{2}$$

(2) *p-Transverse wave*

$$\omega^2 = (B+3A)(1-\cos k)$$

$$C_g = \left(\frac{B+3A}{2}\right)^{1/2} \cos\frac{k}{2}$$

(3) *z-Transverse wave*

$$\omega^2 = 4D(1-\cos k)$$

$$C_g = \sqrt{2D}\cos\frac{k}{2}$$

*Wave Propagation in yz Plane*

$$D(k) = \begin{pmatrix} D^x & 0 \\ 0 & D_p \end{pmatrix}$$

where

$$D^x = (A+3B)(1-\cos k_2) + Ek_3^2$$
$$D^p_{11} = (3A+B)(1-\cos k_2) + Ek_3^2$$
$$D^p_{12} = D^p_{21} = F\sqrt{3}k_3\sin k_2$$
$$D^p_{22} = 4D(1-\cos k_2) + Hk_3^2$$

(1) *x-Transverse wave*

$$\omega_x^2 = (A+3B)(1-\cos k_2) + Ek_3^2$$

$$e_x = [1,0,0]$$

$$C_{gy} = \frac{(A+3B)\sin k_2}{2\omega_x}$$

$$C_{gz} = Ek_3/\omega_x$$

$$C_{gx} = 0$$

(2) *p₁ Wave*

$$\omega_{p1}^2 = R_1(1-\cos k_2) + T_1 k_3^2 + X$$

$$e_{p1} = [0, \sin\theta_p, \cos\theta_p]$$

$$C_{gy} = \frac{\sin k_2}{2\omega_{p1}}\left[R_1 + \frac{P_y}{X}\right]$$

$$C_{gz} = \frac{k_3}{2\omega_{p1}}\left[2T_1 + \frac{P_z}{X}\right]$$

$$C_{gx} = 0$$

### (3) $p_2$ *Wave*

$$\omega_{p2}^2 = R_1(1-\cos k_2) + T_1 k_3{}^2 - X$$

$$\mathbf{e}_{p2} = [0, -\cos\theta_p, \sin\theta_p]$$

$$C_{gy} = \frac{\sin k_2}{2\omega_{p2}}\left[R_1 - \frac{P_y}{X}\right]$$

$$C_{gz} = \frac{k_3}{2\omega_{p2}}\left[2T_1 - \frac{P_z}{X}\right]$$

$$C_{gx} = 0$$

where

$$R_1 = \tfrac{1}{2}(3A+B+4D)$$

$$T_1 = \tfrac{1}{2}(E+H)$$

$$R_2 = \tfrac{1}{2}(3A+B-4D)$$

$$T_2 = \tfrac{1}{2}(E-H)$$

$$X^2 = [R_2(1-\cos k_2)+T_2 k_3{}^2]^2 + 3F^2 k_3{}^2 \sin^2 k_2$$

$$P_y = R_2[R_2(1-\cos k_2)+T_2 k_3{}^2] + 3F^2 k_3{}^2 \cos k_2$$

$$P_z = 2T_2[R_2(1-\cos k_2)+T_2 k_3{}^2] + 3F^2 \sin^2 k_2$$

and $\theta_p$, the angle of polarization with respect to the $z$ axis in the $yz$ plane, is given by

$$\theta_p = \frac{\pi}{2} - \frac{1}{2}\tan^{-1}\frac{\sqrt{3}\,Fk_3 \sin k_2}{R_2(1-\cos k_2) + T_2 k_3{}^2}$$

## 2. Numerical Results

We notice from Eq. (108) that in the present model for a composite with hexagonal symmetry, the force constants are completely determined in terms of the elastic constants. For the elastic constants we have used two different sets of values as given below.

*Set 1.* These values are based on the calculations of Heaton and have been kindly supplied to us by Dr. W. N. Reynolds. They refer to a carbon fiber–epoxy resin composite with 50% fiber loading.

$$\text{Units: } 10^{11} \text{ dynes cm}^{-2}$$

$$c_{11} = 3.157$$

$$c_{12} = 0.959$$

$$c_{13} = 2.058$$

$$c_{33} = 43.720$$

$$c_{44} = 2.273$$

*Set 2.* These values have been obtained by a rough comparison between the calculated and observed values for the phase velocities of long elastic waves (see Section XI) in a carbon fiber–epoxy composite with 50% fiber loading.

$$\text{Units: } 10^{11} \text{ dynes cm}^{-2}$$

$$c_{11} = 2.00$$

$$c_{12} = 0.60$$

$$c_{13} = 1.30$$

$$c_{33} = 29.00$$

$$c_{44} = 0.90$$

For the numerical calculations in this section we have used the elastic constants in Set 1. The results are shown in Figs. 15–24. The frequencies are expressed in units of $\omega_h$ where

$$\omega_h^2 = G/4\pi a^2 \rho$$

and $G = 10^{11}$ dynes cm$^{-2}$, and the velocities are expressed in units of $c_h$ where

$$c_h = \omega_h 2\pi a.$$

The elastic constants in Set 2 will be used in Section XI in an attempt to derive information about the alignment of the fibers.

## XI. Comparison with the Experimental Results for Long Elastic Waves: Calculation of the Fiber Orientation Function

In this section we shall calculate $F(\theta)$, the fiber orientation function as an illustration of the technique described at the end of Section IX for this purpose we have used the preliminary measurements by Reynolds for the angular dependence of the phase velocity of long acoustic waves in a carbon fiber–epoxy composite with 50% fiber loading. The experimental results are shown in Figs. 25 and 26.

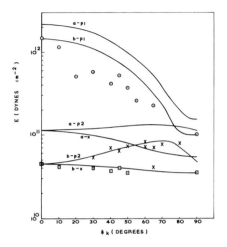

**Fig. 25.** Angular dependence of the effective modulus of elasticity of a hexagonal or tetragonal composite for long elastic waves, using the elastic constants of (a) Set 1 and (b) Set 2. The points denote the experimental results obtained by Reynolds (1970): O $p_1$ wave, × $p_2$ wave, and □ $x$-transverse wave.

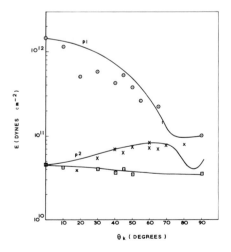

**Fig. 26.** Angular dependence of the effective modulus of elasticity of a hexagonal or tetragonal composite for long elastic waves using the elastic constants in Set 2 and the fiber orientation function [Eq. (102)] with parameters as given in Table IV. The points denote the experimental results as in Fig. 25.

Let us consider a wave traveling in the $yz$ plane. For this plane the form of the dispersion relations will be the same for tetragonal and hexagonal symmetries. Using the Set 1 for the elastic constants (Section X), and $\rho = 2$, $E(\theta_k)$ has been plotted in Fig. 25 using Eq. (95) with $|\mathbf{k}| \rightarrow 0$.

We notice from Fig. 25 that the agreement between the calculated and experimental results for the $p_1$ wave is rather poor. One possible explanation could be that the calculated values of the elastic constants do not agree with the actual elastic constants of the sample used for the experiments. We can of course determine the experimental values of the elastic constants from the measured values of the phase velocities. The elastic constants, thus determined, are given as Set 2 in Section X. The elastic constants $c_{11}$, $c_{66}$, $c_{44}$, and $c_{33}$ have been obtained from the measured phase velocities of the three waves for $\theta = 0$ and 90°. The elastic constant $c_{13}$ has been determined from the measured phase velocity of the $p_2$ wave for $\theta = 45°$. It may be mentioned here that it is not possible to choose a physically acceptable value of $c_{13}$ so as to get an agreement between the measured and calculated values for the velocity of the $p_1$ wave for any value of $\theta$.

The calculated values of the phase velocities for the three waves using the Set 2 elastic constants are shown in the same figure. The agreement between the calculated and the observed values is found to be satisfactory for the $p_2$ and the $x$-transverse waves but is unsatisfactory for the $p_1$ wave. It does not seem to be possible to improve upon this agreement by a different choice of elastic constants.

Let us attribute this discrepancy for the $p_1$ wave to the misalignment of the fibers in the composite. To measure the fiber misalignment we introduce the fiber orientation function as discussed in Section IX. Taking three terms in the series in Eq. (102), and choosing the values for the six parameters $t_n$ and $\phi_n$ as given in the following tabulation, we obtain a reasonable agreement between the calculated and observed values for the $p_1$ wave. The final results for all three waves are shown in Fig. 26. A pleasing feature of these results is that although $F(\phi)$ was calculated for the $p_1$ wave, the agreement between the calculated and experimental values for other two waves is also reasonable. The overall agreement could be improved by taking more terms in the series in Eq. (102) but was not considered to be worthwhile in view of the qualitative nature of the present calculations.

| $\phi_n$ (deg) | $t_n$ |
|---|---|
| −4.6 | 1.26 |
| 1.7 | −0.36 |
| 3.4 | 0.08 |

From this tabulation we infer that all the fibers are within $\pm 5°$ from the $z$ axis, so that the fibers in that particular sample were well aligned. We also notice that the value of $t_n$ for $\phi_n = 1.7°$ is negative. Since $t_n$ denotes the number of fibers, this result is apparently unphysical. However, it is suggested that the negative value of $t_n$ can be interpreted as an indication of cracks or the presence of hollow fibers inclined at the corresponding angle $\phi_n$. In a real composite, the presence of such defects is very normal and therefore this interpretation for the negative values of $t_n$ does not seem to be too drastic. It must be emphasised, however, that the treatment given in this section is for the purpose of illustration only and the results are only qualitative. Since the experimental measurements are preliminary and thus may be somewhat inaccurate, and since the fitting procedure was not accurate enough, it would not be reasonable to associate any quantitative significance to the numbers given in the tabulation.

## XII. Discussion

In the theory of elastic wave propagation in composite materials, as discussed in this paper, we have neglected three effects which may be important—namely, the effects due to (i) anharmonic forces, (ii) long range elastic interactions between the fibers, and (iii) the finite thickness of the fibers. The effect of finite thickness of the fibers has been partially included in the present theory by defining an effective mass of the fibers and using the averaged elastic constants and the density of the whole composite. However, the finite thickness of the fibers will also be manifest through the effects of internal friction, phonon–phonon interaction, and the absorption of elastic waves due to various defects in the atomic structure of the fibers. The contribution due to these effects will increase with an increase in the percentage fiber loading of the composite and a decrease in the wavelength of the elastic wave. A proper determination of the magnitude of these and other corrections has to await a rigorous experimental study of the elastic wave propagation in composites.

Nevertheless, some interesting qualitative inferences can be derived from the numerical results reported in this paper. We notice from Figs. 3–9 and Figs. 15–19 that, as previously asserted, the wave propagation in the $xy$ plane is strongly dispersive and quite nondispersive in the $yz$ plane as seen from Figs. 10–13 and Figs. 20–23. The effect of the fiber loading, and therefore a change in the force constants, on dispersion for the wave traveling in the $xy$ plane can be estimated from Fig. 6. We notice that the change in $\omega^2(\mathbf{k})$ is a maximum when $|\mathbf{k}| = \pi$ or the wavelength is equal to twice the spacing between the fibers in a tetragonal composite. The vanishing of the group velocity at $|\mathbf{k}| = \pi$, predicting a Bragg-type reflection of the elastic waves from a composite is shown in Fig. 7. The group velocity may also vanish at other points in $k$ space such as for example in Fig. 9. This however is accidental. It depends upon the choice of the force constants and is not a basic property of the composite.

We also notice from the figures that the nature of various curves for the elastic wave propagation in the $yz$ plane is quite similar for a tetragonal and a hexagonal composite. This is obvious on physical grounds since the wave traveling along the fibers will not notice the arrangement of the fibers in the $xy$ plane. Similarly a long elastic wave even if traveling in the $xy$ plane will also not be very sensitive to the arrangement of the fiber since it sees only a quasi-homogeneous structure of the composite. The situation is different for a short elastic wave traveling in the $xy$ plane which will be extremely sensitive to the geometry of the arrangement of the fibers in the composite. This can be verified by comparing, for example, Figs. 7 and 16 or Figs. 6 and 15. Thus, information about the arrangement of the fibers can be obtained from a study of the propagation of short elastic waves in the $xy$ plane.

As can be seen from Figs. 12, 13, 21, and 23, the dependence of $C_g$, $\theta_p$, and $\theta_g$ on $\theta_k$ for a wave traveling in the $yz$ plane shows a strong transitional behavior near $\theta_k = 90°$. This occurs because $c_{33}$ is very much larger than $c_{11}$ or $c_{44}$. For most values of $\theta_k$, the wave propagation is determined by $c_{33}$ which has a nondispersive contribution. For $\theta_k \approx 90°$, the contribution of $c_{11}$ or $c_{44}$ becomes comparable to that of $c_{33}$. For short elastic waves it marks the onset of the dispersive effects. This can be easily seen from, for example, Eq. (85), where the contribution of the dispersive term containing $(1 - \cos k_2)$ is appreciable only when $k_3$, i.e., $\cos \theta_k$ is of the order of $(c_{11}/c_{33})^{1/2}$. The denominator in Eq. (85) will become zero when $\theta_k$ satisfies the equation

$$\frac{1 - \cos(k \sin \theta_k)}{k^2 \cos^2 \theta_k} = \frac{c_{33} - c_{44}}{2(c_{11} - c_{44})}$$

On approaching this value the curve for $\theta_p$ shows a steep rise as in Fig. 13; the transitional behavior of other curves near $\theta_k = 90°$ can also be interpreted similarly.

The collimation properties of a tetragonal and a hexagonal composite for elastic waves are shown in Figs. 14 and Fig. 24, respectively. The two curves are quite similar, as expected, in view of the earlier discussion, since we are dealing here with the wave propagation in the $yz$ plane. These figures show a plot of the angular distribution function $g(\theta)$ vs. $\theta$ for the $p_1$ wave; $g(\theta)d\theta$ defines the number of waves emerging between the angles $\theta$ and $\theta + d\theta$ from the $z$ axis, when a set of elastic waves is incident at one end of the composite at angles of incidence ranging from $-90°$ to $90°$ from the $z$ axis. For the calculation of $g(\theta)$ we have assumed that $|\mathbf{k}|$ is the same for all the incident waves and the angles of incidence are uniformly distributed, i.e., the number of waves incident between the angles $\theta$ and $\theta + d\theta$ is same for all values of $\theta$, where the range of $\theta$ is from $-90°$ to $90°$.

We notice from Fig. 14 and Fig. 24 that $g(\theta)$ is very small for $|\theta| > 5°$ compared with the height of the peak at $\theta = 0°$. This means that most of the elastic energy will travel within the angles $\pm 5°$ from the fibers and very little across the fibers. The height and the inverse width of the peak at $\theta = 0°$ increase with a decrease in the wavelength of the incident waves. For $|\mathbf{k}| = \pi$ nearly 93% of the incident energy travels within the angles $\pm 1°$ from the fibers. The distribution function for the other two waves is more or less constant with respect to $\theta$; the $p_2$ waves do show a little accumulation near $\theta = 0$.

For most angles of incidence the $p_1$ wave is polarized in a direction close to the fiber axis (Figs. 13 and 23), and thus on traveling through a composite, the original set of elastic waves will emerge as a highly collimated beam of more or less longitudinally polarized elastic waves. Such a large accumulation of energy seems to be very interesting and significant from an application point of view.

Alternatively, the composite can be used as a filter for elastic waves. We have seen that the amount of elastic energy traveling across the fibers is relatively small. Thus if two structures are joined to each other through a composite in such a way that the line joining the two structures is perpendicular to the fibers, each structure will be more or less unaffected by the stress waves produced in the other one.

Perhaps the most important conclusion of the present paper is that it suggests that from an academic as well as application point of view a study of the propagation of short elastic waves in a composite would be worthwhile. To estimate the frequency of the ultrasonic generator required for this purpose we can use the Eqs. (70)–(72). The maximum frequency for the longitudinal wave is given by

$$\omega^2 = 4(A + 2D)$$

or

$$v^2 = \frac{4 \times c_{11}}{4\pi^2 \times \rho \times 4\pi^2 \times a^2}$$

Taking $c_{11} = 10^{11}$ dynes cm$^{-2}$, $a = 10\mu$, and $\rho = 2$ gm/cc, we get

$$v = 11.3 \, \text{mHz}$$

The corresponding frequency will be even less for the two transverse waves. It should not be too difficult to generate ultrasonic waves of this frequency, and using such short waves, one can measure the dispersion relations in the composite. With the help of such observed dispersion relations, the force constants and therefore the bonding strength of the fibers can be estimated.

Some estimate of the force constants can also be obtained by a measurement of the thermodynamic quantities. The calculation of thermodynamic quantities and the effect of bonding defects on the dispersion relations using the present model will be given in a later paper.

It may be mentioned here that as shown in a recent paper Leigh *et al.* (1971) it may not be possible to determine the force constants uniquely by a measurement of the frequencies alone; one has to measure the polarization vectors as well. This conclusion has been derived for dispersion curves obtained by neutron scattering from a crystal lattice and will also apply to some extent to the present case of the scattering of elastic waves from composites. However, it may be that the uncertainty in the force constants between fibers in a composite due to an uncertainty in the eigenvectors, is much less than the corresponding uncertainty in a crystal lattice. This is because the elastic displacements of fibers can be determined from the theory of elasticity unlike the atomic displacements in a crystal lattice.

Finally, the main results of the present paper are as follows.

(1) The bonding strength of the fibers in a composite can be measured by using short elastic waves. The measurement will be most sensitive to the bonding strength if the wave is traveling in a plane perpendicular to the fibers and its wavelength is of the order of the spacing between the fibers. Such a measurement will also give some information about the arrangement of the fibers in the composite.

(2) A wave of wavelength equal to twice the fiber spacing, traveling in a plane perpendicular to the fibers, will be reflected from the composite, a phenomenon which is analogous to the Bragg reflection of X-rays from crystals. This, however, is a consequence of the assumption of the periodic arrangement of the fibers, and may not be exactly obeyed by a real composite. Certainly for a densely reinforced composite, one should be able to observe an appreciable reduction in the group velocity for a wave of wavelength equal to *twice the average* fiber spacing.

(3) Some information about the alignment of the fibers can be obtained by studying the wave propagation in a plane parallel to the fibers. Long elastic waves may be adequate for this purpose.

(4) If a set of elastic waves is incident at one end of a composite, a large percentage of the incident elastic energy would travel along the fibers and very little across the fibers. Thus, a composite can be used as a collimator and filter for elastic waves.

The qualitative results, as summarized above are quite independent of the details of the model. They can be attributed to the strongly anisotropic and fibrous nature of the composite. As mentioned before, it is not possible at the moment to test the validity of the present model due to a lack of sufficient experimental and previous theoretical work in this field. However, since both the Born–von Karman model and the Debye model have been extremely successful in other fields, it seems unlikely that the present model, which is a mixture of the two, will be far from reality.

## Appendix A: Operations of the Tetrahedral Point Group

On the basis of the Cartesian axes, as shown in Figs. 1 and 2, the matrix representations of the tetrahedral point group for the composite are given below:

1. $\begin{bmatrix} 1 & 0 & 0 \\ 0 & 1 & 0 \\ 0 & 0 & 1 \end{bmatrix}$
7. $\begin{bmatrix} -1 & 0 & 0 \\ 0 & -1 & 0 \\ 0 & 0 & 1 \end{bmatrix}$

2. $\begin{bmatrix} 1 & 0 & 0 \\ 0 & 1 & 0 \\ 0 & 0 & -1 \end{bmatrix}$
8. $\begin{bmatrix} -1 & 0 & 0 \\ 0 & -1 & 0 \\ 0 & 0 & -1 \end{bmatrix}$

3. $\begin{bmatrix} 1 & 0 & 0 \\ 0 & -1 & 0 \\ 0 & 0 & 1 \end{bmatrix}$
9. $\begin{bmatrix} 0 & 1 & 0 \\ 1 & 0 & 0 \\ 0 & 0 & 1 \end{bmatrix}$

4. $\begin{bmatrix} 1 & 0 & 0 \\ 0 & -1 & 0 \\ 0 & 0 & -1 \end{bmatrix}$
10. $\begin{bmatrix} 0 & 1 & 0 \\ 1 & 0 & 0 \\ 0 & 0 & -1 \end{bmatrix}$

5. $\begin{bmatrix} -1 & 0 & 0 \\ 0 & 1 & 0 \\ 0 & 0 & 1 \end{bmatrix}$
11. $\begin{bmatrix} 0 & 1 & 0 \\ -1 & 0 & 0 \\ 0 & 0 & 1 \end{bmatrix}$

6. $\begin{bmatrix} -1 & 0 & 0 \\ 0 & 1 & 0 \\ 0 & 0 & -1 \end{bmatrix}$
12. $\begin{bmatrix} 0 & 1 & 0 \\ -1 & 0 & 0 \\ 0 & 0 & -1 \end{bmatrix}$

13. $\begin{bmatrix} 0 & -1 & 0 \\ 1 & 0 & 0 \\ 0 & 0 & 1 \end{bmatrix}$
15. $\begin{bmatrix} 0 & -1 & 0 \\ -1 & 0 & 0 \\ 0 & 0 & 1 \end{bmatrix}$

14. $\begin{bmatrix} 0 & -1 & 0 \\ 1 & 0 & 0 \\ 0 & 0 & -1 \end{bmatrix}$
16. $\begin{bmatrix} 0 & -1 & 0 \\ -1 & 0 & 0 \\ 0 & 0 & -1 \end{bmatrix}$

## Appendix B: Operations of the Hexagonal Point Group

On the basis of the Cartesian axes, as shown in Figs. 1 and 2, the matrix representations of the hexagonal point group for the composite are given below.

1. $\begin{bmatrix} 1 & 0 & 0 \\ 0 & 1 & 0 \\ 0 & 0 & 1 \end{bmatrix}$
7. $\frac{1}{2}\begin{bmatrix} -1 & \sqrt{3} & 0 \\ \sqrt{3} & 1 & 0 \\ 0 & 0 & 2 \end{bmatrix}$

2. $\begin{bmatrix} 1 & 0 & 0 \\ 0 & 1 & 0 \\ 0 & 0 & -1 \end{bmatrix}$
8. $\frac{1}{2}\begin{bmatrix} -1 & \sqrt{3} & 0 \\ \sqrt{3} & 1 & 0 \\ 0 & 0 & -2 \end{bmatrix}$

3. $\begin{bmatrix} 1 & 0 & 0 \\ 0 & -1 & 0 \\ 0 & 0 & 1 \end{bmatrix}$
9. $\frac{1}{2}\begin{bmatrix} 1 & -\sqrt{3} & 0 \\ \sqrt{3} & 1 & 0 \\ 0 & 0 & 2 \end{bmatrix}$

4. $\begin{bmatrix} 1 & 0 & 0 \\ 0 & -1 & 0 \\ 0 & 0 & -1 \end{bmatrix}$
10. $\frac{1}{2}\begin{bmatrix} 1 & -\sqrt{3} & 0 \\ \sqrt{3} & 1 & 0 \\ 0 & 0 & -2 \end{bmatrix}$

5. $\frac{1}{2}\begin{bmatrix} 1 & \sqrt{3} & 0 \\ \sqrt{3} & -1 & 0 \\ 0 & 0 & 2 \end{bmatrix}$
11. $\frac{1}{2}\begin{bmatrix} -1 & -\sqrt{3} & 0 \\ \sqrt{3} & -1 & 0 \\ 0 & 0 & 2 \end{bmatrix}$

6. $\frac{1}{2}\begin{bmatrix} 1 & \sqrt{3} & 0 \\ \sqrt{3} & -1 & 0 \\ 0 & 0 & -2 \end{bmatrix}$
12. $\frac{1}{2}\begin{bmatrix} -1 & -\sqrt{3} & 0 \\ \sqrt{3} & -1 & 0 \\ 0 & 0 & -2 \end{bmatrix}$

13.
$$\begin{bmatrix} -1 & 0 & 0 \\ 0 & -1 & 0 \\ 0 & 0 & 1 \end{bmatrix}$$

19.
$$\frac{1}{2}\begin{bmatrix} 1 & -\sqrt{3} & 0 \\ -\sqrt{3} & -1 & 0 \\ 0 & 0 & -2 \end{bmatrix}$$

14.
$$\begin{bmatrix} -1 & 0 & 0 \\ 0 & -1 & 0 \\ 0 & 0 & -1 \end{bmatrix}$$

20.
$$\frac{1}{2}\begin{bmatrix} 1 & -\sqrt{3} & 0 \\ -\sqrt{3} & -1 & 0 \\ 0 & 0 & -2 \end{bmatrix}$$

15.
$$\begin{bmatrix} -1 & 0 & 0 \\ 0 & 1 & 0 \\ 0 & 0 & 1 \end{bmatrix}$$

21.
$$\frac{1}{2}\begin{bmatrix} -1 & \sqrt{3} & 0 \\ -\sqrt{3} & -1 & 0 \\ 0 & 0 & 2 \end{bmatrix}$$

16.
$$\begin{bmatrix} -1 & 0 & 0 \\ 0 & 1 & 0 \\ 0 & 0 & -1 \end{bmatrix}$$

22.
$$\frac{1}{2}\begin{bmatrix} -1 & \sqrt{3} & 0 \\ -\sqrt{3} & -1 & 0 \\ 0 & 0 & -2 \end{bmatrix}$$

17.
$$\frac{1}{2}\begin{bmatrix} -1 & -\sqrt{3} & 0 \\ -\sqrt{3} & 1 & 0 \\ 0 & 0 & 2 \end{bmatrix}$$

23.
$$\frac{1}{2}\begin{bmatrix} 1 & \sqrt{3} & 0 \\ -\sqrt{3} & 1 & 0 \\ 0 & 0 & 2 \end{bmatrix}$$

18.
$$\frac{1}{2}\begin{bmatrix} -1 & -\sqrt{3} & 0 \\ -\sqrt{3} & 1 & 0 \\ 0 & 0 & -2 \end{bmatrix}$$

24.
$$\frac{1}{2}\begin{bmatrix} 1 & \sqrt{3} & 0 \\ -\sqrt{3} & 1 & 0 \\ 0 & 0 & -2 \end{bmatrix}$$

ACKNOWLEDGMENT

The authors wish to thank Dr. A. B. Lidiard and Professor E. W. J. Mitchell for their interest in this work, and Dr. W. N. Reynolds for useful discussions and also for supplying his experimental results prior to their publication.

# References

Behrens, E. (1967a). *J. Acoust. Soc. Amer.* **42**, 367.
Behrens, E. (1967b). *J. Acoust. Soc. Amer.* **42**, 378.
Behrens, E., and Kremheller, A. (1969). *Nondestruct. Test.* **2**, 55.
Born, M., and Huang, K. (1954), "Dynamical Theory of Crystal Lattices." Oxford Univ. Press, London and New York.
Fedorov, F. I. (1968). "Theory of Elastic Waves in Crystals." Plenum Press, New York.

Hashin, Z., and Rosen, B. W. (1964). *J. Appl. Mech.* **31**, 223.
Heaton, M. D. (1968). *J. Phys. D.* **1**, 1039.
Heaton, M. D. (1970). *J. Phys. D.* **3**, 672.
Hill, R. (1964), *J. Mech. Phys. Solids* **12**, 199.
Hill, R. (1965), *J. Mech. Phys. Solids* **13**, 189.
Kothari, L. S., and Singwi, K. S. (1957). *Phys. Rev.* **106**, 230.
Kothari, L. S., and Tewary, V. K. (1962). *J. Chem. Phys.* **38**, 417.
Kothari, L. S., and Tewary, V. K. (1963). *Phys. Lett.* **6**, 248.
Langley, M. (1970). *Chartered Mech. Eng.* **17**, 56.
Leigh, R. S., Szigeti, B., and Tewary, V. K. (1971). *Proc. Roy. Soc., Ser. A.* **320**, 505.
Maradudin, A. A., Montroll, E. W., and Weiss, G. H. (1963). *Solid State Phys., Suppl.* **3**.
Musgrave, M. J. P. (1954a). *Proc. Roy. Soc., Ser. A* **226**, 339.
Musgrave, M. J. P. (1954b). *Proc. Roy. Soc. Ser., A* **226**, 356.
Reynolds, W. N. (1969). *Plast. & Polym.* **37**, 155.
Reynolds, W. N. (1970). Private communication. Results quoted from Reynolds (1970) are only preliminary. Further work is in progress.

# Substitutional–Interstitial Interactions in bcc Alloys[†]

D. F. HASSON and R. J. ARSENAULT

*Engineering Materials Group*
*College of Engineering*
*University of Maryland*
*College Park, Maryland*

## I. Introduction

The important effects of interstitial solutes on the properties of bcc metals have been the subject of numerous investigations and discussions. Interstitial atoms have a great effect on the mechanical properties of bcc metals such as the yield stress (Fleischer, 1967). They are also known to influence other properties such as the magnetic properties of silicon iron used for laminated transformer sheets (Leak and Leak, 1957). The effects of substitutional

[†]This work is supported by The Advanced Research Projects Agency at The University of Maryland, College Park, Maryland.

solutes on the properties of bcc metals have not been as thoroughly investigated as have interstitial solutes. However, there is an important and unusual effect of substitutional solute additions to bcc metals—there is a reduction of the yield stress of the alloy in comparison with the unalloyed metal. This phenomenon is not completely understood, but it has been proposed that there is an interaction between substitutional and interstitial solutes.

The dilute addition of the Group VII metal, rhenium, to molybdenum or tungsten increases the ductility and, thus, the workability of these materials. This result, designated as the "Re-effect" by Jaffee and Hahn (1963), was attributed to the interaction of rhenium with trace amounts of interstitial elements. In fact, it has been shown very convincingly by Hasson and Arsenault (1970) that the addition of titanium to vanadim removes oxygen and nitrogen from random solution. However, the question remaining is whether the titanium addition results in solid solution weakening of the vanadium–titanium alloy (Pink and Arsenault, 1971). The importance of studying and clarifying substitutional–interstitial interactions, especially in refractory bcc alloys, is due to an attempt to explain the alloy weakening effect, which modifies the usually expected large temperature dependence of the yield stress at low temperatures.

To be more specific about solution weakening, it is necessary to account for the division of the yield stress into its two components: namely, the thermal component (called the effective stress) $\tau^*$ and the athermal component $\tau_G$. This is illustrated schematically in Fig. 1, where the yield stress-versus-temperature variation for hypothetical solid solutions are compared with the variation for a pure bcc metal. As depicted in Fig. 1, the solute addition always increases the athermal component of the yield stress and can result in either a decrease (i.e., solution weakening) in the thermal component (as for solid solution No. 1) or an increase in the thermal component (as for solid solution No. 2). It is noteworthy that the solution weakening applies only to a decrease in the effective stress due to the addition of a solute.

Early general discussions of the large temperature dependence of the strength of bcc metals and their alloys at low temperatures, and particularly of solution weakening, can be found in papers by Hahn *et al.* (1963), Armstrong *et al.* (1963), and Seigle (1964). As pointed out by Arsenault (1967) and Ravi and Gibala (1969), the initial observations of solution weakening in a number of cases were explained in terms of a "scavenging" mechanism. Scavenging is the reduction by the substitutional solute of the random interstitial concentration. Credence to the scavenging mechanism is substantiated somewhat by the work of Fleischer (1967), who proposed that the thermal stress $\tau^*$ is proportional to the square root of the interstitial concentration, which means that the impurity interstitials are the short range barriers (s.r.b.) to dislocation motion. Thus, if the random interstitial concentration, which are the s.r.b., is

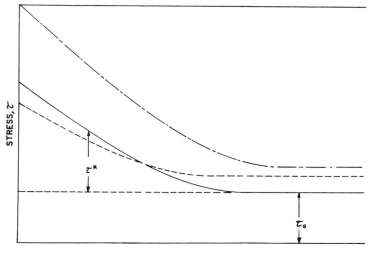

**Fig. 1.** An illustration of the division of yield (or flow) stress into thermal (effective) $\tau^*$ and athermal $\tau_G$ components for a pure bcc metal (——) and hypothetical solid solutions (No. 1, – – –; No. 2 — ·· — ··). $\tau = \tau^* + \tau_G$.

reduced by scavenging in an alloy exhibiting solution weakening, then scavenging is the mechanism responsible for the solution weakening.

There are several mechanisms by which the substitutional atoms may reduce the random interstitial concentration. One possibility is the reaction of solute and interstitial to form a compound; e.g., an oxide, which we define as scavenging. A second hypothesis might be that the solubilities of the interstitial atoms are changed by the introduction of small quantities of substitutional alloying elements. Third, the solute alloying elements, by virtue of their atomic misfit in the solvent lattice, produce preferred sites for the interstitials, thus relieving internal strains.

An explanation of solution weakening, hypothesized first by Weertman (1958) and then by Armstrong *et al.* (1963) and later developed in detail by Arsenault (1967), is that solution weakening can be explained by the localized reduction of the Peierls–Nabarro stress due to alloying. Arsenault's explanation assumes that the effective stress is controlled by the intrinsic lattice, (i.e., Peierls stress).

Thus, it becomes important to determine which substitutional solutes interact with random interstitials to reduce their concentration or to influence their mobility. Such results would identify alloy systems in which significant substitutional–interstitial interactions (s–i) occur. Most of the results which directly manifest s–i interactions are from internal friction studies. Before proceeding, however, it is necessary to note the importance of the early work by Snoek (1941) who, by the internal friction technique, demonstrated the correlation between the anelastic effect and the concentration of randomly distributed interstitial carbon and nitrogen dissolved in alpha iron. For example, a comparison of the internal friction spectra with and without a substitutional addition can indicate the presence of s–i interactions. Thus, the internal friction technique can give insight into the cause of solution weakening. Other methods such as mechanical property tests can also be utilized, but a detailed study of the internal friction spectra gives more explicit information on s–i interactions which might be responsible for some of the unusual properties of bcc alloys.

In order to evaluate the effect of substitutional additions on normal internal friction Snoek peaks, the existing experimental internal friction results for bcc ternary alloys will be discussed. First, experimental internal friction results for the bcc iron solid solutions will be presented, followed by a discussion of the refractory alloys of the Group V and Group VI metals. Applicable experimental mechanical property data, which compliment the internal friction data, will also be considered. Models describing the effect of the substitutional atoms on interstitial mobility will also be presented. Furthermore, theoretical studies, which attempt to identify the s–i complexes responsible for the experimental relaxation spectra, will be presented and discussed.

## II. Background

### A. General

The internal friction technique yields the most explicit information on s–i interactions. Therefore, some general discussion of this technique is in order. Snoek (1941) was the first to recognize that a mechanical relaxation is related to interstitial mobilities and concentration. He found for bcc metals subjected to a periodic strain that interstitially dissolved elements cause a dissipation of elastic energy. Equilibrium is established in the unstrained metal when the solute atoms are randomly distributed, as illustrated in Fig. 2a. When the metal is elastically strained, however, a redistribution takes place, as illustrated in Fig. 2b.

The modes of this redistribution cannot be easily identified, i.e., interstitials can jump from octahedral to octahedral site (O–O), tetrahedral to tetrahedral (T–T), or some combination of octahedral–tetrahedral jumps. Beshers (1965) has examined this problem and on the basis of the lowest activation energy for diffusion, he has proposed the preferred paths for interstitials in various bcc metals. The presence of substitutional atoms, however, further complicates

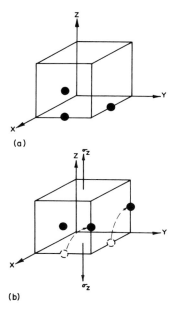

(a)

(b)

**Fig. 2.** Interstitial sites and diffusion paths in a bcc alloy for (a) no applied stress and (b) with an applied stress.

the problem (e.g., the mode can be different for an interstitial in the neighborhood of a solvent atom than for a substitutional solute atom). An illustration of two possible different jumps is given in Fig. 3.

Since there are several excellent comprehensive presentations of the relation between the mechanical relaxation and interstitial mobility and concentration

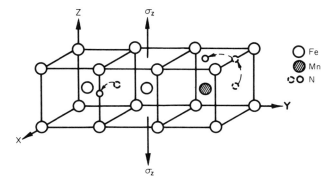

**Fig. 3.** Proposed models for interstitial diffusion with an applied stress for a pure bcc metal (left) and in the vicinity of a substitutional solute atom (right).

(see, e.g., Nowick, 1953, 1965; Zener, 1948), only the salient features are presented in the following discussion. The octahedral interstices of bcc crystals are situated midway along the cube edges and at the face-centered sites, as shown in Fig. 2a. They may be thought to comprise three equivalent sets, coordinated by atoms in the $X$, $Y$, and $Z$ directions, respectively. Ordinarily, interstitial solute atoms occupy all sets with equal probability; however, this random distribution can be altered by the action of external forces as shown in Fig. 2b.

An elastic periodic stress (e.g., biased along one of the cube axes), dilates the interstices lying in that direction at an instant in time, while causing contraction in the other two sets as illustrated in Fig. 4. To minimize the total strain energy, a fraction of the interstitial solute atoms occupying smaller sites move to the larger, lower energy sites. This transition follows an Arrhenius-type rate law which relates the relaxation time to make the jump with the temperature and the activation energy required for the jump. Since the internal friction is related to the frequency of the periodic stress and the relaxation time, the internal friction is dependent on the Arrhenius rate law. Therefore, when the internal friction is measured at a constant frequency as a function

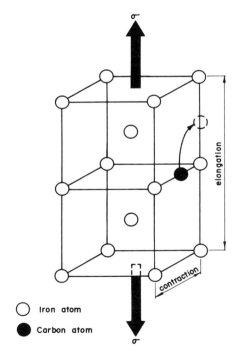

○  Iron atom

●  Carbon atom

*Fig. 4.* Model of the alteration of the bcc lattice due to a periodic stress at some instant in time which shows the probable interstitial jump.

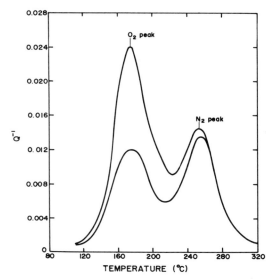

**Fig. 5.** Internal friction of vanadium containing interstitial oxygen and nitrogen for two concentrations of interstitial oxygen at one Hz (Powers, 1954).

of temperature, a single relaxation process [e.g., carbon in iron (Zener, 1948)] is found, and a sharp maximum occurs in the internal friction curve at a well-defined temperature. This maximum, which has been designated the "Snoek peak" for interstitials dissolved in pure bcc metals, occurs when the period of the applied stress is approximately equal to the relaxation time for the interstitial. Here the applied stress cycle is short enough for the anelastic (i.e., time-dependent) strain to have a finite value, while the stress variation is not short enough for equilibrium to be reached. Under this condition, the anelastic strain does not vary linearly with the stress, and the total strain becomes nonlinear. The resultant stress–strain diagram for the cycle is an elliptical loop. The area inside the loop, $\oint \sigma \, d\varepsilon$, has the dimensions of energy and represents the energy loss in the specimen per unit volume during a complete cycle. Thus, when the period of oscillation of the applied stress is comparable to the relaxation time, the dissipation of energy attains a maximum, and the sharp maximum in the internal friction curve is observed. Since the various interstitials in different bcc metals have different jump activation energies, the peak temperatures are different, suggesting that this result can be used to identify which interstitials are present. Also, the height of the internal friction peak is a measure of the concentration of the particular interstitial. To illustrate these points, a representative internal friction curve from the work of Powers (1954) for tantalum containing both oxygen and nitrogen is given in Fig. 5. At high interstitial concentrations clustering can occur, and the Snoek peak becomes altered, usually by broadening. Finally, at sufficiently high interstitial concentrations, new peaks can occur which are associated with the relaxations of pairs, triplets, etc., of the interstitials. These relaxations usually require higher activation energies and consequently, new peaks occur at higher temperatures.

### B. Definitions of Non-Reactive and Reactive Substitutional Solutes

Non-reactive solutes, in general, can cause two effects: (1) no s–i interaction or (2) a strong s–i interaction. Reactive solutes cause a s–i reaction.

A substitutional solute will be termed "non-reactive" if the chemical affinity of the substitutional solute and the solvent are approximately equal (i.e., they both have equal probability of forming compounds with the interstitials). Fast (1951) and Fast and Dijkstra (1951) published the earliest works that reported the effect of a substitutional solute on the interstitial Snoek peak. Fast and Dijkstra (1951) found that the addition of 0.5% manganese to an iron + 0.04% carbon bcc alloy had no effect on the iron–carbon Snoek peak. There was also no effect on precipitation after quenching, which, of course, would indicate the absence of a s–i interaction effect. These investigators, however, also studied the effect on the Snoek peak of the same substitutional addition to an Fe–Mn + 0.02% N bcc alloy. These results were completely different from the results for Fe–Mn–C, namely (1) the internal friction peak broadened and the peak temperature shifted to a higher temperature; and (2) the addition of manganese to the iron–nitrogen alloy almost completely stopped the precipitation of nitrogen after quenching. Fast and Dijkstra (1951) concluded that since the addition of manganese does not remove the peak, the nitrogen atoms are still able to move between the interstitial positions. The widening of the peak, however, signifies that there is now more than one relaxation process operating.

The explanation of Fast and Dijkstra (1951) for the shift of the peak to a higher temperature is based on the assumption that there is an increase in the mean relaxation time, and, therefore, an increase in the mean time that the atoms remain in a particular interstitial position. They further hypothesized that it is more energetically favorable for nitrogen atoms to occupy an interstice next to a manganese atom than it is to be surrounded only by iron atoms. The fact that the interstices are no longer energetically equivalent leads to the occurrence of different relaxation times. Fast and Dijkstra (1951) explained the lack of precipitation as being due to the rapid recapturing of a liberated nitrogen atom by the same or some other manganese atom. This may result from a greater affinity between manganese and nitrogen than for iron and nitrogen.

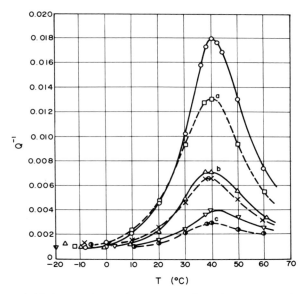

**Fig. 6a.** Internal friction results at about one Hz for α-iron with 0.04% carbon (solid curves) and of iron with 0.5% manganese and 0.04% carbon (broken curves). Results (a) and (b) from Fast and Dijkstra (1951), (c) is from Dijkstra and Sladek (1953).

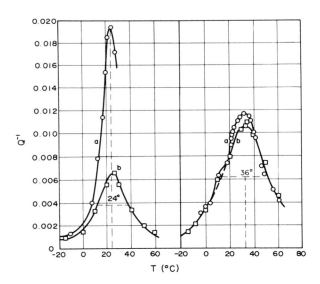

**Fig. 6b.** Internal friction results at about 1 Hz for α-iron with 0.02% nitrogen (left) and iron with 0.5% manganese with 0.02% nitrogen (right). Curves (a) and (b) from Fast and Dijkstra (1951); (c) from Dijkstra and Sladek (1953).

The above results of Fast and Dijkstra (1951) (shown in Figs. 6a and 6b), illustrate the fact that a substitutional solute can or cannot have an effect on the Snoek peak. The former case is defined in this text as a substitutional–interstitial interaction (i.e., a s–i interaction). The damping spectrum (shown in Fig. 6b) is considered in the simplest situation to consist of two processes: (1) the true Snoek relaxation, involving the stress-indiced reorientation of unassociated interstitials, and (2) the s–i relaxation, occurring invariably at a higher temperature and involving the local reorientation of the interstitials among sites in the immediate neighborhood of a substitutional solute atom. An approximate decomposition of the damping spectrum into these two components is shown in Fig. 6c for a Fe–Mn–N alloy from the later results of Dijkstra and Sladek (1953). Figure 3 shows the way in which the reorientation of both types of sites can occur. The data in Fig. 6b, as discussed previously, indicates that the mean-time-of-stay of the nitrogen atom near a manganese atom is greater than if it is only in the vicinity of iron atoms. If the nearest neighbor site is occupied by the interstitial atom, the simplified model of Fig. 3 for a two-jump s–i process accounts for the longer time-of-stay near manganese atoms.

The above results illustrate the existence of nonreactive s–i interactions in a bcc alloy. Since the early work of Fast and Dijkstra (1951) there have been extensive internal friction investigations, primarily in iron alloys. Also, detailed analyses have been performed in an attempt to identify the various s–i interactions which are present in the experimental relaxation spectra. These analyses will be discussed in Section IV.

An intermediate case of an s–i interaction, for which the substitutional solute has a rather high chemical affinity for the interstitial atoms, was found

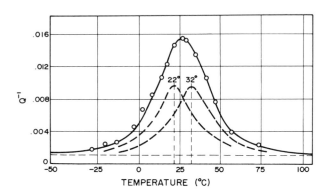

**Fig. 6c.** Internal friction results at about 1 Hz for iron with nitrogen and 0.5% manganese with proposed subsidiary peaks. Curve with open circles: experiment; —— subsidiary peak; — — sum of subsidiary peaks.

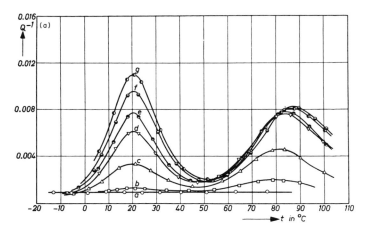

***Fig. 7a.*** Internal friction results at 0.77 Hz for Fe–0.5 at.% V–N alloys showing original internal friction results. (Fast and Meijering, 1953.)

in the Fe–V–N system by Fast and Meijering (1953). Fast and Dijkstra (1951) hypothesized that a substitutional addition, with a higher affinity for $N_2$ than for Mn, accentuates the s–i interaction (i.e., the broadening) effect. What perhaps occurs is a damping spectrum with two separate peaks; i.e., one peak associated with the free nitrogen (the Snoek peak) and another due solely to the s–i interaction. They selected vanadium because vanadium nitride has twice the heat of formation of $Mn_4N$. Their results as shown in Fig. 7a prove their hypothesis of the formation of two peaks. Fast and Meijering (1953), however, neglected to discuss the initial result which indicates that no damping peaks occur for the Fe + 0.5 at.% V + 0.04 at.% N alloy which is comparable to the Fe + 0.5 at.% Mn + 0.02 at.% N alloy previously studied by Fast and Dijkstra (1951). This result indicates that a substitutional–interstitial reaction rather than an interaction has occurred (i.e., the 0.04% N interstitial atoms have been removed from random solution). For higher concentrations of interstitial nitrogen a peak appears at a temperature which corresponds to the normal Fe–N Snoek peak. If the height of this peak is plotted versus nitrogen as in Fig. 7b, the data indicates that a reaction based on a stoichiometric VN formation is saturated at only one sixth of the vanadium available for such a reaction. The Fe + V + N system is, therefore, representative of an intermediate class of s–i interactions. Perry *et al.* (1966) have developed an analytical model for these results which will be discussed in detail in Section IV. The emphasis in the analytic study of Perry *et al.* (1966) is an attempt to explain why the high temperature peak broadens and shifts with temperature as the concentration varies.

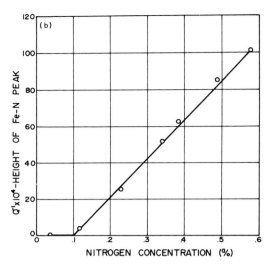

***Fig. 7b.*** Fe–N peak height versus nitrogen concentration (Fast and Meijering, 1953).

Reactive substitutional solutes are best characterized by the Group IV metals: titanium, zirconium and hafnium, all of which have a very high chemical affinity for the most commonly considered interstitial solutes, namely, oxygen, nitrogen, and carbon. Indirect evidence for a reaction can be found from changes in mechanical properties such as ductility. For example, Lundin and Klodt (1961) observed that in V–Y solid solutions increased ductility occurred upon the addition of yttrium. Solomon *et al.* (1969) states that for a Fe–Ti solution, the addition of the titanium scavenges the random interstitials as determined by internal friction. One might conclude, therefore, that the addition of a substitutional solute, which has a strong chemical affinity for the various interstitials, would result in scavenging of the random interstitials, and, hence, the normal bcc solvent–interstitial peak does not appear. This statement should be qualified by noting that the Snoek peak, associated with the free (i.e., random) interstitials, is absent only if the interstitial concentration is less than that of the substutional solute concentration.

One demonstration of scavenging is provided by the internal friction work of Hasson (1970) for the vanadium–titanium–oxygen system. The resultant internal friction spectra are similar in form to those presented in Fig. 7a from the work of Fast and Meijering (1953). Thus, the addition of reactive substitutional solutes results in a substitutional–interstitial reaction as represented by the absence of the normal Snoek peak up to a concentration of interstitials which equals the substitutional concentration. Beyond this concentration, the addition of reactive substitutional solutes gives a behavior similar to the aforementioned Fe–V–N type s–i interactions. The variation with interstitial concentration in the broadness and peak-temperature shift of the high temperature peak is somewhat larger than for the Fe–V–N results. An analytical model for the reactive type of solute (i.e., a scavenger) has not been developed.

In summary, the addition of substitutional solute to a bcc metal–interstitial alloy results in four types of substitutional–interstitial interactions: (1) no interaction, (2) an s–i interaction, (3) a s–i reaction, which becomes saturated when two internal friction peaks occur, one of which is associated with the free interstitial and the other with the s–i interaction, and (4) a s–i reaction which eliminates the interstitial Snoek peak until the concentrations of substitutional and interstitial are equal, and, beyond this concentration, two internal friction peaks occur as in type (3). These four interactions can be characterized by the systems Fe–Mn–C, Fe–Mn–N, Fe–V–N, and V–Ti–O, respectively. Models to describe the internal friction spectra are available only for the first three types of interactions. The model for the Fe–V–N interaction presently involves an attempt to explain the high temperature peak. The applicability and exactness of this model will be discussed later. Next, the experimental results will be presented and discussed for bcc iron ternary alloys (i.e., Fe + S + i), followed by the experimental results for ternary Group V and VI alloys.

## III. Experimental Results

Results from low frequency torsional pendulum experiments are quite extensive for iron ternary alloys (i.e., Fe + S + i); these results will be grouped by substitutional addition with a particular interstitial (e.g., Fe + Mn + C). The available work for Group V or Group VI ternary alloys is not so extensive; these results will be presented individually for each of the Group V and VI solvents. As mentioned previously, the studies of substitutional–interstitial interactions have been primarily internal friction investigations. Thus, the following discussion will be concerned almost entirely with the internal friction spectra.

### A. Iron Ternary Alloys

Internal friction data are available for ternary iron alloys with substitutional elements from Groups IV B, V B, VI B, VIII, I B, III A, IV A, and V A. The alloys investigated are the systems Fe + Ti + N; Fe + V + C; Fe + V + N; Fe + Cr + C; Fe + Cr + N; Fe + Mo + C; Fe + Mo + N; Fe + Mn + C; Fe +

Mn + N; Fe + Ni + C; Fe + Ni + N; Fe + Cu + N; Fe + Al + C; Fe + Si + C; Fe + Si + N; and Fe + P + C. The presentation of the results will follow this order.

### 1. IRON–TITANIUM–NITROGEN ALLOYS

As previously mentioned, Leslie and Sober (1967) and more recently Solomon *et al.* (1969) measured the internal friction of Fe–Ti interstitial ternary alloys and noted the absence of damping spectra. The former authors called their alloys "interstitial free," while the latter authors concluded that the titanium had scavenged the interstitials. Fast (1961), in an attempt to explain his results on Fe–V–N alloys, also investigated a Fe + 0.5 at.% Ti alloy nitrided at 950° C, and stated that although Ti–N forms as a precipitate, free interstitials are present as well resulting in only one damping peak.

Most recently Szabo-Miszenti (1970) investigated in detail the internal friction of alloys of iron + 0.15 to 0.60 wt% titanium + nitrogen. In addition to the normal Fe–N Snoek peak, he found two higher temperature peaks at 1 Hz which occur at about 120° and 227° C (as shown in Fig. 8) with activation energies of 25 and 31 kcal/mole, respectively, when the alloys were nitrided with ammonia at 590° C. The peak temperatures and widths vary with

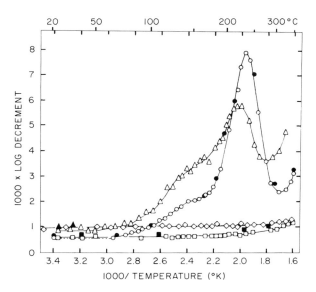

**Fig. 8.** Internal friction for a Fe–0.44 wt% Ti alloy. Open circles: 0.27 wt% N, 1.2 Hz; triangles: 0.19 wt% Ti, 0.7 Hz; squares: treated in moist hydrogen, 1.3 Hz; diamonds: 0.004 wt% N, 1.0 Hz. Closed symbols refer to tests made at double amplitude (from Szabo-Miszenti, 1970).

composition and heat treatment (e.g., both high temperature peaks were suppressed by heating at about 970° C). The peaks were attributed to s–i interactions. The peaks were competitive; that is, the nominal 120° C peak would decrease while the nominal 227° C peak would increase as the nitrogen to titanium atom ratio (designated N/Ti ratio) was varied. This competitive nature was explained on the basis of the following model. Since Ti has a larger atomic diameter than Fe, Ti–Ti pair formation was not considered probable. Consequently, the interstitials were hypothesized to interact mainly with isolated Ti atoms. The highest temperature peak which appears at a low N/Ti ratio was ascribed to nitrogen atoms located in Ti–Fe sites, i.e., the six octahedral sites coordinated with a Ti atom. The difference in activation energy between the 120° and 227° C peaks was interpreted on the basis of two rather different types of interstitials, e.g., jumping in sites differently coordinated with a Ti atom. If, with increasing N/Ti ratio, additional N atoms are progressively located in next-nearest sites (e.g., the cube edge sites coordinated with a body-centered Ti atom) and the inner interstitials are prevented from jumping, the 227° C peak is expected to reach a maximum and then fall off. This was observed experimentally. Thus, the increases in the 120° C peak were ascribed to jumps of interstitials which reside in the outer sites. Several relaxations, related to the actual number of interacting N atoms, are involved and contribute to the broadening and peak-temperature shifting of the two high temperature peaks. Szabo-Miszenti concluded that the tendency of N atoms to reside in preferential sites around Ti atoms is evidenced by the small or negligible normal Fe–N Snoek peak observed when the high temperature peaks are present.

This latter interpretation by Szabo-Miszenti and others is the crux of the problem of discussing s–i interactions in bcc ternary alloys. A s–i interaction is more exactly represented by the broadening of the normal Snoek peak. An example of peak broadening is provided by the Fe–Mn–N system (to be discussed later). The addition of manganese results in interactions between the substitutional atoms and the interstitial nitrogen atoms in addition to the damping associated with interstitial nitrogen in iron. In the case of a substitutional addition which removes the interstitials from random solution at concentrations equal to or greater than the interstitial concentration, there is no indication of a s–i interaction, but there is a s–i reaction. Thus, to study the effect of substitutional additions (especially those for which a strong chemical affinity for the interstitial is known *a priori*), it is best that the alloy originally have an almost negligible concentration of interstitial atoms. The controlled addition of interstitials could then be detected from the internal friction spectra. From the initial absence of peaks and the increase in the concentration of interstitials, the interstitial concentrations where peaks appear allow a more important determination to be made; i.e., the ability of the substitutional addition to remove the interstitials from random solution. Beyond this interstitial concentration, where the free interstitials are now available to take part in the normal Snoek and complex s–i interactions (as manifested by the appearance of internal friction peaks), the significance of the investigation becomes somewhat academic in nature (i.e., an internal friction study of the unusual resultant internal friction spectra). Thus, for substitutional additions which have a high chemical affinity to interstitials, the emphasis should be placed on the determination of the concentration of substitutional solute to remove interstitials from random solution. An example of such an investigation on a Group V solid solution will be discussed later.

## 2. IRON–VANADIUM–INTERSTITIAL ALLOYS

The effect of vanadium on the iron–carbon system was reported by Wert (1952) who found that the addition of 0.5% vanadium had no effect on the iron–carbon internal friction peak, as can be seen in Fig. 9. By measuring the decay of the internal friction peak value versus annealing time for an annealing temperature of 150° C, however, he found that the precipitation rate for this alloy was ten times faster than for other iron ternary alloys (e.g., Fe + 0.5% Mn + C). Wert suggested that vanadium provides more precipitation nuclei for the carbon. His results were somewhat clouded by the fact that for an

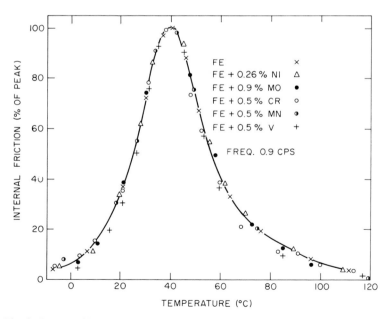

**Fig. 9.** Internal friction of several iron ternary alloys with carbon (Wert, 1952).

iron–carbon alloy the chemical analysis and internal friction results agreed, while the chemical analysis for the Fe + Mn + C alloy showed twice the carbon concentration as measured by internal friction.

Results for a Fe + V + nitrogen alloy were obtained by Dijkstra and Sladek (1953). The resultant internal friction spectrum was composed of the Fe–N Snoek peak and a high temperature peak which they designated an "abnormal" peak and to which they associated a new single relaxation process (i.e., in addition to the Snoek relaxation). The resultant spectrum is similar to that shown in Fig. 7a. Dijkstra and Sladek only indicated the treatment by which the nitrogen was added, but they did not indicate the concentration of additional nitrogen. Also, they found that both peaks were very unstable. That is, the peaks decreased considerably in height with a decrease in temperature during measurements. Furthermore, if they annealed at 700° C for 3 minutes the internal friction was reduced to zero. Thus, Dijkstra and Sladek concluded that vanadium causes the rapid precipitation of nitrides (i.e., VN) in this alloy.

At about this same time Fast and Meijering (1953) also investigated the same substitutional addition (i.e., 0.5 at.% V), but they varied the nitrogen concentration from a very low value of 0.04 at.% up to 0.58 at.%, and also studied various heat treatments. At nitrogen additions up to 0.1 at.%, their results indicate the absence of damping spectra (see Fig. 7a curve a). This clearly indicates that initially vanadium acts as a scavenger of nitrogen in a manner similar to titanium as found by Szabo-Miszenti (1970). The scavenging action, however, is saturated before the nitrogen and vanadium solute concentration are equal. These results indicate that it takes 5 to 6 vanadium atoms to remove the interstitial nitrogen atoms at concentrations where vanadium scavenges the interstitial nitrogen (see Fig. 7b). Fast and Meijering, however, did not discuss the results for low nitrogen concentration, but only emphasized the high temperature peak which they called abnormal as did Dijkstra and Sladek (1953). Fast and Dijkstra (1951) earlier predicted a dual peak in a Fe–Mn–N alloy.

Fast (1970) in a private communication stated that Fe–V alloys which contain 0.5 at.% V (when heated in nitrogen under suitable conditions) absorb one atom of nitrogen per atom of vanadium, and, in addition, dissolve an extra amount of nitrogen corresponding to the solubility of nitrogen in pure iron, which is Fast's original interpretation. He continues to state that there is no significance of the plot shown in Fig. 7b which illustrates that the random interstitial nitrogen Fe–N Snoek peak appears for concentrations much smaller than the substitutional vanadium concentration.

An examination of the variation of the peak temperature for the abnormal peak and an attempt to identify the relaxation processes associated with the abnormal peak in the Fe–V–N system was performed 5 years later by Perry *et al.* (1966). This analysis will be discussed in Section IV.

Jamieson and Kennedy (1966) again studied iron–vanadium–nitrogen alloys with a variation from 0.415 to 0.69% vanadium. They also obtained some results for Fe–V alloys with carbon which also showed the existence of an abnormal peak. This peak, however, was unstable. They concluded that the abnormal peak in Fe–V–N alloys was caused by two relaxation processes, one associated with the nitrogen atom jumps associated with Fe–V sites, and the other with jump processes involving preferred sites in the vicinity of vanadium nitride percipitates. They concluded that the abnormal peak at very high nitrogen concentrations approximates to a single relaxation process due to nitrogen jumps between Fe–V sites. While they considered Fast's later interpretations (Fast, 1961) and Fountain and Chipman's results, (1958), their conclusions, which are not very different from the original interpretations of Fast and Meijering (1953), do not appear to reflect the later thinking of Fast (1961) nor the analytical results of Perry *et al.* (1966). Thus, there remains some doubt as to the validity of their conclusions.

## 3. Iron–chromium–interstitial alloys

The results for an iron + 0.5% chromium + 0.0165% carbon alloy obtained by Wert (1952) show that chromium has absolutely no effect on the height or width of the Fe–C Snoek peak (see Fig. 9). Other results of this investigation on other Fe–X–C alloys (e.g., the Fe–V–C results discussed above) did show a discrepancy between the chemical analysis and internal friction determination of the carbon concentration. But these results were in excellent agreement for the Fe–Cr–C alloy. Thus, the addition of chromium shows absolutely no s–i interaction or scavenging (i.e., a s–i reaction).

Results for the effect of chromium on Fe–N alloys obtained by Dijkstra and Sladek (1953) and Ritchie and Rawlings (1967) show a definite broadening of the Fe–N Snoek peak, and, hence, indicate that a s–i interaction does take place. Dijkstra and Sladek (1953) suggested that the broadened peak for a Fe + 0.5% Cr + N alloy was composed of the Snoek Fe–N component plus a higher temperature relaxation due to a Cr–N (i.e., s–i interaction). Their investigation of the effect of several different substitutional solutes on a particular alloy was not very extensive and, hence, only limited conclusions can be drawn from their study.

Ritchie and Rawlings (1967) investigated 0.94 and 4.2% chromium additions. Their alloys had a finite amount of interstitial carbon. They also noted that in previous works the possibility for relaxation contributions at temperatures lower than the Snoek peak temperature (which is 22° C for 1 Hz) were possible. Hence, they performed their experiments from −20° to +100° C. This is illustrated in Fig. 10 where Dijkstra and Sladek's results are also shown. They synthesized the resultant broad internal friction peak, and reported that six relaxation processes are present starting from a temperature of −7° C: two

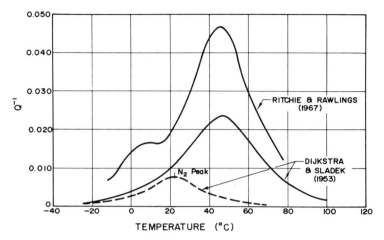

**Fig. 10.** Comparison of internal friction results for Fe–Cr–N alloys with 0.5% Cr (Dijkstra and Sladek, 1953) and 0.94% Cr (Ritchie and Rawlings, 1967).

low temperature relaxations due to nitrogen in the neighborhood of Cr–Cr pairs; the N–Fe Snoek relaxation; a C–Fe Snoek relaxation and a relaxation due to nitrogen in the neighborhood of Fe–Cr; and the highest temperature relaxation at 66° C, probably due to carbon in the neighborhood of Fe–Cr. Similar results were determined by these investigators for the Fe–Mn–N system to be discussed later. One particularly interesting result from their synthesis is the possible existence of two low temperature peaks at −7° and 5° C at 1 Hz. In their discussion the authors make the following important observations. (1) From all of the previous results for iron–nitrogen–ternary systems, there are at least three relaxations for the case of nitrogen atoms near pairs of substitutional atoms. In addition, there are three unknown parameters for each relaxation process, and it is possible that pairs of different types of interstitial atoms (e.g., C and N) can exist. Thus, without more precise data one can still only give a qualitative picture of the solid solutions. (2) The relationship between interstitial concentration and peak height may be different for each relaxation process. (3) The number of interstitials contributing to any one relaxation may change with temperature. It is clear, we feel, that analytical peak synthesis is at best difficult, if not misleading, because there is no *a priori* information on which to base the synthesis.

## 4. IRON–MOLYBDENUM–INTERSTITIAL ALLOYS

Wert (1952) reported results for an iron + 0.9% molybdenum + 0.0155% carbon alloy, and found no effect due to the molybdenum on the Snoek iron–carbon peak (see Fig. 9). Later, Collete (1960) reported quite different results for alloys containing 1.02% Mo with 0.005% C, 1.83% Mo with 0.0075% C, and 3.53% Mo with 0.0120% C. Collette found two peaks at a frequency of 0.217 Hz, one at 18° C, which he attributed to the effect of the distortion of the larger molybdenum atom; and two, the normal Snoek peak at 26° C (see Fig. 11). He verified the existence of the 18° C peak by adding an equal additional concentration of nitrogen which resulted in the absence of the 18° C peak, an increase in the Fe–C peak, and the appearance of the Snoek Fe–N peak. These results for the Fe–Mo–C alloys are the only iron ternary alloys that show a basic disagreement between investigators. From the available results, no explanation can be offered to explain the discrepancy between the Wert (1952) and Collete (1960) results.

The experimental results available for the Fe–Mo–N system are those of Dijkstra and Sladek (1953) for a Fe–0.5% Mo–N alloy. A dual peak results and the findings are very similar to those for the same alloy additions in a Fe–V–N alloy (e.g., after the peaks are measured at ascending temperatures, internal friction measurements at descending temperature show the peaks decreased significantly in height and, thus, a very rapid precipitation of the nitrogen has occurred). The only difference between the Fe–Mo–N and Fe–V–N results is the difference between the Snoek and higher temperature peak heights (i.e., the Snoek peak height is one half of the high temperature peak height for the Fe–V–N alloy, while the Snoek peak is about 1.4 times larger than the high temperature peak for the Fe–Mo–N alloy for the same percentage of added substitutional and interstitial solute). Thus, it appears that vanadium has a greater effect on the random interstitial concentration. This result could be due to the higher chemical affinity of vanadium (e.g., 41.6 kcal/mole for VN as compared to 16.6 kcal/mole for MoN).

## 5. Iron–Manganese–Interstitial Alloys

For a Fe + 0.5% Mn + 0.04% C alloy, Fast and Dijkstra (1951) determined the original results for an Fe + Mn + C carbon alloy and probably the earliest results for the effect of a substitutional addition on the interstitial Snoek peak.

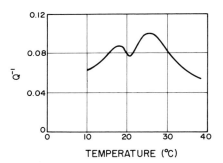

**Fig. 11.** Internal friction for a Fe + 1.83% Mo + 0.0075% C alloy (Collette, 1960).

Later, Wert (1952) reported the results for a Fe + 0.5% Mn + 0.014% alloy, shown in Fig. 9. Both investigators found that this addition of manganese had no effect on the Fe–C Snoek peak and on the rate of precipitation of carbon in $\alpha$-iron.

The bcc ternary alloy system studied most extensively for substitutional–interstitial interactions is the Fe–Mn–N system. The original work of Fast and Dijkstra (1951) showed the addition of 0.5% Mn to a Fe–0.2% N alloy broadened the Fe–N Snoek peak and shifted it to a higher temperature. Moreover, the manganese inhibited almost completely the precipitation of nitrogen in the alloy. These results caused by the affinity between manganese and nitrogen atoms causes (1) Mn sites around which it is easier for the nitrogen atoms to relax, thus producing another relaxation process which broadens the Snoek peak, and (2) nitrogen atoms to be captured by the manganese atoms, thus inhibiting the precipitation of the interstitial nitrogen atoms in the iron solid solution. The explanation of Fast and Dijkstra has not been changed but only examined in more detail by all the later investigators.

Dijkstra and Sladek (1953) next investigated a Fe–Mn–N alloy with a composition of Fe + 0.5 at.% Mn + N. They synthesized the experimental peak with two relaxation processes: (1) a Snoek Fe–N relaxation, and (2) a higher temperature relaxation due to interstitial nitrogen with manganese as a nearest neighbor. They presented a model (to be discussed in Section IV) to explain how a manganese atom may modify the potential energy of a nitrogen atom in the iron lattice. They also derived an expression for the modification in solubility of interstitial nitrogen due to the presence of manganese. From this expression and the peak synthesis, which showed the experimental peak height consists of two equal height relaxation peaks, they derive an expression for the increase in free energy for transferring a mole of nitrogen atoms from Fe–Mn sites to Fe–Fe sites. This value, which is a measure of the substitutional–interstitial interaction, was determined to be 2800 cal/mole. This value will be compared later to the change in tetragonal strain, the importance of which will be discussed in terms of the effect of s–i interactions on the mechanical properties.

Meijering (1961) examined analytically the experimental results first reported by Fast and Dijkstra (1951) and later by Dijkstra and Sladek (1953) for the Fe–Mn–N system. He employed the anelastic effect models of Snoek (1941) and Polder (1946) to determine which relaxation processes are operative and what contributions they make to the broadened peak. He concluded that the presence of the two types of sites (i.e., Fe–N and Mn–N) for the interstitial atoms in a body-centered cubic lattice of Fe–Mn are responsible for the presence of three contributions to the experimental peak for Fe–Mn–N. Meijering stated in principle that the experimental peak could have three contributions without the usual contribution from the tetragonal sites, because the nitrogen atoms are captured by the Mn atoms alone in the ferritic lattice. The existence of a fourth peak contribution was hypothesized which corresponds to the usual Fe–N peak for the case of noncaptured nitrogen atoms. Meijering was the first, however, to point out the fact that the captured interstitial atoms cannot be strictly separated from the noncaptured ones. Furthermore, he noted that the complication introduced by too high a Mn concentration requires a consideration of the possibility of nitrogen atoms being captured by two atoms of Mn. This result would cause another relaxation process to occur and, consequently, would require another hypothesis as to its contribution to the peak synthesis.

Enrietto (1962a) utilized the internal friction technique to study the solubility and precipitation of nitrides in α-iron containing manganese. He found manganese in concentration up to 0.75 wt.% had little effect on the solubility of nitrogen in α-iron at temperatures about 250° C. On the other hand, he found that manganese concentrations as low as 0.15 wt.% inhibited the formation of iron nitrides especially $Fe_4N$, even though it does not form a manganese nitride precipitate. Enrietto found evidence that nitrogen is segregated around manganese in the ferritic lattice, but the precipitation kinetics showed that this effect is not responsible for the inhibition of precipitation.

About this same time, Enrietto (1962b) also questioned the assumptions by Dijkstra and Sladek (1953) and subsequent investigators that there are only two distinct types of interstitial sites in the ternary lattice, and that all of the interaction processes take place with a single relaxation time. These assumptions imply that the influence of the substitutional addition only extends to its nearest neighbors. Enrietto performed an extensive experimental program on the Fe–Mn–N to determine the validity of these assumptions. He studied the effect of very dilute additions (e.g., 0.02 wt.% Mn) through additions (e.g., 0.75 wt.% Mn) which were greater than in most previous investigations while he varied the nitrogen concentration from 0.002 to 0.035 wt.%. Enrietto imposed the following conditions to synthesize his data using the assumptions of the previous investigations: (1) a set of three peak temperatures with some deviation; (2) three activation energies each of whose values was held constant; and (3) the peak temperatures were held independent of the N and Mn concentrations. From his extensive, experimental studies he found that the model of a two site lattice and a single relaxation time for the damping processes constitute too crude an approximation based on the following observations: (1) the relatively poor fit between the theoretical and experimental internal friction curves; (2) the inconsistencies in the behavior of the subsidiary peaks; (3) the large amount of scatter in plots of the subsidiary peak heights versus nitrogen content; and (4) the low value of the binding energy (about 3500 cal/mole for a 0.33 wt.% Mn addition) compared to the value of 5000 cal/mole suggested by the precipitation data. Also Enrietto showed from his results that the assumption that manganese affects only the nearest neighbor interstitial sites is incompatible with the experimental results, and that the influence of the manganese in the ferritic lattice was found to extend over 3 to 4 atomic distances. Enrietto's conclusion was that the usual method of synthesizing the experimentally broadened peaks by the use of a number of subsidiary single Debye peaks becomes questionable. The need is obvious for a different approach either experimentally and/or analytically to determine the exact relaxation processes taking place in bcc ternary alloys.

The first work which considers significantly increased manganese concentration is that of Fast *et al.* (1961). Their results showed that a 2.0% Mn concentration decreased the low temperature peak significantly and increased the high temperature peak which is barely discernable for the 0.5% Mn results just discussed. They hypothesized that the higher temperature peak is due to Mn–N–Mn clusters.

Gladman and Pickering (1965) studied both 0.02 and 2.0 wt.% Mn additions with a nitrogen concentration of 0.02 wt.%. They indicated their Fe + 2.0 wt.% Mn + 0.02 wt.% N alloy was more representative of an actual commercial steel than the alloys. For the 2.0 wt.% Mn addition a single broadened peak occurred at 0° C at 0.7 Hz, which represents a 19° C decrease from the Snoek Fe–N peak temperature. This indicates a more rapid diffusion process for nitrogen in this alloy, which is the only result they obtained that was different from Enrietto's (1962b). However, since Enrietto did not investigate this alloy, his contribution was not negated. They suggested that the dominant relaxation process involves the migration of a nitrogen atom in the neighborhood of a manganese couple. The reason for this conclusion is based on their strain aging results which showed that manganese retards the precipitation of nitrides at temperatures below 300° C, and this result supports the conclusion that long range diffusion of nitrogen was retarded. The mechanism proposed for the effect on the precipitation of nitrogen in ferrite by Gladman and Pickering is quoted as follows:

The formation of manganese atom pairs provides low energy sites for nitrogen atoms and these are then occupied in preference to the normal interstitial sites. This effectively pins the nitrogen and restricts its long range diffusion.

To reproduce their experimental internal friction results, they also presented a model for the vibration of the nitrogen atom between contiguous favored sites. This model will be presented and discussed later in the text. Another important observation they made was that the activation energy from the experimental peak, which was approximately the proper value for nitrogen in iron (i.e., 20 kcal/mole), caused the peak height to be less than it should have been for the concentration of nitrogen present. Their model is based on three relaxations associated with just the Mn couple, and also additional relaxations were proposed. These additional relaxations were associated with the effect of the lattice distortion of the couple at increasing distance on either a single Mn atom or another Mn couple or cluster. Thus, they concluded that a virtually continuous distribution rather than two or three relaxation processes occur in ternary bcc alloys.

Later, Ritchie and Rawlings (1967) investigated Fe–Mn–N alloys in an attempt to determine the exact relaxations responsible for the new low temperature contribution to the internal friction spectrum as found by Enrietto (1962b), and Gladman and Pickering (1965). They investigated additions of 0.2, 0.9, and 2.0 wt.% Mn, and their results showed at least two peaks of considerable height below room temperature instead of the single peak at 0° C as found by the previous investigators. The proposed mechanisms, however, were the same as those proposed previously (i.e., nitrogen atoms jumping around pairs of Mn atoms). Ritchie and Rawlings, through their synthesis and analysis, showed that these two peaks could not be due to jumps around single Mn atoms, since the heights of these two peaks did not vary the same as those due to N atoms around a single Mn atom. Some evidence was presented which showed that the ratio of the heights of these two peaks below room temperature varies with the Mn concentration. This suggests that the two peaks represent two distinct sites or two types of jumps from the same site. Ritchie and Rawlings, however, indicated they needed a larger variation in nitrogen concentration and more accurate determination of the activation energies to quantitatively prove this conclusion. In this same investigation, Ritchie and Rawlings also studied Fe–Cr–N alloys (note that the results were previously discussed in Section III, A, 3), and their important observation as to the difficulty of describing the complex relaxation processes in a ternary alloy is again most applicable.

About this same period in time Couper and Kennedy (1967) also investigated Fe–Mn–N alloys to verify Fast's (1961) suggestion of a three site model as opposed to Enrietto's two site model for the low temperature peak in the higher concentration Fe–Mn–N alloys. Alloys of 0.12, 0.7, and 1.6 wt.% Mn were studied experimentally and their experimental peaks were synthesized with the assistance of a digital computer. Couper and Kennedy noted that there is disagreement in the reported work of the several investigators on the detailed shape of internal friction profile for apparently the same Fe–Mn–N alloy. They compared their results for the 0.7 wt.% Mn alloy to those of Enrietto (1962b) for a 0.75 wt.% Mn alloy. However, Enrietto's profiles showed a higher damping at low temperatures (i.e., around 0° C) and were much broader than Couper and Kennedy's. They noted that aging reduces and broadens the peak profile. According to them, Enrietto had aged his specimens before testing, which could explain the differences noted above and contribute to the marked scatter in Enrietto's plots of the constituent peak heights against nitrogen concentration. This scatter helped Enrietto to reach his conclusions on the nonapplicability of a single relaxation process for s–i interaction processes in iron–ternary alloys. Couper and Kennedy went on to show, nevertheless, their results for the activation energies for those separate peaks at 7°, 23°, and 34° C. The low temperature peak at 7° C was suggested to be a hybrid of two peaks—one due to nitrogen atoms associated with pairs of manganese atoms, and another due to nitrogen atoms jumping from Fe–Fe sites to Fe–Mn sites, which was the reverse of the mechanism proposed for the higher temperature peak at 34° C. Moreover, they concluded that if both of these processes contribute to the low temperature peak, both processes must be characterized by almost identical relaxation parameters (i.e., activation energies and relaxation times).

Couper and Kennedy's (1967) and Ritchie and Rawlings' (1967) analyses of the experimental results and conclusions are similar. The analyses and conclusions of these two works, however, differ from Enrietto's and Gladman and Pickering, but Enrietto and Gladman and Pickering differ from each other. The internal friction profiles from these various investigators are shown in Fig. 12 for comparison. The analysis of the low frequency internal friction spectrum for Fe–Mn–N ternary alloys best illustrates the problem associated with attempting to fit the experimental data with various simple Debye single relaxation peaks, without any *a priori* information on the possible contributions to the experimental internal friction spectrum.

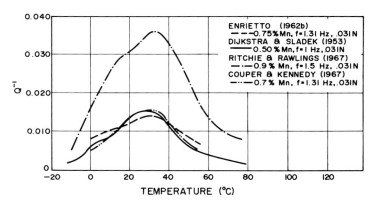

*Fig. 12.* Comparison of internal friction spectrums for Fe–Mn–N bcc ternary alloys from several investigators.

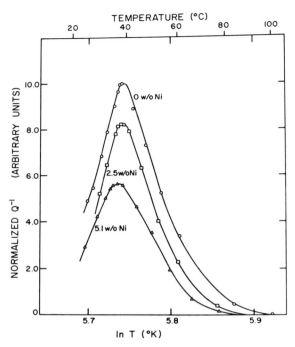

**Fig. 13.** Effect of nickel content on the Fe–N Snoek peak at 1.05 Hz. (Jackson and Winchell, 1964).

### 6. IRON–NICKEL–INTERSTITIAL ALLOYS

One of the earliest results is that for a Fe+0.26% Ni to 0.015% carbon alloy from Wert's (1952) survey of the effect of various substitutional solutes on the internal friction and precipitation of iron–carbon alloys. Wert found no effect on the internal friction or precipitation in this alloy as is shown in Fig. 9. Later, Heller and Brauner (1964), and Jackson and Winchell (1964) investigated the effect of nickel on the diffusion and precipitation behavior of carbon for nickel concentrations from 0 to 5 wt.% and 0 to 16.5 wt.%, respectively. In the latter study only the 0 to 5 wt.% nickel additions produce alpha alloys (i.e., ferritic alloys). Since the results of both investigations for these nickel concentrations were very similar, only the more extensive work of Jackson and Winchell will be discussed. Basically the effect of nickel additions on the iron–carbon Snoek peak was not drastic (as can be seen in Fig. 13). For a concentration of 0.26% Ni, which corresponds to Wert's nickel addition, the effect on the Snoek peak was almost negligible and agrees with Wert's conclusion. Upon increasing the nickel content there were three effects on the iron–carbon internal friction as noted by Jackson and Winchell; (1) a decrease in peak height by about 8% of a Fe–C peak height for each percent nickel; (2) a small increase in the half width of about 11% of the Fe–C Snoek peak width at a concentration of 5.1 wt.% Ni; and (3) a slight shift of the peak maximum by about −2.7° C at a concentration of 5.1 wt.% Ni. The first two effects indicate that nickel has some small influence on the interstitial carbon mobility (i.e., nickel has some short range interaction on the carbon atoms which prevents them from participating in their normal Snoek relaxation process). Jackson and Winchell concluded that the third effect (i.e., the peak temperature shift) indicates a long range influence by the nickel atoms on the movement of carbon atoms so that at 300° K the diffusiveness of carbon in ferrite is increased approximately 35% by the addition of 5.1 wt.% nickel. Jackson and Winchell's study shows that the conclusions drawn for the effect of small substitutional additions may have to be modified as the concentration of nickel is increased (i.e., Wert's result indicated no effect due to nickel, where in reality there is an effect of the substitutional addition on the interstitial mobility and concentration). Most recently, Fischer (1971) has privately communicated that a decrease in the Snoek peak height with increasing nickel concentration depends on whether the alloys are quenched from the ferrite or austenite region. If quenched from the austenite region the peak height decreases, and Fischer indicates that the decrease in the Snoek peak height is due to Snoek ordering or Cottrell atmosphere formation. He supports his conclusions by the fact that in his alloys as the nickel concentration increased, the dislocation density increased after quenching from the austenite region. Fischer then concluded that the decrease in the Snoek peak height is due to the removal of the random interstitials into Cottrell atmospheres. Jackson and Winchell's results suggest that a reexamination of the many systems in this paper may be in order (specifically Fe–Mo–C, Fe–Cr–C, Fe–Mn–C, Fe–V–C, etc.).

Results to determine the effect of nickel on the nitrogen–iron peak were reported by Ritchie and Rawlings (1967). The addition of 2.29% nickel showed absolutely no effect on the Snoek Fe–N peak. It was concluded, therefore, that Ni does not attract N atoms as might have been expected, since Ni does not form a stable nitride at room temperature, whereas, both Mn and Cr do.

## 7. Iron–Copper–Nitrogen Alloys

Fast and Verrijp (1961) reported some results for this system which they investigated because copper has no affinity for nitrogen. After studying copper additions of 0.7–1.5 at.% with 0.1 at.% nitrogen, they studied the influence of copper in solution by charging the wire with nitrogen at 950° C and quenching it to avoid interstitial precipitation. They then heated the wire for 1 hour at 550° C and 0.5 hour at 600° C, and then quenched it to determine if any precipitates were formed. Fast and Verrijp's internal friction curves for a Fe + 1.5 at.% Cu + 0.1 at.% N alloy (see Fig. 14) show that the copper in solution case obtained by the heating to 950° C and quenching produces a somewhat broad single peak; but the curve, after annealing, actually increases and approaches a normal Fe–N Snoek peak. Thus, they concluded that one does not find any indication of the appearance of a dual peak due to the presence of possible precipitates.

## 8. Iron–Aluminum–Interstitial Alloys

Internal friction of low carbon steels, containing from 0.04 to 2.36 wt.% aluminum, has been studied by Laxar *et al.* (1961). Their resultant internal friction spectra are shown in Fig. 15. As the figure shows, the aluminum additions up to 0.36 wt.% had very little effect on the normal Fe–C Snoek peak where the carbon concentration was 0.004 wt.%. The addition of 0.61 wt.% aluminum, however, reduced the peak height by about 50% and drastically broadened the internal friction curve. Higher concentrations resulted in dual peaks. Laxar *et al.* synthesized an analytical spectrum for the 1.83 wt.% Al alloy from four Debye peaks. The four peaks are the normal Snoek peak for Fe–C, and three higher temperature peaks which were hypothesized to be due to jumps between Fe–Al interstices, jumps from Fe–Fe to Fe–Al interstices, and jumps from Fe–Al to Fe–Fe interstices. Again, there

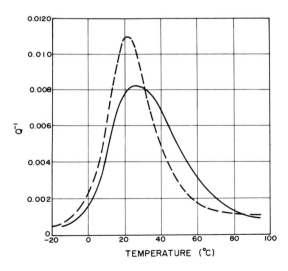

**Fig. 14.** Effect of copper on the Fe–N Snoek peak —— Quenched from 950° C; ——
after anneal. (Fast and Verrijp, 1961).

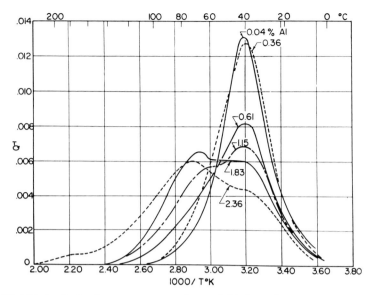

***Fig. 15.*** Effect of aluminum content on internal friction of low carbon steel brine-quenched from 700° C (Laxar *et al.*, 1961).

is the same type of study as for the other iron alloys (i.e., the emphasis has been on the interpretation of the resultant dual peak spectrum rather than the effect of the substitutional solute on the interstitial per se). As shown in Fig. 15, an increasing aluminum concentration produces a decrease in the Fe–C Snoek peak with the same nominal carbon concentration in these alloys. Thus, it appears that the addition of aluminum causes a scavenging action similar to that of titanium and vanadium.

Wolf and Hanlon (1961) reviewed the work of Laxar *et al.* (1961) and disagreed with their synthesization of the experimental peaks for the 1.83 wt.% Al alloy into four contributions. Wolf and Hanlon systematically analyzed the data from the point of view of the activation energies required for the various processes and concluded only two relaxations were responsible for the experimental results, namely, the normal Fe–C Snoek peak and the jumps of interstitials from Fe–Fe to Fe–Al interstices. On the basis of the two relaxation processes being responsible for the experimental dual peak, Wolf and Hanlon estimated that the difference in activation energies is 1700 cal/mole for a carbon jump near an aluminum atom as compared to a jump near only iron atoms.

Fast (1961) reported that addition of 0.5 at.% $N_2$ at 950° C, by the same technique as explained for the Fe–Ti–N alloy, to an Fe + 0.5 at.% Al alloy

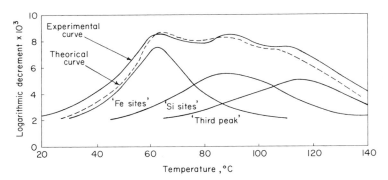

**Fig. 16.** Internal friction of an Fe + 2.83 wt% Si + 0.02 wt% C alloy at 7.125 Hz (Leak and Leak, 1958).

resulted in the precipitation of AlN. However, he did not observe a second peak even up to a temperature of 150° C and did not show his damping spectra; thus, an evaluation of his result is not possible.

### 9. IRON–SILICON–INTERSTITIAL ALLOYS

Leak and Leak (1958) have studied the internal friction of iron + 5.5 at.% silicon + 0.005 to 0.015 wt.% carbon alloys. Two peaks were always observed experimentally (see Fig. 16) and a third higher temperature peak had to be inferred from the analysis of some of the experimental curves. The lower temperature peak was attributed to the normal Fe–C Snoek peak (i.e., the one caused by jumps of the interstitial atoms with iron atoms as nearest neighbors). The second peak was shown to be due to jumps of the interstitial to and from sites where one silicon atom was the nearest neighbor. The activation energy for this later process was found to be 14.7 kcal/mole which is less than the value of 20.8 kcal/mole for the Fe–C peak. This result is explained in terms of a contraction of the lattice (Wert, 1950) caused by the silicon atoms. As a result, the combination of a carbon atom with a silicon atom as one of its nearest neighbors presents less distortion to the surrounding lattice than the two separate atoms would (e.g., carbon by itself distorts the Fe lattice). These correlations are illustrated in Fig. 17. Thus, carbon atoms remain associated with silicon atoms. Since the distortion of this carbon–silicon pair is less than for the carbon–iron complex, the activation energy for jumping of the carbon atom in the neighborhood of the silicon atom is less. From solubility measurements and the relative heights of the Fe–C and s–i peaks for various carbon concentrations, Leak and Leak (from their previous work on Fe–Si–N alloys to be discussed below), determined that the interaction between silicon and carbon atoms was much less than between silicon and nitrogen atoms. They pointed out, however, that use of high frequency damping internal friction measurements were required to resolve the peaks more accurately. From these same measurements, Leak and Leak concluded that the third peak corresponded to carbon atoms jumping from iron to silicon sites and back again (i.e., from a silicon to an iron site). Also, the high frequency technique was suggested as a method of checking this conclusion. Thus, the addition of silicon to iron with carbon causes s–i interactions, which are more complex than those associated with just broadening as represented by the Fe–Mn–N system.

Earlier, Leak *et al.* (1955) made a study of the solubility of nitrogen in an iron + 5.5 at.% silicon alloy. They found that the internal friction spectrum and relaxation processes for Fe–Si–N and Fe–Si–C are similar and that the activation energies for the first two peaks were 18 and 12 kcal/mole. They also found a third unstable peak and hypothesized that it was due to nitrogen atoms in the vicinity of nitride precipitate particles. This corresponds with Fast and Meijering's (1953) original interpretation of the Fe–V–N system results. Furthermore, they stated that sites with two or more Si neighbors for the interstitials would be very rare and would not contribute to the measured peak.

Rawlings and Robinson (1961) performed a similar investigation to check

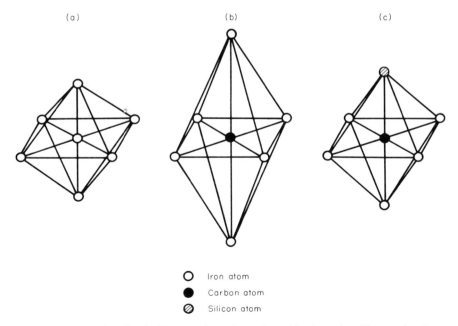

○  Iron atom
●  Carbon atom
⊘  Silicon atom

*Fig. 17.* Octahedral sites in (a) pure α-iron plus carbon, (c) α-iron plus silicon and carbon.

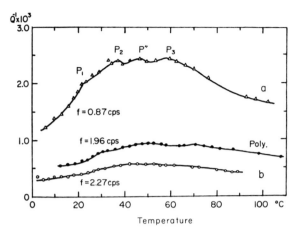

**Fig. 18.** Internal friction of single and polycrystalline Fe–Si–N specimens from Hashizume and Sugeno (1967). Curves (a) and (b) are for single crystals of $\langle 111 \rangle$ and $\langle 100 \rangle$ orientations, respectively.

the work of Leak *et al.* (1955), utilizing silicon additions of 0.5%, 1.1%, and 1.0%. They quenched from temperatures in the gamma and alpha range, and found three as did Leak *et al.* (1955), and in some cases five internal friction peaks. Starting from the lowest temperature, the first peak was identified as the usual Snoek Fe–N peak; the second was associated with N atoms with a neighboring Si atom; the fourth was associated with N–N pairs; and the fifth could not be identified. Rawlings and Robinson's identifications agree with those of Leak *et al.* (1955) except for the number five peak which they found to be present irrespective of the presence of any precipitates to which the earlier authors had attributed the peak's existence.

An iron—silicon—nitrogen alloy was also investigated by Hashizume and Sugeno (1967). Both poly- and single crystal (especially $\langle 100 \rangle$ and $\langle 111 \rangle$ orientations) specimens of Fe + 5.4 at.% Si + 150 ppm nitrogen were investigated at frequencies of about 2 Hz. As in the previous investigation, two abnormal peaks, in addition to the Fe–N Snoek peak, were observed for the polycrystalline specimen and the single crystal specimen oriented nearly parallel to the $\langle 111 \rangle$ axis. These results are shown in Fig. 18. It was concluded that tetragonal or $\langle 100 \rangle$ orthorhombic are the symmetries of the strains produced by the defects which correspond to the additional higher temperature abnormal peaks. The relaxation mechanisms proposed for these peaks were (1) jumps of solute nitrogen atoms to and from octahedral sites nearest to silicon atoms and (2) reorientation of N–N pairs along the $\langle 100 \rangle$ orientation with a silicon atom at its center. Their second proposed mechanism differs from that of Leak *et al.* (1955) described above. Furthermore, the probability of sufficient N–N pairs for a fairly low concentration of 150 ppm nitrogen to cause the highest temperature peak does not seem reasonable. The usual experimental spectrum in Fe–Si interstitial (i.e., for both C and N) alloys, where three peaks are actually measured, is unexplainable at present, and further investigation should be pursued to determine the mechanism responsible for the new third high temperature peak.

## 10. IRON–PHOSPHOROUS–CARBON ALLOYS

Dickenscheid and Seeman (1958) have studied the effect of phosphorous on the iron–carbon Snoek peak. They varied the phosphorous concentration from 0.02 to 0.18%. They found no effect of phosphorous on the Snoek peak height, but phosphorous did broaden the Snoek peak at the higher temperatures. The peak broadening and skewing on the high temperature side also increased as the phosphorous concentration increased. Dickenscheid and Seeman did not perform any peak synthesis, and they only remarked that the phosphorous altered the activation energy for carbon in iron.

## 11. SUMMARY OF EXPERIMENTAL INTERNAL FRICTION RESULTS FOR bcc IRON TERNARY ALLOYS

The internal friction results from low frequency torsional pendulum experiments for the various bcc iron ternary alloys are summarized in Table I. The four effects of the substitutional addition on the solvent–interstitial Snoek peaks, namely (1) the production of no effects on the interstitial Snoek peaks, (2) the broadening of the Snoek peak usually associated with a s–i interaction, (3) absence of Snoek peaks due to s–i reaction, and (4) the resulting dual peak, which for a substitutional concentration (i.e., $C_s$) is greater than the interstitial concentration (i.e., $C_i$), produce scavenging showing a direct correlation to the heat of formation of carbides or nitrides of the substitutional additions and also atomic size factor. For example, titanium, a Group IV metal, which acts as a scavenger of interstitial nitrogen, produces a double peak (Note: This occurs for $C_i > C_s$) and has a larger atomic radius than iron and the heat of formation of TiN is $-80.7$ kcal/mole as compared to $-2.55$ kcal/mole for $Fe_4N$. On the other hand, nickel, which forms no stable nitrides or carbides and has almost the same atomic radius as iron, has almost no effect on the iron–interstitial Snoek peaks. The effect of silicon is somewhat different than those for other substitutional additions because here the strong s–i interaction manifested by the complex internal friction spectra is due to the fact that the radius of silicon is smaller than that of iron, and hence, the interstitial atoms are found in the lower energy silicon sites. From the summary in Table I, one might conclude that there are some discrepancies in the results

## TABLE I

SUMMARY OF EXPERIMENTAL INTERNAL FRICTION CHARACTERISTICS FOR bcc
IRON–SUBSTITUTIONAL–CARBON AND
IRON–SUBSTITUTIONAL–NITROGEN TERNARY ALLOYS

| Alloy | Solute group no. | Solute atomic radius (Å) | Stable carbide or nitride | Heat of formation of stable carbide or nitride (kcal/mol) | Type of damping peak | Remarks |
|---|---|---|---|---|---|---|
| Fe–C | — | — | $Fe_3C$ | Metastable | Snoek | $C_{Fe} = 1.24$ Å |
| Fe–V–C | V B | 1.31 | V C | $-28.0$ | Snoek | |
| Fe–Cr–C | VI B | 1.25 | $Cr_4C$ | $-16.4$ | Snoek | |
| Fe–Mo–C | VI B | 1.36 | $Mo_2C$ | $-4.2$ | Snoek | For Mo < 0.5% |
| Fe–Mo–C | VI B | 1.36 | $Mo_2C$ | $-4.2$ | Double | For Mo > 0.5% additional low temperature peak |
| Fe–Mn–C | VII B | 1.12 | $Mn_3C$ | $-3.6$ | Snoek | |
| Fe–Ni–C | VIII | 1.25 | $Ni_3C$ | Not stable | Snoek | For Ni < 0.26% |
| Fe–Ni–C | VIII | 1.25 | $Ni_3C$ | Not stable | Broadened | For Ni > 0.26% peak is slightly broadened and shifted to higher temperature |
| Fe–Al–C | III A | 1.43 | $Al_4C_3$ | $-39.9$ | Double | |
| Fe–Si–C | IV A | 1.17 | SiC | $-13.0$ | Triple | |
| Fe–P–C | V A | 1.08 | — | — | Broadened | |
| Fe–N | — | — | $Fe_4N$ | $-2.55$ | Snoek | |
| Fe–Ti–N | IV B | 1.44 | Ti N | $-80.7$ | Snoek | For 0.5% Ti + 0.5% N nitrided at 950° and quenched |
| Fe–Ti–N | IV B | 1.44 | Ti N | $-80.7$ | Triple | See Fig. 8 |
| Fe–V–N | V B | 1.31 | V N | $-40.8$ | Double | |
| Fe–Cr–N | VI B | 1.25 | CrN | $-29.5$ | Broadened | |
| Fe–Mo–N | VI B | 1.36 | $Mo_2N$ | $-16.6$ | Double | |
| Fe–Mn–N | VII B | 1.12 | $Mn_4N$ | $-30.3$ | Broadened | |
| Fe–Mn–N | VII B | 1.12 | $Mn_4N$ | $-30.3$ | Double | For Mn > 0.5% additional low temperature peak |
| Fe–Ni–N | VIII | 1.25 | | Not stable | Snoek | |
| Fe–Cu–N | I B | 1.28 | CuN | $-60.23$ | Broadened | |
| Fe–Al–N | III A | 1.43 | AlN | $-64.0$ | Snoek | |
| Fe–Si–N | IV A | 1.17 | $Si_3N_4$ | $-179.0$ | Triple | |

for particular substitutional solutes (e.g., the effect of molybdenum on the Fe–C peak). However, this is explained easily by the fact that a more complex spectrum results as the substitutional concentration increases. As the substitutional concentration is increased, there is a higher probability of substitutional–substitutional pairs, which introduces new possible relaxation processes for the interstitial in the neighborhood of the pair. These relaxations were manifested by the appearance of peaks at low temperatures (e.g., in most cases about 0° C).

The exact mechanisms responsible for the various experimental internal friction spectra have been postulated for almost all of the results from analytic syntheses utilizing single relaxation Debye-type peaks. In cases where at least two investigators studied the same system there is usually some disagreement as to the relaxation processes responsible for the experimental spectrum. These problems arise because there is no *a priori* information (e.g., activation energy, peak temperature, relaxation time, relaxation strength–concentration relation) for these subsidiary peaks. More extensive experiments, along with some modified peak synthesis methods, appear to be required in order to resolve this problem.

## B. Group V Alloys

The results of Group V alloys are divided into those for vanadium alloys, niobium alloys, and tantalum alloys. The discussion follows this grouping.

### 1. VANADIUM ALLOYS

Keith (1969) examined the effect of adding titanium and molybdenum to vanadium in terms of the mechanical properties. These alloys contained primarily interstitial oxygen but with an appreciable concentration of nitrogen. He also reported on the variation of the interstitial Snoek peak values as a function of the concentration of substitutional addition. Keith stated that the effect of molybdenum on the internal friction of the alloys had no measurable effect at all, and, furthermore, only the oxygen and nitrogen peaks occurred in the internal friction spectra. In his plot of $Q_{max}^{-1}$ vs. Mo concentration for commercial bomb-reduced vanadium alloys, however, there is reduction of about 25% in $Q_{max}^{-1}$ for oxygen only at a concentration of 0.2 at.% Mo. From 0.5 to 4.75 at.% Mo the $Q_{max}^{-1}$ for oxygen is constant as is the $Q_{max}^{-1}$ for nitrogen which is constant from 0.0 Mo to 4.75 at.%. Since Keith did not present his internal friction spectra, it is difficult to negate his statement about the lack of a substitutional–interstitial interaction. Thus, Keith concluded that the Group VI solute–molybdenum does not show a s–i interaction.

Keith's vanadium–titanium results showed that titanium had a very strong effect on the concentration of the random interstitials. For example, a titanium concentration of only 0.1 at.% reduced the maximum oxygen damping value by 97%. The nitrogen peak changed in a different manner in that there was no effect on the maximum damping value until about 0.7 at.% titanium when the peak height was reduced in a linear manner to zero at one atomic percent titanium. At this same titanium concentration, the oxygen peak height was also zero. In addition to the oxygen and nitrogen peaks, an extra broad peak was found at a high temperature (for which he did not mention a variation in the peak temperature) which increased in magnitude from 0.0 to 0.75 at.% titanium, but this peak also vanished at one atomic percent titanium. Keith concluded that the addition of titanium to vanadium causes the removal of the interstitials, oxygen and nitrogen, from solution. Furthermore, he stated that approximately five times the stoichiometric quantity of titanium is required to accomplish this. His result is based on the use of Powers and Doyle's (1959) relaxation strength per atomic percent interstitial solute for oxygen in vanadium. Furthermore, this result is similar to Fast and Meijering's (1953) results for a Fe + 0.51 at.% V + N alloy (i.e., it takes about six vanadium

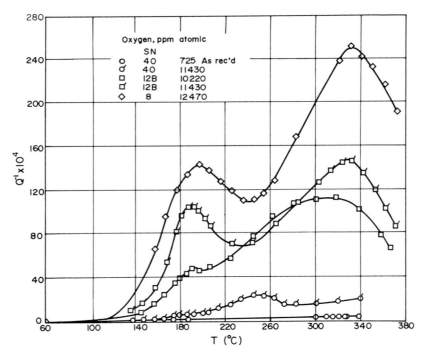

**Fig. 19.** Internal friction for V + 1.31 at.% Ti + O alloys at 1.73 Hz (Hasson, 1970).

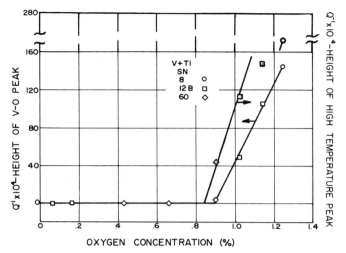

***Fig. 20.*** V–O peak and high temperature peak heights versus oxygen concentration for V + 1.31 at. % Ti + O alloys (Hasson, 1970).

atoms to remove one nitrogen atom from solution). Keith utilized internal friction data, precision lattice parameter measurements, and strain aging results to conclude that the titanium is partitioned between compound formation and solid solution formation below the titanium concentration required to remove the interstitials from solution.

The titanium reaction with oxygen in the V–Ti–O system was studied in more detail by Hasson (1970) and reported in the literature by Hasson and Arsenault (1970). Oxygen was added incrementally to an "as received" vanadium + 1.31 at. % titanium alloy from a starting concentration of 222 ppm atomic oxygen to a concentration of 1.25 at. %. The results are shown in Fig. 19. The Snoek peak, however, associated with interstitial oxygen in vanadium, is not detectable until the level of added oxygen is nearly equal to the concentration of the substitutional solute. Also, the peak temperature associated with the substitutional–interstitial interaction does vary. A plot of the oxygen peak height and the substitutional–interstitial peak height versus oxygen concentration (see Fig. 20) indicates that it takes 1.45 titanium atoms to remove one oxygen atom from solution. These data, however, are from specimens which were heated in vacuum several times in order to add the oxygen incrementally to the same specimen. Electron microprobe analysis of the specimen showed a depletion of the titanium due to repeated heating in a vacuum of $10^{-8}$ Torr, therefore, the ratio of titanium atoms to remove one oxygen atom from solution is one. This result is quite different than that of Keith (1969) for the same system. The apparent discrepancy, however, can be explained partially by Hasson and Arsenault's utilization of a relaxation strength per atomic percent interstitial solute value of 0.168 (Hasson and Arsenault, 1971) as compared to Keith's use of the Powers and Doyle (1959) value of 0.55, which is 3.24 times larger. Keith's rather high interstitial nitrogen concentration, which was equal to approximately half of the oxygen concentration value, would explain the remaining discrepancy (i.e., 30–40% of the titanium atoms are required to remove the interstitial nitrogen atoms). In the work of Hasson and Arsenault, this initial nitrogen concentration was 52 ppm atomic, and the experimental apparatus and procedures were designed to avoid further nitrogen contamination. From their results, Hasson and Arsenault concluded that titanium is reactive with interstitial oxygen. They also stated that titanium is a scavenger as was predicted by Hahn *et al.* (1963) for the Group IV metal solutes in the Group V metals. Beyond the point where the ability of titanium to scavenge is exceeded (i.e., the substitutional and interstitials have equal concentration), a two peak internal friction spectrum occurs. The low temperature peak is the classic V–O Snoek peak, and the high temperature peak represents several complicated relaxation processes associated with the substitutional solute. The variation in peak shape and temperature makes the interpretation or even the proposing of simple substitutional–interstitial complex models very difficult, and, therefore, Hasson and Arsenault proposed none. The application of the analysis and models of the similarly behaving Fe–V–N system due to Perry *et al.* (1966) will be presented later.

## 2. NIOBIUM ALLOYS

The earliest reported work on the effect of a substitutional addition on a Group V metal–interstitial alloy is that done by Powers and Doyle (1959) for Nb–Zr–N alloys. Their interest was to illustrate a source of distribution of relaxation times. For a 0.02 Zr alloy, there was no effect on the Nb–N damping curve, while a 0.3% Zr addition broadened the Nb–N Snoek peak, and the addition of 0.8% Zr produced higher temperature peak in addition to the Nb–N Snoek peak. Powers and Doyle said the latter two cases were reminiscent of the previous work on Fe–Mn–N (Fast and Dijkstra, 1951) and Fe–V–N (Fast and Meijering, 1953), respectively. Other than noting from their results that nitrogen atoms preferred to spend a larger fraction of their time in the neighborhood of zirconium atoms than niobium atoms, Powers and Doyle apparently did no further investigation of the Nb + Zr alloy system.

A few years later Bunn *et al.* (1962) examined in greater detail the effect of zirconium on internal friction in niobium. Since this investigation, Szkopiak and Ahmad (1969), Thurber *et al.* (1966), Mosher *et al.* (1970), Miner *et al.*

(1970), and Hasson (1970) have also studied the effect of zirconium on internal friction in niobium. The study of Bunn *et al.* on the effect of 1.0 at.% Zr on the Nb–O peak showed that the oxygen peak did not appear until an oxygen concentration of 0.8 at.%. Furthermore, a new peak, corresponding to the substitutional–interstitial interaction or a Zr–O peak, was present for an oxygen concentration of just 0.029 at.%. Bunn *et al.* noted that this peak did not change when the oxygen concentration was increased and from this result, they indicated that the diffusion of oxygen about a zirconium atom is a single relaxation process. A nitrogen peak, however, was present from the start and increased as the level of oxygen was increased. This latter result indicates nitrogen contamination must have taken place. Their results at the oxygen concentration where the oxygen peak appears are plotted with those of Hasson in Fig. 21. Hasson's results indicate the appearance of the Snoek Nb–O peak at an oxygen concentration equal to the zirconium concentration. In Hasson's experiments the concentration of nitrogen was only 30% of that of Bunn *et al.* (1962) and the concentration did not increase measurably after several additions of oxygen. Putting these results aside temporarily, Bunn *et al.* (1962) did not observe the Zr–N peak found by Powers and Doyle (1959). They were successful, however, in demonstrating the existence of the Zr–N peak by degassing a specimen of oxygen which originally contained 0.58 at.% of oxygen with 0.1 at.% of nitrogen. They observed the Zr–N peak, which increased as the oxygen decreased, and also a decrease of the Nb–N peak. From this experiment they found that as the concentration of oxygen is reduced, zirconium attracts nitrogen away from niobium atoms, altering the height of both peaks, and that the Zr–N peak also shifted its position to

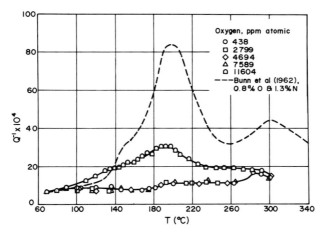

*Fig. 21.* Internal of Nb+1 at.% Zr–O alloys at 0.64 Hz (Hasson, 1970).

higher temperatures. This experiment can be used to explain the difference with Hasson's results, because the lower nitrogen concentration in Hasson's specimens would allow the zirconium to remove more oxygen from solution; thus, it would take a higher concentration of zirconium atoms to remove the oxygen (i.e., a ratio of one Zr to one $O_2$ atom as opposed to that of Bunn *et al.* 1.25 Zr/one $O_2$ atom). Bunn *et al.* concluded that there was a strong attraction between zirconium and oxygen atoms. Gibala and Wert (1971) called this result an extreme case of high binding of s–i pairs. Hasson (1970) supports the idea that for this system the substitutional (Zr)–interstitial (O) interaction is of the reactive type, and he puts the Group IV solute Zr for the Nb + Zr + O ternary system in the same category (i.e., Zr is a "scavenger") as the Group IV solute Ti in the V + Ti + O ternary system. Bunn *et al.* did not hypothesize any models for the s–i complexes which might be responsible for the peaks designated as Zr–O and Zr–N.

Szkopiak and Ahmad (1969) investigated zirconium alloys of 0.82 and 2.04 at.% with interstitial concentrations of 2600 ppm atomic oxygen with 1000 ppm atomic nitrogen, and 3600 ppm atomic oxygen with 730 ppm atomic nitrogen, respectively. Szkopiak and Ahmad's internal friction spectrum agrees with the results of Bunn *et al.*, especially with regard to the height and presence of the Zr–O peak before the niobium–oxygen peak appears. Again the high concentration of nitrogen can partially explain this result (i.e., some of the zirconium is associated with the nitrogen). Their internal friction spectrum for the 2.04 at.% zirconium addition shows a reduction of 67% of the Zr–O peak height as compared to the 0.82 at.% alloy, but more significantly, the Zr–O and Zr–N peak heights have been shifted higher by 40° C. This result would indicate that the energy associated with the substitutional–interstitial pair is increased by increasing the substitutional solute concentration.

An investigation by Thurber *et al.* (1966) on a Nb + 1.09 at.% Zr alloy containing 3100 ppm oxygen and 120 ppm nitrogen was also found to be in agreement with the work of Bunn *et al.* (1962). The main emphasis of Thurber *et al.*, however, was on the effect of aging; and they found that both oxygen and nitrogen were precipitating during aging at a temperature lower than the solution annealing temperature. They did not compare the kinetics of the precipitation with that of the aging process to determine which interstitial impurity was responsible for the aging in the alloy.

The most recent investigation of the Nb–Zr–O ternary system was performed by Miner *et al.* (1970) who employed the high frequency internal friction technique on oriented single crystals of Nb and Nb + 1 at.% Zr. Their internal friction spectra were similar to the earlier works (i.e., the initial absence of the oxygen peak and the presence of higher temperature peaks). They then indicated that the interstitial oxygen may be present in several different arrangements; (1) as isolated interstitial atoms, (2) or as part of an isolated s–i defect, and (3) at positions not detectable by internal friction (i.e., a substantial portion of the oxygen is not detectable because of the precipitation of the very stable $ZrO_2$ oxide phase). Their alloys also contained a significant residual of interstitial nitrogen as an impurity. By the use of a large variation in oxygen interstitial concentration and computer techniques, they performed a desynthesis of the measured internal friction spectra to speculate on five processes which they concluded were responsible for the experimental peak. In their computer model they assumed each process could be represented by a single Debye-type peak. One of their assumptions to support their model (namely, that the nitrogen peak height is independent of the oxygen concentration) is in conflict with the experimental results of Bunn *et al.* (1962) who found that the nitrogen peak did not even appear until the oxygen concentration was reduced significantly. Bunn *et al.* (1962), as noted previously, stated that as the concentration of oxygen is reduced, zirconium attracts nitrogen away from the Nb atoms altering the height of the Zr–N and Nb–N peaks. The investigation of Miner *et al.*, nevertheless, was very extensive, and their orientation and interstitial variation appear adequate to make their conclusions. They found that the internal friction spectra of the Nb + 1.0 at.% Zr + O alloys could be synthesized on the basis of five single relaxation time theoretical Debye peaks—two corresponding to the oxygen and nitrogen Snoek peaks, and three extraordinary peaks resulting from the relaxation of s–i defects. From the oriented single crystal specimens, they concluded that s–i defect symmetry was either tetragonal or ⟨100⟩ orthorhombic. From these results and the compositional dependence of the five processes, they further concluded that the defects responsible for the three extraordinary peaks were (1) an oxygen atom, and (2) a nitrogen atom on nearest neighbor octahedral sites to a substitutional zirconium atom, and (3) a pair of oxygen atoms on opposite nearest neighbor octahedral sites around a zirconium atom. Thus, the interpretation of the Zr–O peak by Miner *et al.* differs from that of Bunn *et al.* The analyses of Bunn *et al.* and Miner *et al.* are complicated by the effect of the residual interstitial nitrogen concentration, but the latter's work is certainly more extensive and may be more valid. The single relaxation Debye peak synthesis, however, is somewhat questionable in light of the remarks by Meijering (1961) and Ritchie and Rawlings (1967) all of whom indicate that the complexity of the s–i interactions makes the single relaxation synthesis somewhat questionable.

Mosher *et al.* (1970) also studied the Nb + Zr system, but they selected nitrogen as the interstitial to be studied for the s–i interaction. They also studied zirconium alloys with a concentration variation from 0/1 wt.% Zr to 1.0 wt.% Zr. Experimentally, they examined the Zr–N interaction peak and synthesized the experimental peak on the basis of the single relaxation Debye

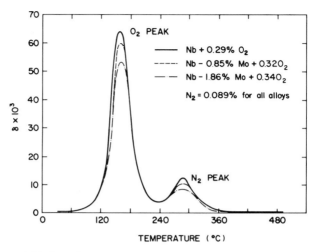

**Fig. 22.** Internal friction of Nb + Mo + O alloys at 1.0 Hz (Szkopiak and Ahmad, 1969).

peak analysis, which was utilized in the other Nb + Zr investigations. On the basis of these results, they derived expressions for the concentrations of Zr–N pairs and single N atoms with zirconium pairs (i.e., Zr–N–Zr) in niobium alloys containing small concentrations of both zirconium and nitrogen. They also determined the activation energies and relaxation times for these two processes which they hypothesized were responsible for Zr–N extraordinary peak. Their analytical approach was exactly similar to the work of Miner

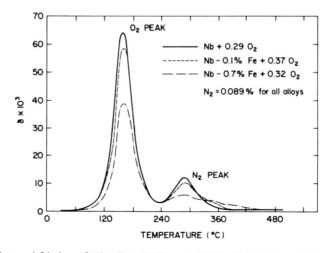

**Fig. 23.** Internal friction of Nb + Fe + O alloys at 1.0 Hz (Szkopiak and Admad, 1969).

*et al.* (1970); and, thus, the same earlier remarks as to its validity in the interpretation of the extraordinary peak would apply.

A niobium alloy with molybdenum as the substitutional addition has been investigated in the study of Szkopiak and Ahmad (1969) discussed in the preceding paragraphs. They investigated 0.85 and 1.86 at.% molybdenum additions where only the oxygen and nitrogen Snoek peaks were present, and found reductions of 6 and 15.7% in the oxygen Snoek peak height (Fig. 22), respectively. They felt this slight reduction in Snoek peak height was indicative of a s–i interaction. This result is similar to the reduction in the maximum damping reported by Keith (1969) for a V + 0.2 at.% Mo + O alloy. In the V–Mo–O alloy, this effect occurred at only one molybdenum concentration and Keith stated molybdenum had no effect on the oxygen Snoek peak. Szkopiak and Ahmad (1969) also investigated as part of this same study niobium–iron alloys of 0.1 at.% Fe and 0.7 at.% Fe which produced reductions in the oxygen Snoek peak height of 12 and 30% (Fig. 23), respectively. They noted that the greater effect of iron could be attributed to the larger size factor percentage between iron and niobium than between molybdenum and niobium, and also the slightly higher chemical affinity. On this basis the substitutional solute tungsten should have an effect similar to molybdenum. Mosher (1969) stated that tungsten additions of 0.5, 1.5, and 3 at.% had no

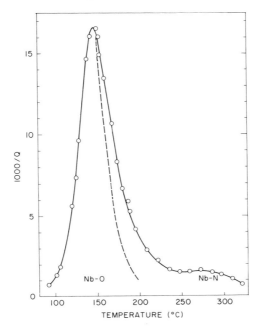

**Fig. 24.** Internal friction of Nb + 3 at.% W + 0.3 at.% O; f = 0.95 Hz (Mosher, 1969).

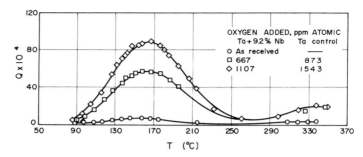

**Fig. 25.** Internal friction of Ta + 9.2 at.% Nb–O alloys at 2.32 Hz (Hasson and Arsenault, 1970).

effect on the niobium–oxygen Snoek peak height for oxygen concentrations to 0.3 at.% (Fig. 24), but he did not perform independent chemical analysis. Mosher did observe that there was an increase in the peak width due to the tungsten addition. This peak broadening was interpreted by Mosher as being due to an increase of di- and tri-oxygen complexes induced by the tungsten addition. Thus, the addition of tungsten, molybdenum, and iron have an effect on the interstitials present in niobium. However, these solutes do not totally remove the interstitials from solution, i.e., there is no apparent scavenging.

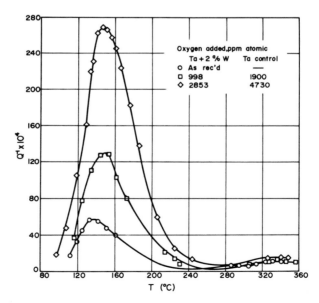

**Fig. 26.** Internal friction of Ta + 2 at.% W + O alloys at 1.15 Hz (Hasson, 1970).

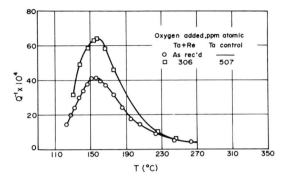

**Fig. 27.** Internal friction of Ta + 3.55 at.% Re + O alloys at 4.0 Hz (Hasson, 1970).

## 3. TANTALUM ALLOYS

The experimental investigations of tantalum alloys are not extensive at present. Results for Ta + 9.2 at.% Nb + O, Ta + 2 at.% W + O, and Ta + 3.55 at.% Re + O ternary alloys are available from the work of Hasson (1970). The experimental data for these systems are shown in Figs. 25, 26, and 27, respectively. The addition of niobium reduces the apparent concentration of interstitial oxygen by 25%, while both tungsten and rhenium reduce the apparent concentration of oxygen by 40%. These values were determined from a comparison of the interstitial oxygen added to a pure tantalum specimen (designated Ta control) and to the alloy specimens. The addition of niobium also resulted in a 40% reduction in the experimental Ta–O Snoek peak activation energy, while the tungsten and rhenium additions only reduced the Ta–O Snoek peak half width by 10%. The niobium addition produces definite peak broadening which is manifested by the reduction in activation energy mentioned above and is similar to the Fe–Mn–N results of Fast and Dijkstra (1951). Therefore, one could hypothesize that a s–i interaction is operative and its different relaxation characteristic contributes to the peak broadening. The slight decrease in the peak width for the W and Re additions indicates that the substitutional addition does not interact (i.e., a s–i interaction occurs) significantly with the interstitials to affect their mobility. Furthermore, the situation for the tungsten and rhenium is complicated somewhat in that reduction in random interstitial concentration must be explained. Examinations of the heat of formation of the various oxides in Table II for niobium, tungsten, and rhenium from Goldschmidt (1967) does not provide a clue as to why W and Re have a greater ability to reduce the random interstitial concentration (e.g., the heat of formation of the dioxides shows $NbO_2$ has a higher heat of formation than $WO_2$ and $ReO_2$ in that order). If one examines the size factor of these three solutes, niobium is

TABLE II

HEATS OF FORMATION OF THE OXIDES OF GROUPS IV THROUGH VII METALS[a, b]

| | Group | | |
|---|---|---|---|
| IV | V | VI | VII |
| Ti | V | | |
| TiO  $-135$ | VO  $-106$ | | |
| Ti$_2$O$_3$  $-375$ | V$_2$O$_3$  $-300$ | | |
| Ti$_3$O$_5$  $-587$ | VO$_2$  $-176$ | | |
| TiO$_2$  $-219$ | V$_2$O$_5$  $-383$ | | |
| Zr | Nb | Mo | |
| ZrO$_2$  $-259$ | NbO  $-116$ | MoO$_2$  $-141$ | |
| | NbO$_2$  $-194$ | MoO$_3$  $-180$ | |
| | Nb$_2$O$_5$  $-463$ | | |
| | Ta | W | Re |
| | Ta$_2$O$_5$  $-499$ | WO$_2$  $-135$ | ReO$_2$  $-101$ |
| | | WO$_3$  $-200$ | Re$_2$O$_7$  $-300$ |
| | | | ReO$_3$  $-147$ |

[a] Values in kcal/mole.
[b] Goldschmidt (1967).

almost exactly the same as tantalum, while both W and Re are about 4% smaller. Thus, it is energetically favorable for interstitial oxygen atoms to be neighbors of W and Re atoms rather than solvent tantalum atoms and the solute niobium atoms, which are essentially the same size. The explanation of the different behavior of the substitutional solutes Nb, W, and Re, with interstitial oxygen, therefore, is not availabe at this time, and a systematic variation of the substitutional solute concentration in two of these alloys (e.g., Ta–Nb and Ta–W) would be most useful.

The results of the addition of tungsten to tantalum discussed above may be different from those for Nb–W–O alloys investigated by Mosher (1969) who found no effect on peak height (i.e., interstitial concentration) and only peak broadening (i.e., activation energy). The reason that a definitive statement cannot be made concerning the results of Nb–W–O in comparison to Ta–W–O is due to the fact that Mosher did not measure the oxygen concentration independently so there is no way of knowing whether there was a reduction in peak height or not.

## 4. SUMMARY OF EXPERIMENTAL INTERNAL FRICTION RESULTS FOR bcc GROUP V TERNARY ALLOYS

A summary of the results is shown in Fig. 28 where lines from the solvent indicate whether scavenging occurs, an s–i interaction is present and responsible for peak broadening (for Group V alloys the definition of s–i interactions assumes a broader definition, which includes the possibility of interstitial clustering induced by the substitutional addition), or no s–i interaction or scavenging occurs. The Group IV elements, Ti and Zr, both act as scavengers of interstitial oxygen and nitrogen in Group V solutes. Group VI elements in a higher series (e.g., Mo in V) produce no effect on the solvent–interstitial Snoek peak, while solutes of the Group VI elements of the same series produce some evidence of a s–i interaction (e.g., Mo in Nb). For the cases of Nb in Ta (i.e., a Group V solute in a Group V solvent) and Fe in Nb (i.e., the solute is three groups higher and one series lower) definite peak broadening occurs, which is indicative of s–i interactions occurring. For the case where only slight evidence is shown for the existence of s–i interaction (e.g., Ta–W), the solute does reduce the solubility of interstitials more than in the cases which show a significant s–i interaction. An explanation of these apparent opposite effects is not presently available. Further extensive research, such as has been performed in the iron bcc alloys, is required to provide information on the effect of substitutional solutes on the solubility and mobility of interstitial solutes in the Group V, bcc alloys.

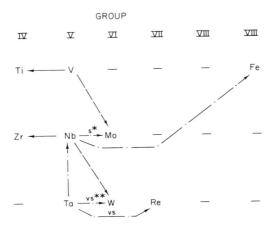

*Fig. 28.* Summary of internal friction results for Group V solid solutions with interstitial oxygen. —— Scavenging; —·— S–i interaction; —— no interaction. (Note: s*—slight; vs**—very slight).

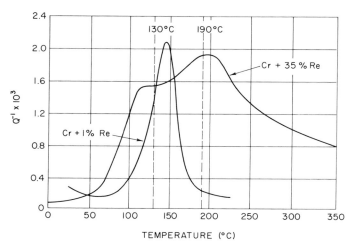

**Fig. 29.** Internal friction of Cr + Re + N alloys at 0.8 Hz (Klein and Clauer, 1964).

## C. Group VI Alloys

The measurement of internal friction for unalloyed Group VI metals is in itself very sparse because the solubility of interstitials in these metals is very low. Data is available, however, for nitrogen in chromium (DeMorton, 1962; Klein and Clauer, 1965.); carbon in molybdenum (Schnitzel, 1964); and carbon in tungsten (Schnitzel, 1965). Klein and Clauer (1964) have investigated the internal friction spectra of chromium with 1% rhenium and 35% rhenium with nitrogen as the interstitial. For the Cr–1% Re alloy, the nitrogen peak did not appear to differ significantly from nitrogen peak in unalloyed chronium. On the other hand, the internal-friction spectrum induced by nitrogen in Cr + 35Re alloy resulted in a complex dual peak. These results are shown in Fig. 29. Furthermore, neither of the dual peaks occurred at the Cr–N peak temperature. The unusual peaks indicate that a number of relaxation processes make up the spectrum induced by nitrogen. It is apparent that research on alloys of intermediate concentration of rhenium is required to understand the mechanisms responsible for the unusual internal friction spectrum of the 35% rhenium alloy.

Although there are no known internal friction data for molybdenum and tungsten alloys, some effects of substitutional additions on the interstitial solubility or mobility have been inferred from the result of mechanical property tests. Jaffee and Hahn (1963), for example, state that the "Re-effect" (i.e., the inhibiting of brittleness in W due to an addition of rhenium) is due to the influence of rhenium on the dissolved interstitial atoms, which are usually carbon. Furthermore, in molybdenum–titanium–carbon alloys Semchyshen and Barr (1961) have noted a TiC dispersoid exists. This TiC dispersoid is similar to the titanium oxide produced in the V–Ti–O system. Thus, one could hypothesize that titanium acts as a scavenger in Mo–T–C ternary alloys.

A concerted attempt to perform definitive internal friction studies of Group VI metals and alloys should be made to determine interstitial solubilities and mobilities in these metals. The initial effort should be with the Mo–Ti and W–Re alloys which display the unusual effects mentioned above (e.g., the Re-effect).

## IV. Analyses and Models

The experimental evidence of substitutional–interstitial interactions has been divided into two groups—those due to the nonreactive substitutional solutes and those due to reactive substitutional solutes. The case of the reactive solutes before $C_i > C_s$ is rather straightforward (i.e., a chemical compound forms between the substitutional addition and the interstitial addition). Therefore the direct effect of the substitutional or interstitial atom is no longer present. However, the case of the nonreactive substitutional solute addition is not as straightforward because a possibility of an interaction exists and the question arises as to what degree the effect of the substitutional and interstitial atoms on the properties of the base metal have been affected. The magnitude of this effect is manifested experimentally by internal friction peak broadening and a shift in the peak temperature. In addition, for the reactive substitutional solutes for the case of $C_i > C_s$, the peak width and temperature of the resultant separate high temperature peak varies with interstitial concentration (i.e., $C_i$). An understanding of these latter two cases is a little more complicated. The available analyses and models for these cases will be discussed in the following text.

Most investigators [notably, Dijkstra and Sladek (1953), Ritchie and Rawlings (1967), and Nacken and Kuhlmann (1966) for iron ternary alloys; and Miner *et al.* (1970), Mosher *et al.* (1970), and Gibala and Wert (1971) for the Group V ternary alloys] have utilized the addition of several single relaxation time peaks of the Debye type to synthesize the experimental internal friction spectrum. These peak contributions are defined by the Debye equation as follows:

$$Q_i^{-1} = \Delta_i \frac{\omega \tau_i}{1 + \omega^2 \tau_i^2} \tag{1}$$

where $Q_i^{-1}$ = internal friction of the $i$th component of the spectrum, $\omega$ = frequency of the test, $\tau_i$ = relaxation time of the $i$th component, and $\Delta_i$ = relaxation strength of the $i$th component

One of the earliest such syntheses was performed by Dijkstra and Sladek (1953) for the experimental results on an Fe + 0.5 at.% Mn + N ternary alloy. This alloy which is representative of a nonreactive type, produces a definite s–i interaction. They synthesized the experimental peak with two single relaxation contributions (see Fig. 6c). Later both Nacken and Kuhlmann (1966) and Ritchie and Rawlings (1967) synthesized the Fe + Mn + N internal friction peak using seven relaxation contributions. Nacken and Kuhlmann's synthesis is shown in Fig. 30. Enrietto (1962b) had previously investigated these alloys by systematically varying both the Mn and N concentrations. He concluded that the broad internal friction peaks observed could not be approximated by the sum of a number of subsidiary single relaxation peaks. Couper and Kennedy (1967) made a more recent investigation (similar to Enrietto's study) of Fe–Mn–N alloys. The former synthesized their experimental results with just three single relaxation Debye peaks. The identification of the precise relaxation processes responsible for the experimental broadening and peak temperature shifting in this one bcc ternary alloy system is obviously quite confused because they reported at least four conflicting interpretations. With the exception of Enrietto's conclusions, however, all the investigators have used the approach of synthesizing several single relaxation Debye peaks.

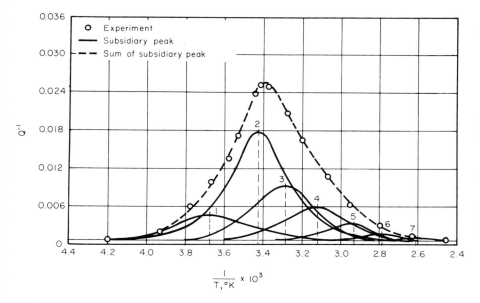

**Fig. 30.** Synthesis of experimental internal friction results of an Fe + 0.27 wt.% Mn + 0.05 wt.% N alloy from single relaxation time Debye subsidiary peaks (Nacken and Kuhlmann, 1966).

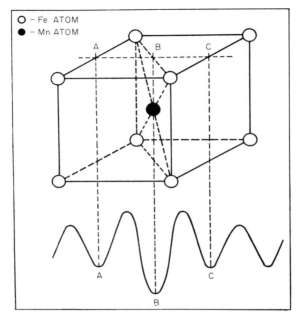

*Fig. 31.* Model depicting the modfication by a manganese atom of the potential energy of a nitrogen atom in the iron lattice (Dijkstra and Sladek, 1953).

It is noteworthy that this same approach has been employed in the interpretation of some Group V bcc ternary alloy results [e.g., Miner *et al.* (1970) for Nb+Zr+O, and Mosher *et al.* (1970) for Nb+Zr+N]. The individual investigators apparently had success matching their experimental results with their calculated curves because they can vary the parameters (e.g., the relaxation strength and time) and essentially curve-fit the experimental results due to the immense capacity of digital computers.

There have been several models, i.e., the arrangement of the substitutional and interstitial atoms, and the jump positions, that have been proposed to account for the assumed number of single Debye peaks. The main features of some of these models will be discussed below. Determination of the number of single Debye peaks assumed to account for the peak broadening is very important in determining the magnitude of the change in the tetragonal strain of the defect. Dijkstra and Sladek (1953) are responsible for the model for these low manganese concentration analytical results for the effect of a manganese atom on the potential energy of a nitrogen atom in the iron lattice. This model, shown in Fig. 31 was based on two assumptions. First, every Mn atom in substitutional solid solution creates around it six interstices located between an iron and a manganese atom as shown in Fig. 31. These interstices for nitrogen, called Fe–Mn interstices, have a free energy level lower than the normal Fe–Fe interstices between two iron atoms. This means that there is a binding energy between the Mn and the $N_2$ atoms. For the 0.5 at.% Mn addition that Dijkstra and Sladek studied, the possibility that two Mn atoms will be nearest neighbors is disregarded because of the low Mn concentration. Second, assuming that in addition to the normal relaxation time $\tau_{Fe}$, at least one new relaxation time $\tau_{Mn-Fe}$ enters into the phenomenon which is associated with the elementary diffusion transitions between a Fe–Mn interstice and an adjacent Fe–Fe interstice. In Fig. 31 the free energy for a nitrogen atom as a function of its position along the line ABC is shown schematically. This free energy distribution suggests why it is more favorable for a nitrogen atom to stay near a Mn site (i.e., the depth of the wall at B). Thus, a new relaxation process with the relaxation time $\tau_{Mn-Fe}$, which is greater than $\tau_{Fe}$, is possible. Other models for the low concentration of Mn additions have been developed by Meijering (1961) and Gibala and Wert (1971). The latter model is shown in Fig. 3.

Gladman and Pickering (1965) investigated a higher concentration manganese alloy (i.e., a 2% Mn addition), and observed that the peak shifted to a lower temperature (i.e., $T_p = 0°$ C at 0.7 Hz) and that the precipitation rate of the 2% Mn alloy was lower than that of the 0.5% Mn alloys. They developed a model to explain these results, as shown in Fig. 32. The diffusion transition of the nitrogen atoms, which are remote from Mn atoms, requires an activation energy $Q$ (see Fig. 32) and it is this jump which causes the normal Snoek peak in binary Fe–N alloys at 10° C for the frequency of 0.7 Hz. If the nitrogen atom is in an interstitial site adjacent to a single Mn atom, the energy level would be decreased by $E_{Mn}$, and the activation energy changed to $Q_{Mn}$ (see Fig. 32). Thus, a nitrogen atom jumping from a site of energy $E$ to one of energy $E_{Mn}$ encounters a smaller activation energy barrier than it would in making a normal transition in the iron lattice. This result, they remarked, could produce an internal friction peak at a lower temperature; but this low temperature peak would always be associated with a high temperature peak, and this was not observed in the 2% Mn alloy. They went further to propose that for a 2% Mn alloy, the probability of a Mn couple occurring becomes quite large. Gladman and Pickering's explanation is excerpted in the following text. The binding energy for nitrogen atoms, associated with a Mn–Mn pair (indicated by the curves in Fig. 32) should be larger. The dotted lines show connecting paths between similar sites; each is associated with a Mn couple. All sites are $a/2$ from the nearest Mn atom and $a/\sqrt{2}$ from the second nearest Mn atom. Assuming a binding energy between the Mn couple and nitrogen, the energy of a N atom $E$ is these sites will be lower than in other sites because of the elastic expansion of the surrounding ferritic lattice by the Mn. The activation energy for the jump process between similar Mn couple sites has a value of $Q$ (see Fig. 32). The absence of a high temperature peak, corresponding to the energy barrier $Q$ (see Fig. 32) for a nitrogen atom leaving the Mn couple, is due to the prohibitively high activation energy and may also be due to the lower positional energy of the nitrogen atom in the Mn couple sites than in Fe–Fe sites, which are distended by the internal friction stress.

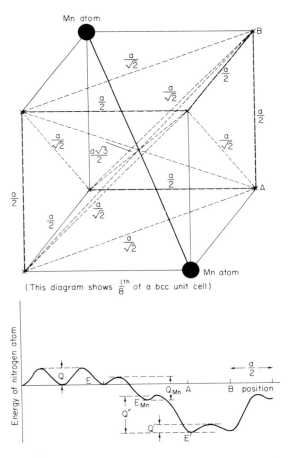

**Fig. 32.** Model for the interaction between nitrogen and manganese solute atoms in iron (Gladman and Pickering, 1965).

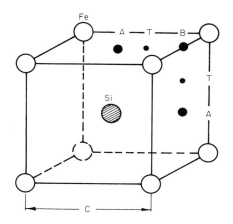

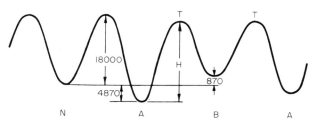

**Fig. 33.** Octahedral sites and strain energies for a Fe–Si–N alloy. A and B are octahedral sites in neighborhood of a silicon atom; T is a tetrahedral site; and N indicates the normal octahedral sites. The strain energy-differences are in cal/mole (Hashizume and Sugeno, 1967).

Another interesting observation made by Gladman and Pickering was that the activation energy of the nitrogen jump process may vary according to the nature of the path and the distance between the two low energy positions. Three distinct types of jumps (shown in Fig. 32) having lengths $a/2$, $a\sqrt{2}$, $a\sqrt{3}/2$, may have subtly varying activation energies which could account for the observed experimentally broadening of the peak.

The other available models for nonreactive substitutional solutes are those of Leak *et al.* (1955) and Hashizume and Sugeno (1967) for iron–silicon–nitrogen alloys. Leak *et al.* proposed a model exactly like that shown in Fig. 17. This simple model was proposed to explain the effect of silicon to broaden the Fe–N peak. However, the model of Hashizume and Sugeno is more detailed; it is shown in Fig. 33. This model was proposed to explain the two higher temperature peaks which occurred in addition to the normal Snoek peak. The presence of silicon was hypothesized to create a different strain field for the nitrogen atoms (e.g., interstitial nitrogen favors the octahedral sites around the silicon atoms rather than the normal octahedral sites of the iron solvent lattice). The strain energies of a first neighbor octahedral A site and B site, which corresponds to a second octahedral site (see Fig. 33), were calculated and found to be $\Delta U_A = -4870$ cal/mole and $\Delta U_B = 870$ cal/mole, respectively.

The definitions of these quantities they provided were, namely,

$$\Delta U_A = 2(C_{11} - C_{12})(E_1 - E_2) e_{\Theta\Theta} C^3 \tag{2}$$

and

$$\Delta U_B = -(C_{11} - C_{12})(E_1 - E_2) e_{\Theta\Theta} C^3 \tag{3}$$

where $C_{11}$ and $C_{12}$ are the elastic constants of the alloy ($C_{11} = 2.37 \times 10^{11}$ and $C_{12} = 1.41 \times 10^{11}$ dyn/cm$^2$), $E_1$ and $E_2$ are the principal values of the strain tensor characterizing a nitrogen atom ($E_1 = 0.44$ and $E_2 = 0.049$) in the normal octahedral site, and $C$ is the lattice constant of the alloy. In these equations

$$e_{\Theta\Theta} = \tfrac{1}{3}\frac{(r_{Si}^3 - r_{Fe}^3)}{r_{Fe}^3}$$

The values of $r_{Si}$ and $r_{Fe}$ used were 1.211 Å and 1.239 Å, respectively. Their estimate indicates that a nitrogen atom is more stable in an A-site than the normal octahedral site but less in a B site (see Fig. 33). Their explanation of the jump process is as follows. If a N atom associated with a silicon atom diffuses along the same path as in alpha iron, the saddle point for the diffusion is the tetrahedral position (marked T in Fig. 33). Unfortunately the strain energy calculation of site T is not feasible, because no experimental data is available on the strain produced by a nitrogen atom occupying a tetrahedral site. From Fig. 33, however, it is suggested that the strain energy of site T is reduced by the presence of the silicon atom. Therefore, the energy barrier H for the jump A → B is lower than the 18 (the activation energy for the Fe–N jump) + 4.87 kcal/mole energy barrier. If the diffusive jump occurs in an A–T–B–T–A sequence, the process might not be described as a single relaxation process, and the activation energy for this jump is probably higher. These two paths were advanced as explanations for the various relaxation processes responsible for the two additional experimental peaks.

Perry *et al.* (1966) utilized the technique of adding single Debye peaks to synthesize the separate experimental high temperature peak found in Fe + V + N alloys originally reported by Fast and Meijering (1953). This peak results in a reactive solute after $C_i > C_s$. Although they obtained a reasonable fit of the experimental data curve, Perry *et al.* were the first to indicate that there were problems associated with this approach. One of the problems they found was that only particular site locations were allowed for interstitials in the neighborhood of a substitutional atom. Another more important problem they found was the difficulty in the interpretation of the experimental data, because not all the interstitials in the various types of sites are needed to give rise to unique observable relaxation processes. This fact they exemplified with two cases; namely (1) if more than one interstitial "type" contributes to a given relaxation peak, then the processes are experimentally indistinguishable; and (2) if, during the buildup of interstitial shells of the inner atoms are blocked jumping, a given relaxation can pass through a maximum value and fall with increasing fraction of interstitials. Inherent in these conclusions is that there is no *a priori* information to assign values for the relaxation strengths and times in Eq. (1) for the various $i$ components which contribute to the complex experimental relaxation spectrum.

More recently, Perry and Boon (1969) have suggested a combined analytical–experimental method to furnish information about the origin of the extraordinary internal friction peaks which result from s–i interactions. They noted that the low frequency method has the disadvantage that the spectrum analysis is difficult, inasmuch as the peak relaxation strengths and relaxation times depended on temperature. They pointed out that when the frequency is varied and the temperature held constant, this difficulty is somewhat alleviated. One of the drawbacks, however, was that usually only a limited range of frequency was available and, thus, only part of the damping spectrum could be analyzed isothermally. For the Fe–V–N system which they used as a model, the frequency range required was 0.5 to 50 Hz. Perry and Boon, nevertheless, set out to overcome this difficulty, and their method appears to be useful. For details of this approach one can consult the original paper. The strength of their approach results from the fact that the apparent distribution of interstitials can be derived from the variable frequency experiment, and, thus, a specific s–i model does not have to be hypothesized to derive interaction energies. Their analysis shows that for suitably precise experimental damping data (e.g., a precision of $\pm 3 \times 10^{-5}$ in the damping measurement) both the relaxation parameters and number of subsidiary peaks which are required to synthesize the experimental damping curve can be determined. At present, however, an experimental verification of their technique and analysis has not been performed. If a verification of their approach was positive, this would increase the understanding of s–i interactions manifold. In a private communication, however, Perry (1970) again pointed out that the above approach has restrictions (e.g., the available experimental frequency). Perry stated that he was inclined to regard a great deal of the existing analyses as dubious, but not the experimental results.

The only additional models available are those for the high temperature peak of the reactive Nb–Zr–N alloys by Mosher *et al.* (1970) and Miner *et al.* (1970) who utilized single crystal specimens with orientations of $\langle 100 \rangle$, $\langle 110 \rangle$, and $\langle 111 \rangle$. They concluded that the s–i defect symmetry was either tetragonal or $\langle 100 \rangle$ orthorhombic. Their defect models are shown in Fig. 34. The processes (i.e., processes 2, 3, and 5) denoted in Fig. 34 are the three separate relaxations responsible for a separate high temperature peak that occurs in these alloys. This has been previously discussed in Section III, B, 2 of this text. These are models primarily for the defect symmetry; when combined with their internal friction results, they were used solely to identify their hypothesized processes for the s–i interactions in the separate high temperature peak of the Nb + Zr + O alloys.

There is an alternate explanation of the peak broadening which is not directly the result of a substitutional–interstitial interaction. Mosher (1969) has proposed that broadening observed in the Nb–W–O system is due to the clustering of interstitials induced by the substitutional tungsten atoms. Also, there has been no consideration given to the possibility that a distorted lattice (due to substitutional addition) will in itself give rise to peak broadening.

As has been stated several times in this text, the present available analysis technique for the explanation and identification of s–i interactions (namely, the superposition of single Debye peaks) has several drawbacks. It should also be indicated that the difficulty is not in the superposition technique itself,

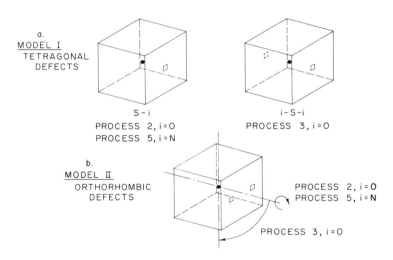

**Fig. 34.** Schematic illustration of the two defect models for Nb + Zr + O alloys that are symmetrically consistent and produce the proposed number of relaxation processes (Miner *et al.*, 1970).

but in the assignment of parameters in the individual Debye equation [see Eq. (1)] without any experimental information. Perhaps another approach that was suggested by Perry and Boon (1969), or a new analysis method, would be most timely in light of the extensive experimental results available. The available models, however, are consistent with the present method of analysis, and new models will become available only as the analytical method improves.

## V. Discussion

Most of the available experimental data relating to substitutional–interstitial interactions or reactions in bcc ternary alloys can only be obtained from internal friction investigations. The trends in the experimental internal friction results for the various bcc ternary alloys are very consistent; for example, reactive substitutional solutes from the Group IV metals remove the interstitial atoms from random solution for both iron and Group V ternary alloys. Also, for reactive substitutional solutes, the normal solvent–interstitial Snoek peak does not appear until the interstitial concentration is greater than the substitutional concentration (i.e., $C_i > C_s$). At these concentrations, a dual peak is formed with the new peak corresponding to complex s–i interactions. The nonreactive substitutional solutes were found to have either no effect or produced a broadening of the Snoek peak. In nonreactive iron ternary alloys the tendency for strong s–i interactions as attested to by peak broadening can be correlated very well with the heat of formation of the substitutional–interstitial carbide or nitride as compared to the pure metal, and also to a size effect (i.e., atomic radii). For example, the addition of manganese, whose nitride has a heat of formation of $-30.3$ kcal/mole, to Fe–N alloys causes definite peak broadening; while the addition of nickel, which has no stable nitride, does not effect the Fe–N Snoek peak.

Correlation between the heat of formation of substitutional–interstitial oxides and the solvent oxides in nonreactive Group V ternary alloys is, however, not as straightforward as that found in the iron ternary alloys. There is at least some evidence of s–i interactions or clustering of interstitials taking place in all of the substitutional solutes added to the Group V metals (that is, peak broadening was observed experimentally for all of the Group V ternary alloys). The nature of these s–i interactions has not been determined and more extensive investigations must be performed in the future to obtain an insight as to why they even occur in some cases (e.g., in the Ta–W–O system).

The method presently available to analyze the existing internal friction data, e.g., peak broadening, has many difficulties associated with it. The assignment of relaxation strengths and times and activation energies for the subsidiary peaks, which are utilized to synthesize the experimental spectrum, without *a priori* information is the greatest drawback to the present method of summing individual subsidiary peaks. Furthermore, the method to determine these parameters experimentally is not straightforward. Confusion even arises, as is illustrated by the results for the Fe–Mn–N alloys, where at least four different sets of subsidiary peaks have been hypothesized. The combined experimental–analytical method of Perry and Boon (1969) could be one possible solution, but the truncation of complete internal friction spectrum due to the limits of the available experimental frequency could also provide difficulties. The exact explanations and models for the substitutional–interstitial interactions which most often occur in the nonreactive substitutional solutes are, therefore, presently not available, and hence the importance of these s–i interactions from the viewpoint of mechanical properties will be indicated in the following discussion.

The possible effect of a reduction in the random interstitial concentration by alloying on the temperature dependence of the yield stress (specifically solution weakening) has been discussed in the Introduction. Recent studies of this temperature dependence, which is found commonly in bcc alloys, include work on iron alloys (Leslie and Sober, 1967; Solomon *et al.*, 1969); the Group V alloys of niobium (Ravi and Gibala, 1969; Szkopiak and Ahmad, 1965); vanadium (Keith, 1969; Pink and Arsenault, 1971) and tantalum (Arsenault, 1966, 1967; Das and Arsenault, 1968; (Mitchell and Raffo, 1967; Smialek *et al.*, 1970); and alloys of the Group VI metal rhenium (Stephens and Witzke, 1970). In the studies of these alloys where the solution weakening was attributed to the reduction of the stress necessary for the formation of a double kink, the internal friction results generally are in agreement. That is, the internal friction results do not show a drastic reduction in the interstitial concentration due to a substitutional addition. An example of the last statement is provided by the internal friction (Hasson, 1970) and mechanical property (Arsenault, 1969) results for tantalum + niobium + oxygen alloys. One system where the internal friction and mechanical property results indicate a different mechanism is responsible for the solution weakening, however, is that proposed by Hasson and Arsenault (1970) and Pink and Arsenault (1971) using V–Ti–O alloys. The addition of the reactive substitutional solute titanium to a vanadium–oxygen alloy removes the interstitial oxygen from random solution (i.e., titanium scavenges the interstitial oxygen). Thus, one might conclude that the solution weakening observed in the mechanical property results is due to a "scavenging mechanism." Pink and Arsenault (1971), however, noted that the addition of 0.1 at.% titanium to their alloys from the internal friction results (Hasson and Arsenault, 1971) was sufficient to remove the interstitial concentration present in their alloys. Thus, they concluded that observed solution weakening for their 4 at.% titanium addition alloy could not be explained, therefore, in terms of the scavenging mechanism. Furthermore, from their strain rate analysis they found that the rate controlling mechanism responsible for the solution weakening was the double kink mechanism. The important conclusion from these results is that both internal friction and mechanical property tests are required to identify the exact mechanism responsible for solution weakening which can be observed in the temperature dependence of the yield stress.

The analysis of mechanical property data, with respect to internal friction data in the case of the nonreactive substitutional solutes which produce a peak broadening, is not as straighforward. In the nonreactive case as was pointed out previously, the peak broadening due to a s–i interaction could be due to a binding energy between the interstitial and substitutional solutes. Thus, if the binding energy of the s–i pair is significant, the interstitials associated with these pairs would be effectively removed from random solution (i.e., one could say they were pseudoscavenged). Since the random interstitials could be the short range barriers to dislocation motion and hence, control the temperature dependence of the yield stress, the pseudoscavenging nature due to s–i interactions could result in a reduction of the yield stress at low temperatures.

In order to assess the significance of the effect of a s–i pair, a comparison of the strain of the s–i pair and the tetragonal strain due only to interstitials is required. If the strain associated with the s–i pair relative to the tetragonal strain is a significant value (e.g., 15%), the temperature dependence of the yield stress in an alloy with the strong s–i interaction may have to be interpreted in terms of an interstitial mechanism (e.g., the scavenging mechanism). The strain associated with the s–i pair can be assumed to be proportional to the binding energy of the s–i pair. An estimate of the binding energy and the resultant strain associated with the s–i pair is provided from the analysis presented below.

If one assumes that the s–i pairs are in equilibrium with unassociated interstitial and substitutional atoms, one can write the reaction

$$\text{s-i(Metal)} \rightleftarrows \text{i(Metal)} + \text{s(Metal)} \tag{4}$$

The brackets are added to note that these are not pure components but pseudocomponents in the solvent metal. For this reaction the respective atom fractions (ratio of the number of defects to the number of occupiable sites) $C_i$, $C_s$, and $C_{s-i}$ for a dilute ternary solution are related by the expression

$$C_{s-i} = \frac{m_{s-i}}{m_i m_s} C_i C_s \exp(\Delta G_{s-i}/RT) \tag{5}$$

where $C_{s-i}$ = atom fraction of s–i pairs, $C_i$ = atom fraction of unpaired interstitials, $C_s$ = atom fraction of unpaired substitutionals, $m$ = multiplicity of orientations of respective component, and $\Delta G_{s-i}$ = molar free energy of the binding of the s − i pair. Since $m_i = 3$ and $m_s = 1$, the term $m_{s-i}/m_i m_s$ is equal to $Z_{s-i}/3$, where $Z_{s-i}$ is the number of independent positions the interstitial can assume about the substitutional atom in the s–i pair.

The application of Eq. (5) to the results of Dijkstra and Sladek (1953) shown in Fig. 6c with the substitutional–interstitial configuration shown in Fig. 3 yields a binding energy $\Delta G_{s-i}$ of 2800 cal/mole. The value of $Z_{s-i}$ used in Eq. (5) was 6, since there are six Fe–Mn interstitials per Mn atom. The concentration of unassociated substitutional atoms $C_s$ is the value for the particular ternary alloy. For the Fe–Mn–N results of Dijkstra and Sladek, this value was 0.5 atomic percent Mn. The concentrations of s–i pairs and unassociated interstitials were assumed equal, and the corresponding temperature of 27° C was determined from the point where the two subsidiary peaks intersect in fig. 6c. Inherent in this assumption is the fact that the relaxation parameters of the subsidiary peaks are the same. Since the subsidiary peaks shown in Fig. 6c seem to yield the experimental curve, this assumption appears to be valid. The general validity of this assumption, however, is somewhat questionable as has been pointed out several times in this text. Nevertheless, until a more exact analysis for peak synthesization becomes available, the above results are useful at least for comparative purposes. The results of Hasson and Arsenault (1970) for the Ta + 2% W alloy with 0.285 at.% oxygen added were also synthesized into two peaks, and the binding energy for the s–i pair was determined to be 2780 cal/mole. Peak syntheses of the high temperature peaks of the Fe 0.5%–V–N alloy by Perry et al. (1966) and the Nb–Zr–N alloy by Mosher et al. (1970) yielded values of 6200 cal/mole and 7200 cal/mole, respectively, for the s–i pairs in these alloys. It should be remembered that these two alloys represent reactive solutes.

To assess the effect of the strain (or its counterpart, the binding energy) associated with the s–i pair on the temperature dependence of the yield stress, one must determine the change in the tetragonal strain due to the binding energy of the s–i pairs. To determine this strain value we note the following relationship between the binding and strain energies—specifically, the binding energy of the s–i pair $\Delta G_{s-i}$ is proportional to the strain energy of the s–i pair minus the strain energy of the substitutional atom in solvent and the tetragonal strain of interstitial. This relation is written explicitly as

$$\Delta G_{s-i} = GV[\varepsilon_{s-i} - \varepsilon_s - \varepsilon_i]^2 \qquad (6)$$

where $G$ = shear modulus of solvent and $V$ = volume of solvent lattice. Now, we assume $\varepsilon_{s-i}$ is composed of $\varepsilon_s$ and $\varepsilon_i'$ which is the strain energy of an interstitial atom near a substitutional atom. Thus, if we replace $\varepsilon_{s-i}$ in Eq. (6) by these terms, we have

$$\Delta G_{s-i} = GV[\varepsilon_i' - \varepsilon_i]^2 \qquad (7)$$

and if we denote $[\varepsilon_i' - \varepsilon_i]$ as $\Delta(\Delta_\varepsilon)$ we have

$$\Delta(\Delta_\varepsilon) = \left(\frac{\Delta G_{s-i}}{GV}\right)^{\!\!\frac{1}{2}} \tag{8}$$

where $\Delta(\Delta_\varepsilon)$ is defined, therefore, as the change in the tetragonal strain due to the s–i pair.

Calculations of the change in the tetragonal strain due to the s–i pair [i.e., $\Delta(\Delta_\varepsilon)$] utilizing Eq. (8) for the available binding energy estimates mentioned previously were performed, and the results are summarized in Table III. The value of $\Delta(\Delta_\varepsilon)$ for all of the systems is significant and is also larger for the reactive solute systems (e.g., Zr in Nb) as one might expect. The importance of the present calculations, however, is the change in the tetragonal strain of the nonreactive systems (e.g., Fe–Mn–N and Ta–W–O). The value of $\Delta(\Delta_\varepsilon)$ for the Ta–W–O alloy is large enough so that the tetragonal strain associated with just the interstitial oxygen in the tantalum is reduced significantly by the addition of the substitutional solute. Thus, in nonreactive bcc ternary alloys which exhibit strong s–i interactions, the tetragonal strain can be reduced by the binding energy associated with these s–i interactions. This reduction in the tetragonal strain is similar to a reduction in the random interstitial concentration due to scavenging, and hence, ternary alloys which exhibit strong s–i reactions may be called pseudoscavengers. These alloys should have as similar effects as those alloys which exhibit scavenging on the temperature check dependence of the yield stress (e.g., solution weakening possibly associated with the effective reduction of the random interstitial concentration). However, it must be kept in mind that for scavenging to lead to a solid solution weakening, it has to be demonstrated that the interstitials are the short range barriers to dislocation motion.

In the above analysis it was assumed that the peak could be represented by only two subsidiary peaks. If more than two subsidiary peaks are assumed,

**TABLE III**

Summary of Changes in Tetragonal Strain Due to Binding Energy of s–i Pairs

| Alloy | Tetragonal strain | Binding energy (cal/mole) | $\Delta(\Delta_\varepsilon)$ | Percentage of tetragonal strain (%) |
|---|---|---|---|---|
| Fe–0.5Mn–N | 0.4175 | 2800 | 0.1056 | 25.3 |
| Ta–2.0W–O | 0.2540 | 2780 | 0.0880 | 34.7 |
| Fe–0.5V–N | 0.4175 | 6200 | 0.1568 | 37.5 |
| Nb–0.3Zr–N | 0.2810 | 7200 | 0.1922 | 68.4 |

the value of the binding energy quoted above will change. In order to provide a direct comparison to the above results, specific experimental results for the Fe–Mn–N system are selected, namely, the results of Nacken and Kuhlmann (1966). Nacken and Kuhlmann's synthesis (see Fig. 30) shows four contributions to the experimental peak at the experimental peak temperature. The equilibrium reaction equation for the four contributions which Nacken and Kuhlmann have hypothesized is as follows:

$$4(S-i) \rightleftarrows S + 4i + (S\text{-}S) + (S\text{-}s) \qquad (9)$$

where (S–S), i, (S–i), and (S–s) refer to the subsidiary peaks 1, 2, 3, and 4, respectively (see Fig. 30). The expression for the free energy of binding of the s–i pair would then be

$$\Delta G_{s-i} = RT \ln \frac{(C_{s-i})^4}{2C_S C_{SS} C_i C_{Ss}} \qquad (10)$$

where $C_{s-i}$ = atom fraction of s–i pairs, $C_i$ = atom fraction of unpaired interstitials, $C_S$ = atom fraction of unpaired substitutionals, $C_{SS}$ = atom fraction of substitutional pairs, and $C_{Ss}$ = atom fraction of solvent–substitutional pairs. (Note: The factor two in the denominator of Eq. (10) comes from the multiplicity factor.) The value of $\Delta G_{s-i}$ calculated from Eq. (10) for the Fe + 0.27 wt.% Mn + N results of Nacken and Kuhlmann is 4780 cal/mole, and this value corresponds to a $\Delta(\Delta_\varepsilon)$ value of 0.1375 and the corresponding percentage of tetragonal strain is 32.8. This calculation represents the overall effect of increasing the number of subsidiary peaks from 2 to 4. In order to see the effect of reducing the s–i peak due to the four peak synthesis, a calculation of the resulting binding energy utilizing Eq. (5) for this case yields a value of 2600 cal/mole. This binding energy value corresponds to a $\Delta(\Delta_\varepsilon)$ value of 0.101 and the corresponding percentage of tetragonal strain is 24.3. This latter calculation indicates that the effect is small on the change in tetragonal strain due to the binding energy of the s–i pair which in turn is due to increasing the number of subsidiary peaks. The change in tetragonal strain, nevertheless, for the case of two versus four subsidiary peaks is not significant enough to negate the previous conclusions.

The peak broadening which occurs as a result of a substitutional solute addition, as proposed by Mosher (1969), can be due to a larger degree of clustering of the interstitial atoms induced by the substitutional solute. Still another possible reason for the peak broadening is due to the fact that the lattice is distorted by the solute addition, and this could result in a decrease in the activation energy, i.e., peak broadening.

## VI. Summary

The experimental internal friction results for bcc ternary alloys including iron alloys, Group V alloys, and Group VI alloys have been examined. The addition of Group IV substitutional solutes to bcc metals which contain interstitials was found to remove the interstitials from random solution, and these solutes were categorized as reactive solutes. The other substitutional solutes were categorized as nonreactive solutes. The nonreactive solutes were found to have no effect on the normal solvent–interstitial Snoek peak or they caused broadening and usually a shift in the temperature of the experimental peak. This broadening can be associated with substitutional–interstitial interactions designated s–i interactions. The extent of the s–i interaction, that is, the propensity to occur as manifested by the amount of experimental broadening of the normal Snoek peak is related to the comparative heat of formations between the solvent–interstitial atoms and the substitutional solute–interstitial atoms (e.g., for a Fe–Mn–N ternary alloy the heat of formation of $Fe_4N$ comparted to $Mn_4N$). This correlation is especially accurate for iron ternary alloys. For Group V ternary alloys, peak broadening is present in all the results, however, a comparison of the heat of formations does not indicate why one substitutional solute broadens the experimental internal friction spectrum more than another substitutional solute with a low heat of formation of the substitutional–interstitial atoms.

If a comparison is made of the change in peak height due to a substitutional addition, it appears that for the iron ternary alloys there is no change in peak height; while for the Group V ternary alloy there is, in general, a reduction in the peak height. This difference in apparent behavior is probably due entirely to method of testing. In iron ternary alloys controls were not run and therefore it was not possible to determine the change in peak height.

The peak broadening observed in iron ternary alloys has been interpreted in terms of a substitutional–interstitial interaction. The exact s–i interaction responsible for these results have in most cases not been unambiguously identified. This result was found to be due to the fact that the presently available analysis methods cannot yield unique results. The present analytic method which utilizes single Debye peaks to synthesize the experimental results depends on the arbitrary selection of relaxation strengths, times, and activation energies. The analytical results for Fe–Mn–N alloys show, at present, at least four different interpretations and consequent s–i interactions of the experimental peak. The broadening observed in the Group V metals can be explained in terms of several possible mechanisms. Substitutional–interstitial interactions are a distinct possibility, However, there are at least two alternate explanations. One, clustering of interstitials induced by the substitutional addition, and two, a distortion lattice which can cause broadening. The need for definitive experiments to provide information on the relaxation parameters of the individual s–i interactions has been indicated. Testing at constant frequency with a similar analytical approach has been proposed as one possible approach to provide more exact information on these s–i interactions.

The magnitude of the effect of a s–i interaction on the tetragonal strain of the interstitial was estimated for Fe–Mn–N and Ta–Nb–O. In both cases the strain associated with the binding energy of the s–i defect was found to be a maximum of about 20% of the tetragonal strain which is associated with only the interstitial in the solvent lattice. It was suggested that this effect is somewhat significant in reducing the random interstitial concentration, and the s–i interaction could be classed as a pseudoscavenger. However, if more than two peaks are used in the synthesis, then the change in the tetragonal strain becomes smaller.

As stated in the Introduction one of the more unusual aspects of bcc solid solutions is that solution weakening can occur. Also that there are alternative explanations of this phenomenon, i.e., a reduction in the interstitial concentration "scavenging," or a change in stress necessary to form a double kink. In cases where the substitutional addition has no effect on the Snoek peak, the choices of explanations are limited to the change in the double kink mechanism. If there is a strong reaction between the substitutional and interstitial atoms, then it is possible that scavenging is the cause of the weakening. However, it must first be proved that the interstitial atoms are the short range barriers to dislocation motion. A very good example of this are the V–Ti–O systems; the titanium does a very effective job of scavenging the interstitials, but solution weakening continues to occur at concentrations far in excess of that necessary to scavenge the interstitials. Therefore, the solution weakening is due to a change in the double kink mechanism.

Finally, the third possibility, peak broadening which can be due to substitutional–interstitial interactions, or clustering of the interstitials induced by the substitutional atoms, cannot be analyzed in a straightforward manner. If the peak broadening is due to substitutional–interstitial interactions with a concurrent reduction of the tetragonal strain of the interstitial atoms, then the situation could be very similar to a case of scavenging. However, if the substitutional addition induces clustering of the interstitial, then it is not certain whether weakening or strengthening will occur. If the individual interstitial atoms are the short range barrier to dislocation motion then weakening would probably result due to clustering. However, if di- or tri-interstitials are the short range barriers to dislocation motion, then strengthening would result due to the increased density of di- and tri-interstitials. The reason for considering di- and tri- interstitials is due to the fact that it is very difficult to propose that single interstitials are the short range barriers. For example, the tetragonal strain of nitrogen in tantalum is $< 0.2$, whereas the strain required to explain the mechanical properties data has to be $\approx 1.0$. It is for this reason that clusters have been proposed to explain effect of interstitials on the effective stress.

In conclusion, it is evident from an examination of the data that the effect on the interstitial atoms in bcc alloys due to presence of substitutional atoms is complex. Unfortunately, this complexity makes the correlations between internal friction and mechanical property data rather difficult to interpret. However, there are a few cases where the correlation is straightforward.

### ACKNOWLEDGMENTS

The authors wish to acknowledge several important discussions with Professor D. Beshers of Columbia University, Professor R. Gibala, Case Western Reserve University, and numerous other investigators. The authors also gratefully acknowledge the support of the Advance Research Projects Agency through the Center of Materials Research of The University of Maryland.

# References

Armstrong, R. W., Bechtold, J. H., and Begley, R. T. (1963). *In* "Refractory Metals and Alloys-II" (M. Semchyshen and I. Perlmutter, eds.), Vol. 17, pp. 159–190. Wiley (Interscience), New York.

Arsenault, R. J. (1966). *Acta Met.* **14**, 831.

Arsenault, R. J. (1967). *Acta Met.* **15**, 501.

Arsenault, R. J. (1969). *Acta Met.* **17**, 1291.

Beshers, D. (1965). *J. Appl. Phys.* **36**, 290.

Bunn, P. M., Cummings, D. G., and Leavenworth, H. W., Jr. (1962). *J. Appl. Phys.* **33**, 3009.

Collette, G. (1960). *C. R. Acad. Sci.* **215**, 2017.

Couper, G. J., and Kennedy, R. (1967). *J. Iron Steel Inst., London* **205**, 642.

Das, G. C., and Arsenault, R. J. (1968). *Scr. Met.* **2** 495.

DeMorton, M. E. (1962). *J. Apply. Phys.* **33**, 2768.

Dickenscheid, W., and Seeman, H. J. (1958). *Rev. Met. (Paris)* **55**, 872.

Dijkstra, L. J., and Sladek, R. J. (1953). *Trans AIME* **197** 69.

Enrietto, J. F. (1962a). *Trans. AIME* **224**, 43.

Enrietto, J. F. (1962b). *Trans. AIME* **224**, 1119.

Fast, J. D. (1951). *Philips Tech. Rev.* **13**, 175.

Fast, J. D. (1961). *Metaux* **36**, 431.

Fast, J. D. (1970). Private communication.

Fast, J. D., and Dijkstra, L. J. (1951). *Philips Tech. Rev.* **13**, 172

Fast, J. D., and Meijering, J. L. (1953). *Philips Tech. Rev.* **8**, 1.

Fast, J. D., Meijering, J. L., and Verrijp, M. B. (1961). *Metaux* **36**, 112.

Fast, J. D., and Verrijp, M. B. (1961). Metaux **36**, 447.

Fischer, J. J. (1971). Private communication.

Fleischer, R. L. (1967). *Acta Met.* **15**, 1513.

Fountain, R. W., and Chipman, J. (1958). *Trans. AIME* **212**, 737.

Gibala, R., and Wert, C. A. (1971). *In* "Energetics of Metallurgical Phenomena." Gordon & Breach, New York (to be published).

Gladman, T., and Pickering, F. B. (1965). *J. Iron Steel Inst.*, **203**, 1212.

Goldschmidt, H. J. (1967). "Interstitial Alloys," Plenum Press, New York.

Hahn, G. T., Gilbert, A., and Jaffee, R. I. (1963). *In* "Refractory Metals and Alloys-II" (M. Semchyshen and I. Perlmutter, eds.), Vol. 17, pp. 23–64. Wiley (Interscience), New York.

Hashizume, H., and Sugeno, T. (1967). *Jap. J. Appl. Phys.* **6**, 367.

Hasson, D. F. (1970). Ph.D. Dissertation. University of Maryland.

Hasson, D. F., and Arsenault, R. J. (1970). *Con Proc., Int. Con. Strength Metals Alloys. 2nd, 1970* Vol. 1, p. 267.

Hasson, D. F., and Arsenault, R. J. (1971). *Scr. Met.* **5**, 75.

Heller, W., and Brauner, J. (1964). *Arch. Eisenhüettenw.* **35**, 1105.

Jackson, J. K., and Winchell, P. G. (1964). *Trans AIME* **230**, 216.

Jaffee, R. I., and Hahn, G. T. (1963). DMIC Rep. No. 182. Battelle Memorial Institute, Columbus, Ohio.

Jamieson, R. M., and Kennedy, R. (1966). *J. Iron Steel Inst.*, *London* **204**, 1208.

Keith, G. H. (1969). U. S., *Bur. Mines, Rep. Invest.* **7262**. **RI**.

Klein, M. J., and Clauer, A. H. (1964). ARL 64–88 AFSC/WPAB AD603506.

Klein, M. J., and Clauer, A. H. (1965). *Trans AIME* **233**, 1771.

Laxar, F. H., Blickwede, D. J., and Frame, J. W. (1961). *Trans. Amer. Soc. Metals* **53**, 683.

Leak, D. A., and Leak, G. M. (1957). *J. Iron Steel Inst. London* **187**, 190

Leak, D. A., and Leak, G. M. (1958). *J. Iron Steel Inst.*, *London* **189**, 256.

Leak, D. A., Thomas, W. R., and Leak, G. M. (1955). *Acta Met.* **3**, 501.

Leslie, W. C. and Sober, R. J. (1967). *ASM (Amer. Soc. Metals) Trans. Quart.* **60**, 99.

Lundin, C. E., Jr., and Klodt, D. T. (1961). *Trans. Amer. Soc. Metals* **13**, 735

Meijering, J. L. (1961). *Metaux* **36**, 107.

Miner, R. F., Gibbons, D. F., and Gibala, R. (1970). *Acta Met.* **18**, 419.

Mitchell, T. E., and Raffo, P. L. (1967). *Can. J. Phys.* **45**, 1047.

Mosher, D. (1969). Ph.D. Dissertation. University of Denver.

Mosher, D., Dollins, C., and Wert, C. (1970). *Acta Met.* **18**, 797.

Nacken, N., and Kuhlmann, U. (1966). *Arch. Eisenhüettenw.* **37**, 235.

Nowick, A. S. (1953). *Progr. Metal Phys.* **4**, 1.

Nowick, A. S. (1965). *Book ASTM Stand.* **STP** 378.

Perry, A. J., and Boon, M. H. (1969). *J. Mater. Sci.* **4**, 873.

Perry, A. J. (1970). Private communication.

Perry, A. J., Malone, M., and Boon, M. H. (1966). *J. Appl. Phys.* **37**, 4705.

Pink, E., and Arsenault, R. J. (1971). *Radiat. Effects* **10**, 27.

Polder, D. (1946). *Philips Res. Rep.* **1**, 5.

Powers, R. W. (1954). *Acta Met.* **2**, 604.

Powers, R. W., and Doyle, M. V. (1959). *J. Appl. Phys.* **30**, 514.

Ravi, K. W., and Gibala, R. (1969). *Scr. Met.* **3**, 547.

Rawlings, R., and Robinson, R. M. (1961). *J. Iron Steel Inst.*, *London* **197**, 211.

Ritchie, I. G., and Rawlings, R. (1967). *Acta Met.* **15**, 491.

Schnitzel, R. H. (1964). *Trans. AIME* **230**, 609.

Schnitzel, R. H. (1965). *Trans. AIME* **233**, 186.

Seigle, L. J. (1964). *In* "The Science and Technology of Tungsten, Molybdenum, Niobium and their Alloys" (N. E. Promisel, ed.), p. 63. Macmillan, New York.

Semchyshen, M., and Barr, R. Q. (1961). *In* "Refractory Metals and Alloys" (M. Semchyshen and J. J. Harwood, eds.), Vol. 11, p. 283. Wiley (Interscience), New York..

Smialek, R. L., Webb, G. J., and Mitchell, T. E. (1970). *Scri. Met.* **4**, 33.

Snoek, J. L. (1941). *Physica* **8**, 711.

Solomon, H. D., McMahon, C. J., Jr., and Leslie, W. C. (1969). *ASM* (*Amer. Sec. Metals*) *Trans. Quart.* **62**, 886.

Stephens, J. R., and Witzke, W. R. (1970). *NASA Tech. Note* **NASA TN D-7000**.

Szabo-Miszenti, G. (1970). *Acta Met.* **18**, 477.

Szkopiak, Z. C., and Ahmad, M. S. (1969). *High Temp. Mater.*, *6th Plansee Semin.* p. 345.

Thurber, W. C., Distefano, J. R., Kollie, T. G., Inouye, H., McCoy, H. E., and Stephenson, R. L. (1966). *High Temp. Refract. Metals* **34**, 59.

Weertman, J. (1958). *J. Appl. Phys.* **20**, 1685.

Wert, C. A. (1950). *Trans. AIME* **188**, 1242.

Wert, C. A. (1952). *J. Metals* 602.

Wolf, J. D., and Hanlon, J. E. (1961). *Trans. Amer. Soc. Metals* **53**, 955.

Zener, C. (1948). "Elasticity and Anelasticity of Metals." Univ. of Chicago Press, Chicago.

# The Dynamics of Microstructural Change

R. T. DeHOFF

*Department of Metallurgical and Materials Engineering*
*University of Florida*
*Gainesville, Florida*

## I. Introduction

The inducement and control of microstructural change is a central objective of physical metallurgy. Much of metallurgical literature is devoted to qualitative discussions of microstructures and the processes that cause their development. A much smaller literature is concerned with the quantitative characterization of the path and rate of evolution of microstructures. Recently, the discovery and acceptance of new methods in quantitative microscopy have made it possible to greatly expand the set of tools for the quantitative characterization of microstructure. The objective of the present article is to show how the tools provided by quantitative microscopy, making full use of all of the information that can be obtained, may be applied to the analysis of microstructural change.

A microstructure is an array of points, lines, surfaces, and volumes; thus a microstructural change involves the formation and motion of points, lines, surfaces, and volumes in space. The viewpoint taken here is that the description of the "dynamics of microstructural change" is essentially a problem in geometry; physical aspects of the problem (e.g., what diffusional flows govern the local velocity of the interface) are outside the scope of this treatment.

Thus, we shall seek to present the geometric concepts necessary to describe the dynamics of microstructural change. To accomplish this, the following topics will be explored.

(1) The description of the microstructural state of a system

(2) The concept of the path of microstructural change, which constitutes a process through which the system passes

(3) Geometric equations which govern the rate of traverse of this path

(4) Applications of these equations to geometrically simple processes, including nucleation and growth and redistribution processes

(5) An approach to modeling microstructural change in geometrically complex systems

(6) A discussion of problems that remain to be solved in the field.

The entire development is limited to the description of systems that contain only two distinguishable kinds of volumes, separated by an identifiable interface. Thus, for example, one could treat two-phase structures, or recrystallization in a single-phase material, or the growth of cellular nodules (such as pearlite) with the concepts and relations that follow. Extension to multiphased structures is, in most cases, made straightforward by simply considering the phases of interest one at a time, and imagining the remainder of the multiphase structure to be the "matrix."

## II. The Description of Microstructural State

A given microstructure is a collection of geometric features, points, lines, surfaces, and volumes, of specific types. The qualitative state of a micro-structure may be specified by listing the kinds of geometric features that it contains. The quantitative microstructural state is specified by assigning values to the geometric properties associated with each of the features in the structure. A complete list of the microstructural features that may exist in two-phase structures is presented in part A of this section. This is followed by a listing and description of the geometric properties that may be associated with each kind of feature. To maintain a viewpoint that is simultaneously general and quantitative, it is necessary to concentrate upon those geometric properties that have unambiguous meaning for structures of arbitrary shape, size, and distribution. The underlying quantitative microscopic measurements that permit the determination of the properties are also briefly reviewed.

### A. Geometric Features that May Exist in Two-Phase Structures

The zero-, one-, two-, and three-dimensional features that may exist in a two-phase structure are listed in Table I. In this listing it is assumed that there

**TABLE I**

LIST OF FEATURES THAT MAY EXIST IN A TWO-PHASE STRUCTURE

| Dimensionality | Feature | Designation of kind of feature[a] |
|:---:|:---:|:---:|
| 3 | Volumes | $\alpha$ |
|   |         | $\beta$ |
| 2 | Surfaces | $\alpha\alpha$ |
|   |          | $\alpha\beta$ |
|   |          | $\beta\beta$ |
| 1 | Lines | $\alpha\alpha\alpha$ |
|   |       | $\alpha\alpha\beta$ |
|   |       | $\alpha\beta\beta$ |
|   |       | $\beta\beta\beta$ |
| 0 | Points | $\alpha\alpha\alpha\alpha$ |
|   |        | $\alpha\alpha\alpha\beta$ |
|   |        | $\alpha\alpha\beta\beta$ |
|   |        | $\alpha\beta\beta\beta$ |
|   |        | $\beta\beta\beta\beta$ |

[a] Each kind of feature is designated by the kinds of volumes (particles of grains) whose incidence upon each other produce the feature.

exists sufficient randomness in the system so that all linear features are triple lines, i.e., are delineated by the conjunction of three microscopically different volumes (e.g., three grains). All zero-dimensional entities are quadruple points, i.e., exist because four volumes meet at a point in three-dimensional space. These assumptions will generally be valid for microstructures; a very high degree of geometric order is required to violate them, as exists, for example, in a regular stack of uniform size cubes.

To specify the qualitative microstructural state, it is necessary to select from Table I those features that exist in the structure. For example, a single-phase structure contains grains ($\alpha$), grain boundaries ($\alpha\alpha$), edges ($\alpha\alpha\alpha$), and quadruple points ($\alpha\alpha\alpha$). A structure with small particles of a second phase which tend to avoid the grain boundaries would contain, in addition to the features existing in a single-phase structure, $\beta$ volumes, and $\alpha\beta$ surfaces. If these particles were on the matrix grain boundaries, then triple lines of the type $\alpha\alpha\beta$ would also exist, as well as quadruple points of the type $\alpha\alpha\alpha\beta$. The most complex two-phase structure contains all 14 of the features listed in Table I.

## B. General Geometric Properties of Microstructural Features

### 1. GLOBAL PROPERTIES

Each of the features listed in Table I has associated with it one or more properties. These geometric properties may be classified into two categories: topological and metric (Table II). A property of a geometric feature is topological if its value is invariant under smooth deformation of the feature, so long as the deformation does not break the feature into additional parts, or produce new connections. The basic topological properties of interest for describing the microstructural state are (1) the *number* of separate features of each specific type in the system; and (2) their *connectivity* (i.e., degree to which the features are multiply connected). For detailed definitions of these properties, and examples of their measurement, see DeHoff (1968), Kronsbein *et al.* (1965), Aigeltinger and DeHoff (1971), and Barrett and Yust (1969).

The metric properties, on the other hand, are those that report the dimensions of the features in the system. These are also listed in Table II and are briefly described in the following. These properties primarily describe the extent of a particular feature in the structure. Thus, the three-dimensional features possess a volume, specified by the normalized parameter $V_V$ which is the volume fraction of the phase in the system. Two-dimensional features have an area, specified by the surface area per unit volume $S_V$. Triple lines have a length, evaluated as the length of the feature per unit volume of micro-

## TABLE II

SUMMARY OF GEOMETRIC PROPERTIES THAT MAY BE DEFINED FOR THE
FEATURES LISTED IN TABLE I.

| | Metric properties | | | | Topological properties | |
|---|---|---|---|---|---|---|
| | Extensive | | Average | | | |
| Feature | Property | Designation | Property | Designation | Property | Designation |
| Volumes | Volume fraction | $V_V^\alpha, V_V^\beta$ | Mean chord intercept | $\bar{\lambda}^\alpha, \bar{\lambda}^\beta$ $\bar{\lambda}_g^\alpha, \bar{\lambda}_g^\beta$ | Connectivity | $C_V^\alpha, C_V^\beta$ |
| | | | Mean area intercept | $\bar{A}^\alpha, \bar{A}^\beta$ | Number | $N_V^\alpha, N_V^\beta$ |
| Surfaces | Area | $S_V^{\alpha\alpha}, S_V^{\alpha\beta}, S_V^{\beta\beta}$ | Average mean curvature | $\bar{H}^{\alpha\beta}$ | Genus | $G_V^{\alpha\beta}$ |
| | Total curvature | $M_V^{\alpha\beta}$ | | | Number | $N_V^{\alpha\alpha}, N_V^{\alpha\beta}$ $N_V^{\beta\beta}$ |
| Lines | Length | $L_V^{\alpha\alpha}, L_V^{\alpha\beta}$ $L_V^{\alpha\beta}, L_V^{\beta\beta}$ | — | — | Connectivity Number | $C_V$ $N_V$ |
| Points | — | — | — | — | Number | $N_V$ |

structure $L_V$. All of these parameters may be defined for features of arbitrary shape, size, and distribution, and thus satisfy the conditions that they are general and quantitative.

The *total curvature* of surface is an additional property that satisfies these requirements. [For a detailed description of curvature, see DeHoff, (1967, 1968) and Cahn, (1967a).] The local curvature of an arbitrarily curved surface is defined by two parameters, called the principal normal curvatures $\kappa_1$ and $\kappa_2$. Two simple combinations of these parameters have found utility in the geometry of surfaces (Struik, 1950): the local *mean curvature*

$$H \equiv \tfrac{1}{2}(\kappa_1 + \kappa_2) \tag{1}$$

and the *Gaussian curvature*

$$K \equiv \kappa_1 \kappa_2 \tag{2}$$

The integrals of these two quantities over the area of the surfaces on which they are defined are also useful properties. The integral of the mean curvature, which we define as the *total curvature of the surface* for the present purpose,

$$M_V \equiv \iint_{S_V} H dS \tag{3}$$

provides an additional global parameter with which to specify the geometry of a surface, which is independent of its area. The integral of the Gaussian curvature, called the *integral curvature*, may be shown to be simply related to the topological properties of the surface, provided the surface is smooth and closed:

$$\iint_{S_V} K dS = 4\pi(N_V - G_V) \tag{4}$$

where $N_V$ is the number of separate parts of the surface, and $G_V$ is its *genus* (Kronsbein *et al.*, 1965; Aigeltinger and DeHoff, 1971; Barrett and Yust, 1969) (i.e., connectivity) per unit volume of structure. Since to perform the integration in Eqs (3) and (4) it is necessary to be able to unambiguously assign the principal curvatures as positive or negative, these integrals may be defined only for surfaces that have an inside and an outside. In the present context, this limits their significance to closed surfaces, and hence, to interfaces of the type $\alpha\beta$, see Table II.

## 2. AVERAGE PROPERTIES

In addition to these global or total properties of the geometric features in the system, it is possible to define some additional, intensive properties. These are not independent of the global properties, but provide convenient visualizations of specific structural aspects which, although contained in the global properties, are not explicitly presented. These derived properties include the mean intercepts $\bar{\lambda}$ of each phase, and of grains in the structure, and the average mean curvature $\bar{H}$ of the $\alpha\beta$ interface. The mean intercept is the average surface-to-surface distance in a particular three-dimensional structural feature. It can be defined separately for the $\alpha$ phase, the $\beta$ phase, grains in the $\alpha$ phase, and grains in the $\beta$ phase. It can be shown that

$$\bar{\lambda}^\alpha = 4(V_V^\alpha / S_V^{\alpha\beta}) \tag{5a}$$

$$\bar{\lambda}^\beta = 4 V_V^\beta / S_V^{\alpha\beta} \tag{5b}$$

$$\bar{\lambda}_g^\alpha = 4 V_V^\alpha / (S_V^{\alpha\beta} + 2 S_V^{\alpha\alpha}) \tag{5c}$$

$$\bar{\lambda}_g^\beta = 4 V_V^\beta / (S_V^{\alpha\beta} + 2 S_V^{\beta\beta}) \tag{5d}$$

In a one-phase structure, where $S_V^{\alpha\beta}$ is zero, this quantity is the familiar mean grain intercept used to specify grain size (ASTM Designation E 112-60T).

The average mean surface curvature is given by Cahn (1967) and DeHoff (1967).

$$\bar{H}^{\alpha\beta} \equiv \iint_{S_V} H dS \bigg/ \iint_{S_V} dS = M_V^{\alpha\beta} / S_V^{\alpha\beta} \tag{6}$$

This quantity has been used as an index of the average driving force for processes, such as sintering, that are driven by surface tension.

Other averages could also be defined by taking ratios of the extensive properties to the number of particles in the structure (e.g., average particle volume or surface area), but the topological properties are, in general, difficult to determine experimentally (Kronsbein et al., 1965; Aigeltinger and DeHoff, 1971) so that the measurement of these properties is very limited in practice.

At the present state of development of applications of geometric properties to microstructural characterization, Table II is a complete list of the concepts available. Thus, the quantitative specification of the state of a microstructure is complete when values have been assigned to all of the independent properties listed in Table II. However, the topological parameters have been measured in only a few cases (Kronsbein et al., 1965; Aigeltinger and DeHoff, 1971), and then only those properties associated with the $\alpha\beta$ interface were determined. Most quantitative studies of microstructural change report the volume fraction with some auxiliary parameters (Hull et al., 1942; Pellesier et al., 1942; Anderson and Mehl, 1945). Rigorous measurements and analyses of additional extensive properties are limited (Vandermeer and Gordon, 1959; Speich and Fisher, 1966; Rhines et al., 1969; Vedula and Heckel, 1970; Vandermeer, 1970). Because the quantitative determination of the topological properties involves a great deal of effort, it may be expected that their application in the development of a dynamics of microstructural change will be severely limited. On the other hand, all of the metric properties listed in Table II (columns 3 and 5) may be determined by applying the counting measurements of quantitative microscopy to a representative microsection (DeHoff, 1968). Accordingly, a set of dynamical equations formulated in terms of the metric properties may be expected to find broad application as the use of quantitative microscopic techniques becomes more widespread.

## 3. Size Distributions

An additional kind of information about a structure is made available through the determination of the particle size distribution. This function reports explicitly information that is only implied by the global properties. Unfortunately, in order to specify and measure a size distribution in the usual sense, it is necessary to assume that all of the particles in the structure are the same shape, and that this shape is simple, and known (Underwood, 1968). Thus, size distributions lack the characteristic of generality that the global properties have; this deficiency is compensated by the additional information contained, which is indispensible in specific applications, such as the study of redistribution processes. In any event, this concept is most useful in developing models for microstructural change in geometrically complex structures (Section VI), which are based upon the evolution of size distribution.

## C. Quantitative Microscopy

The quantities in Table II, as well as the particle size distribution, can all be determined in principle by techniques of quantitative microscopy. The measurement of topological properties requires a synthesis of the three-dimensional structure from a series of closely spaced microsections (serial sections). Readers interested in this technique are referred to Kronsbein *et al.* (1965) and Aigeltinger and DeHoff (1971). The metric properties are accessible from a single plane of observation, provided that it is representative of gradients and anisotropies in the structure. The four kinds of properties listed in Table II, volume fraction, surface area, total curvature, and line length, may be determined from four simple counting measurements made upon the plane of observation, as follows:

(1) The *volume fraction* is determined from the point count, obtained by superimposing a grid on the microstructure and counting the number of points in the phase of interest.

(2) The *surface area* of any given kind of interface is obtained by counting the number of intersections that lines in the grid make with traces of that type of interface on the microsection.

(3) The *total curvature* of $\alpha\beta$ interface is obtained by sweeping a line across the structure and counting the number of points of tangency with particle outline.

(4) The *length* of each kind of triple line is determined by counting the number of points of emergence (triple points) of that type of line in the microstructure.

For a complete discussion of the relationships between these counts and corresponding properties, see DeHoff (1968), or the review in Aigeltinger and DeHoff (1971). For the present purpose, it is only necessary to point out that the nine independent[†] metric properties listed in Table II may be unambiguously determined by applying these counting measurements. The five derived properties in column 5 of Table II may then be calculated by applying Eqs. (5) and (6).

The particle size distribution may be determined if it is reasonable to assume that the particles are all triaxial ellipsoids of the same shape (DeHoff and Bousquet, 1970) (note that the sphere is a special case of the ellipsoid of revolution). This determination requires the measurement of the distribution of sizes of linear intercepts produced by superimposing a random line on the structure. For a complete description of this and other techniques for estimating particle size distributions, see Underwood (1968) and DeHoff and Bousquet (1970).

---

† $V_V{}^\alpha$ and $V_V{}^\beta$ are not independent.

## III. The Path of Microstructural Change

The transition in geometric state from some initial condition to some final condition is defined as a *process*. The sequence of geometric states through which a system passes in changing from the initial to the final condition is defined as the *path* of microstructural change.

The geometric state of a microstructure may be defined qualitatively by listing the kinds of geometric features it contains, selecting pertinent features from Table I. With this viewpoint, the *qualitative path* of microstructural change for a process may be defined in terms of the sequence of qualitatively defined geometric states through which the system passes. This description may be further amplified by a qualitative description of the changes in one or more of the geometric properties of the system. For example, consider a precipitation reaction, in which the new precipitate particles form at grain boundaries in the matrix phase. The following sequence of structures may be expected to form.

(1) The initial state contains $\alpha$ volumes, $\alpha\alpha$ surfaces, $\alpha\alpha\alpha$ triple lines, and $\alpha\alpha\alpha\alpha$ quadruple points (Fig. 1a).

(2) The first particles of the $\beta$ phase form (Fig. 1b). The system passes to a state which, in addition to the features contained in (1), contains $\beta$ volumes, $\alpha\beta$ interfaces, $\alpha\alpha\beta$ triple lines, and $\alpha\alpha\alpha\beta$ quadruple points, the volume fraction of the $\beta$ phase increases, as does the area of $\alpha\beta$ interface and $\alpha\alpha\beta$ triple line length. The number of $\alpha\beta$ interfaces increases and the genus (connectivity) remains zero.

(3) The growing $\beta$ particles impinge upon each other (Fig. 1c). This introduces $\beta\beta$ grain boundaries, $\alpha\beta\beta$ triple lines, and $\alpha\alpha\beta\beta$ quadruple points. The qualitative changes in the properties listed in (2) continue, with the exception that the number of $\alpha\beta$ interfaces may begin to decrease as a result of impingement. The area of $\beta\beta$ interface and length $\alpha\beta\beta$ triple lines increases. The connectivity of the triple line networks may become increasingly complex.

Since no values are assigned to any of the properties in the system, this description is completely qualitative. The qualitative path may be determined by inspection of the microstructures alone.

The path of microstructural change for a process may be made quantitative by assigning measured values to two or more of the geometric properties of features in the system.[†] The accuracy and detail with which the path is defined depends upon the number of geometric properties that are determined for the

---

†Because the path of microstructural change is defined as the variation of geometric properties with each other (e.g., surface area variation with volume fraction), a minimum of two properties must be determined to define a path.

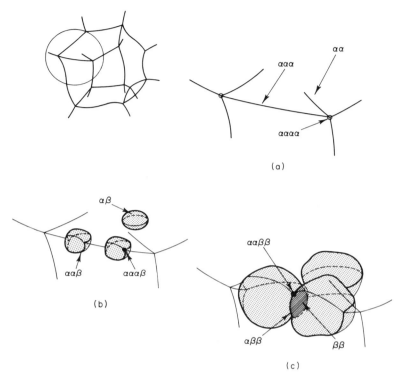

**Fig. 1.** Qualitative path of microstructural change for the precipitation of a phase ($\beta$) at the grain boundaries ($\alpha\alpha$) and edges ($\alpha\alpha\alpha$), illustrating the various features that exist at a sequence of states.

system. For nucleation and growth transformations, for example, the volume fraction is often determined without additional information. This permits an investigation of kinetics, i.e., variation of state with time; however, no insight into the path of change is revealed by a single property determination. (DeHoff and Rhines, 1968). In an investigation of sintering, the volume fraction of porosity, and the area and total curvature of the pore–solid interface have been measured to define the path in some detail. Studies of redistribution kinetics [e.g., coarsening of carbides in steel (Vedula and Heckel, 1970), and swelling in uranium (Bierlein *et al.*, 1962)] occasionally supply the information required to define the path in terms of the evolution of particle size distribution in the system. However, a complete quantitative characterization of a single microstructure has not been reported, much less a complete quantitative path of microstructural change.

Investigations of the dynamics of microstructural change are often undertaken without reference to a path of microstructural change. In such cases, some path is assumed, either explicitly, or, more often perhaps, implicitly. Thus, the rate of change of volume fraction of a growing phase may be reported without any additional information, either qualitative or quantitative, about other features in the structure, or the geometric properties associated with them. The formalization of the concept of the path of microstructural change, due to Rhines *et al.* (1969) may prove to be a significant contribution to the development of a general approach to the analysis of microstructural change.

## IV. The Dynamics of Microstructural Change

The *path* of microstructural change is formulated in terms of the changes that occur in the geometric properties of the system relative to *each other*. Thus, for example, one might conveniently specify the path as the change of $\alpha\beta$ interfacial area and total curvature with the volume fraction of the $\beta$ phase (Rhines *et al.*, 1969). The *dynamics* of microstructural change is formulated in terms of the changes of these properties with *time*.

The description of the dynamics is conveniently divided into two categories: (1) rates of topological change, e.g., rates of formation (or annihilation) of volumes, surfaces, etc., and (2) rates of metric change, e.g., rates of change of volume fraction, surface area, total curvature, etc. These categories correspond to generalizations (in a geometric sense) of the concepts of rate of nucleation and rate of growth. Many of the changes in the toplogical properties listed in Table II are directly related to the rate of formation of new volumes of the phase that is forming. The rates of change of the metric properties are in turn related to the distribution of interface velocities in the system.

### A. Rates of Topological Change

The two basic subclasses of topological properties pertinent to the present discussion are those that report number, and those that report connectivity. The rate of change of the number of some specific class of microstructural features is an essential element in the description of corresponding specific aspects of the dynamics of the overall structural evolution. Thus, the rate of formation of new volumes of the $\beta$ phase, which is equal to the rate of formation of new $\alpha\beta$ interface, and related to the rate of formation of $\alpha\alpha\beta$ triple lines, is necessary to the description of the early stages of precipitation reactions. Rates of annihilation of the same features are important in describing the redistribution (coarsening) process that may follow precipitation. The rate of formation of new $\beta\beta$ grain boundaries (and the associated $\alpha\beta\beta$ triple lines) is

basic to the description of the impingement process that occurs when growing
$\beta$ particles meet. Changes in connectivity of the $\beta$ phase, or of $\alpha\beta$ interface,
also reflect the impingement process. Changes in the genus of the pore–solid
interface during sintering appear to be intimately connected with the kinetics
of the process (DeHoff and Rhines, 1968).

The experimental determination of rates of nucleation, or other topological
rates, has been limited to a few, isolated studies (Hull *et al.*, 1942; Anderson
and Mehl, 1945; Rhines *et al.*, 1969; Vandermeer, 1970; Vedula and Heckel,
1970), In most cases, the number of particles is not determined by direct serial
sectioning, but by indirect methods of quantitative microscopy, e.g., by
determining the size distribution of particles. The experimental view provided
is perhaps oversimplified.

A critical complication of observation of rates of topological change lies in
the difficulty of detecting or even defining the event. The topological changes
occur only when a new feature forms, or two or more features meet. Usually a
relatively small number of atoms is involved in the event, so that its detection
taxes the resolving power of instruments used to detect the change. Second,
the definition of the point in time at which the event is complete is often
ambiguous even if the complete local sequences of changes that produce it may
be observed. If the formation of a nucleus involves a composition change
without a change in crystal structure, for example, the new interface that
defines the separate volume may materialize gradually as a sharpening of a
composition fluctuation. A change in crystal structure may first appear as a
stacking fault, which may form from already present dislocations, or be
present in the initial structure (Hren and Thomas, 1963). "Nuclei" for recrystallization are often present as subgrains in the cold-worked structure
(Hsun Hu, 1963).

As a consequence of these difficulties, it is often necessary to accept a
practical definition of rates of topological change, defined as the rate at which
they become detectable by the method of observation being used in the study.
This reduces the problem to workable from the geometric point of view, but
makes the physical interpretation of the observation less meaningful. Because,
in the present context, the geometric viewpoint is primary, this practical or
experimental definition of rates of topological change will be adopted.

## B. Rates of Metric Change

Motions of the geometric features contained in Table I produce changes in
the geometric properties associated with these features. The growth (or
shrinkage) of any feature is best visualized and described in terms of the
motion of its boundary. Thus, the growth of particles of the $\beta$ phase is formulated in terms of displacements of the $\alpha\beta$ interface. In a grain boundary

reaction, it may be useful to describe the kinetics in terms of the rate of consumption of grain boundary area by the growing phase; in this case, the displacement of the $\alpha\alpha\beta$ triple line, which is the boundary of the growing $\beta$ region in the $\alpha\alpha$ grain boundary, is a useful measure of growth rate.

## 1. DISPLACEMENT EQUATIONS FOR MOVING INTERFACES

Consider a closed $\alpha\beta$ interface bounding a $\beta$ particle (Fig. 2). In some time interval $dt$ this interface moves to a new configuration and position. In the general case, some elements of the interface may be displaced toward the $\beta$ phase, and some away from it. Let it be assumed that this closed surface is

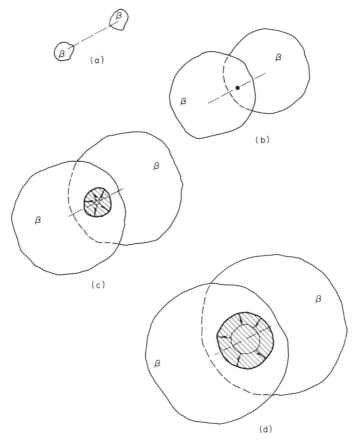

**Fig. 2.** The impingement of two growing $\beta$ particles is equivalent to the nucleation of a $\beta\beta$ grain boundary, and its growth by the motion of its $\alpha\beta\beta$ grain edge.

everywhere smooth and continuous, i.e., it has no sharp ridges or cusps. This assumption is necessary so that each element of surface on the original interface can be unambiguously associated with an element on the displaced surface. This one-to-one association is accomplished by constructing normals to the original surface, and considering the point of intersection of the normal with the new surface as the corresponding point. The local displacement of the surface may then be defined as the distance between corresponding points along the normal, $dn$. The local velocity of the interface is then defined by

$$v \equiv dn/dt \tag{7}$$

This velocity has, for any given system, some distribution of values upon the surface, and some average value. Rates of change of all of the metric properties associated with the interface are related to this velocity distribution. It can be shown that for smooth continuous surfaces

$$\frac{dV_V^\beta}{dt} = \iint_{S_V^{\alpha\beta}} v \, dS \tag{8}$$

$$\frac{dS_V^{\alpha\beta}}{dt} = 2 \iint_{S_V^{\alpha\beta}} v \, H dS \tag{9}$$

$$\frac{dM_V^{\alpha\beta}}{dt} = \iint_{S_V^{\alpha\beta}} v \, K dS \tag{10}$$

where $H$ and $K$ are local values of the curvature functions defined in Eqs. (1) and (2). These relationships are derived in Appendix A. The first two have been adapted from the metallurgical literature; the third is presented here for the first time. Equations (8)–(10) are the basic equations relating the total properties of a moving interface to the distribution of velocities on the interface.

If the interface has singularities, sharp ridges or cusps (e.g., triple lines or quadruple points), then the one-to-one mapping of the initial and displaced interface may break down in the vicinity of these discontinuities. Equations (8)–(10) are not valid in these regions. The errors introduced are of second order in Eq. (8), and may be neglected; however, the effects upon the validity of Eq. (9) and (10) have not been investigated.

*a. Interface Velocity Averages.* Three velocity averages, each weighted differently, may be obtained from the interface displacement equations, by multiplying numerator and denominator of the righ-hand sides by $\iint_{S_V} dS$, $\iint_{S_V} H dS$, $\iint_{S_V} K dS$, respectively, and applying Eqs. (3) and (4)

$$\bar{v}_S \equiv \iint v \, dS \Big/ \iint dS = \frac{1}{S_V^{\alpha\beta}} \frac{dV_V^\beta}{dt} \tag{11}$$

$$\bar{v}_H \equiv \int\int v\,HdS \Big/ \int\int HdS = \frac{1}{2M_V^{\alpha\beta}}\frac{dS_V^{\alpha\beta}}{dt} \tag{12}$$

$$\bar{v}_K \equiv \int\int v\,KdS \Big/ \int\int KdS = \frac{1}{4\pi(N_V^{\alpha\beta}-G_V^{\alpha\beta})}\frac{dM_V^{\alpha\beta}}{dt} \tag{13}$$

The experimental determination of these three dynamical parameters for any structure provides some insight into the distribution of interface velocities in the system, and, in particular, into any correlations that may exist between velocity and curvature.

*b. Dynamical Curvature Averages.* Alternative manipulations of the displacement equations with Eqs. (3) and (4) yield

$$\bar{H}_g \equiv \int\int Hv\,dS \Big/ \int\int v\,dS = \frac{1}{2}\left(\frac{dS_V^{\alpha\beta}}{dt}\Big/\frac{dV_V^{\beta}}{dt}\right) = \frac{1}{2}\frac{dS_V^{\alpha\beta}}{dV_V^{\beta}} \tag{14}$$

$$\bar{K}_g = \int\int Kv\,dS \Big/ \int\int v\,dS = \left(\frac{dM_V^{\alpha\beta}}{dt}\Big/\frac{dV_V^{\alpha\beta}}{dt}\right) = \frac{dM_V^{\alpha\beta}}{dV_V^{\beta}} \tag{15}$$

The "growth rate average mean curvature" $H_g$ has been briefly discussed by Cahn (1967); the "growth rate average Gaussian curvature" $\bar{K}_g$, is a new parameter. If the path of metric change is reported as the change in metric properties with volume fraction of $\beta$ phase, these dynamical curvature averages represent the slope of the paths of surface area and total curvature change, respectively.

For the special case that the interface velocity distribution is very narrow, so that $v$ may be treated as essentially constant in the system, then, from Eq. (11)–(13),

$$\bar{v}_S = \bar{v}_H = \bar{v}_K = v \tag{16}$$

while Eqs. (14) and (15) become

$$\bar{H}_g = \int\int HdS \Big/ \int\int dS \equiv \bar{H}^{\alpha\beta} = \frac{1}{2}\frac{dS_V^{\alpha\beta}}{dV_V^{\beta}} \tag{17}$$

$$\bar{K}_g = \int\int KdS \Big/ \int\int dS \equiv \bar{K}^{\alpha\beta} = \frac{dM_V^{\alpha\beta}}{dV_V^{\beta}} \tag{18}$$

Thus, a simple test to determine the breadth of the distribution of interface velocities is to compare the slope of the surface area versus volume fraction plot with $2\bar{H}$, determined from Eq. (6). Similar comparisons involving Eq. (18) are not as straightforward, because the evaluation of $\bar{K}$ requires a knowledge of the topological properties of the surface, since

$$\bar{K}^{\alpha\beta} \equiv \int\int KdS \Big/ \int\int dS = \frac{4\pi(N_V^{\alpha\beta}-G_V^{\alpha\beta})}{S_V^{\alpha\beta}} \tag{19}$$

Conversely, if Eq. (17) is valid, Eq. (18) could be used to obtain information about the topological properties of the interface, since

$$N_V^{\alpha\beta} - G_V^{\alpha\beta} = \frac{S_V^{\alpha\beta}}{4\pi} \frac{dM_V^{\alpha\beta}}{dV_V^{\beta}} \tag{20}$$

Thus, for the case that $v$ is "position independent" in the structure, it is possible to deduce that the topology of the system is most complicated (i.e., $N_V^{\alpha\beta} = G_V^{\alpha\beta}$) when the total curvature reaches a maximum or minimum, so that $dM_V/dV_V$ is zero.

## 2. DISPLACEMENT EQUATIONS FOR MOVING TRIPLE LINES

If the growing phase is confined to the grain boundaries of the parent phase, its growth may be described in terms of the motion of the $\alpha\alpha\beta$ triple line that bounds the $\beta$ occupied regions in the grain boundary surface, and the changes in metric properties associated with this motion. With this viewpoint, the structural evolution is determined by the rate of topological changes in the boundary, and by the distribution of velocities along the $\alpha\alpha\beta$ triple lines.

This velocity may be defined by analogy with that for moving surfaces. In some time $dt$, the triple line is displaced from some initial position to some infinitesimally removed final position. Each point on the displaced curve is associated with a point on the initial curve by erecting a normal to the initial curve that lies in the grain boundary. If this displacement is locally $dn$, the local velocity of the line is given by

$$v = dn/dt \tag{21}$$

Displacement equations for smooth, continuous triple lines analogous to Eq. (11) for surfaces and volumes are derived in Appendix $B$.[†]

$$\frac{-dS_V^{\alpha\alpha}}{dt} = \int_{L_V^{\alpha\alpha\beta}} v \, d\lambda \tag{22}$$

$$\frac{dL_V^{\alpha\alpha\beta}}{dt} = \int_{L_V^{\alpha\alpha\beta}} v \, k d\lambda \tag{23}$$

where $k$ is the local curvature of the $\alpha\alpha\beta$ triple line, measured in the plane of the boundary, and the integration is carried out over the triple line length $\alpha\alpha\beta$. By multiplying and dividing the right side of Eq. (22) by the total length of

---

†Equation 22 assumes that the matrix grain boundaries in the $\alpha$ phase do not move during the phase transformation, i.e., that grain growth does not accompany precipitation.

triple line in the structure, it is possible to define the average velocity of $\alpha\alpha\beta$ triple line:

$$\bar{v}_L \equiv \int v \, d\lambda \Big/ \int d\lambda = \frac{-1}{L_V^{\alpha\alpha\beta}} \frac{dS_V^{\alpha\alpha}}{dt} \tag{24}$$

A curvature-averaged velocity, analogous to that defined for surfaces in Eq. (12) and (13) may also be defined:

$$\bar{v}_k \equiv \int v \, kdl \Big/ \int kdl \tag{25}$$

Also, a growth-rate-averaged curvature may be defined for triple lines:

$$\bar{k}_g \equiv \int k \, vdl \Big/ \int vdl = -\frac{dL_V^{\alpha\alpha\beta}}{dS_V^{\alpha\alpha}} \tag{26}$$

Unfortunately the denominator in Eq. (25) cannot generally be evaluated from global properties of the elements of the microstructure. It can be shown that, for the special case that the $\alpha\alpha$ grain boundaries are flat

$$\int_{L_V^{\alpha\alpha\beta}} kdl = 2\pi[N_V^{\alpha\alpha\beta} - C_V^{\alpha\alpha\beta}] \tag{27}$$

provided that the connectivity is properly defined. Then

$$\bar{v}_k = \frac{1}{2\pi[N_V^{\alpha\alpha\beta} - C_V^{\alpha\alpha\beta}]} \frac{dL_V^{\alpha\alpha\beta}}{dt} \tag{28}$$

Further, for the special case that $v$ is everywhere constant along the triple line,

$$\bar{k}_g = \int kdl \Big/ \int dl = \frac{2[N_V^{\alpha\alpha\beta} - C_V^{\alpha\alpha\beta}]}{L_V^{\alpha\alpha\beta}}$$

In this case

$$N_V^{\alpha\alpha\beta} - C_V^{\alpha\alpha\beta} = -\frac{L_V^{\alpha\alpha\beta}}{2\pi} \frac{dL_V^{\alpha\alpha\beta}}{dS_V^{\alpha\beta}} \tag{29}$$

so that information about the topology of the $\alpha\alpha\beta$ network may be obtained from measurements of the metric properties. If under these conditions, the $\beta$ particles in the grain boundaries are simply connected, isolated regions, $C_V^{\alpha\alpha\beta}$ is zero in Eq. (29). The number of isolated circuits of triple line is equal to the number of $\beta$ particles. Thus, for the case of simple, isolated particles growing with uniform velocity on flat grain boundaries, the number of particles per unit volume $N_V^{\beta}$, which is a topological property, may be estimated from the path of metric change.

The displacement equations for moving triple lines in grain boundaries may also be applied to the study of the impingement of growing particles. Particle impingement occurs when two moving $\alpha\beta$ surfaces meet. In general this results in either a decrease of one in the number of isolated $\beta$ regions, or an increase in the connectivity of the $\beta$ region. The event itself involves the nucleation of a $\beta\beta$ grain boundary, which grows by the motion of its bounding $\alpha\beta\beta$ triple line (Fig. 2). The dynamics of the impingement process may be described quantitatively with equations analogous to (21)–(29), with appropriate superscript changes.

## C. Evolution of Particle Size Distribution

The equation relating the time variation of the particle size distribution in microstructures to the rate of formation of new surfaces, and the distribution of interface velocities in the system is developed in this section. It is necessary to restrict the development to systems for which the concept of particle size distribution is meaningful. This requires that the structure consist completely of separate, convex particles, distributed in a matrix. This requirement, in turn, effectively limits the range of structures that may be analyzed to those for which the volume fraction of particles is less than about 0.10; systems with larger volume fractions may generally be expected to exhibit increasing numbers of impinged particles and irregular shapes, so that the meaning of the "particle size distribution" losses its precision.

For such small volume fraction systems, it will be shown that the particle size distribution at any time is simply related to the rate of topological change (nucleation rate, or rate of annihilation in coarsening), and to the rate of metric change (interface velocity) as a function of particle size and time. Methods of graphically deducing the nucleation rate and growth rate distribution from experimentally determined size distribution curves, and conversely, for predicting the evolution of particle size distribution from a model of the nucleation and growth rates, are developed in detail.

The equations governing size distribution evolution also form the basis for constructing and testing models for the dynamics of microstructural change in systems that are geometrically more complex. An approach to deriving dynamical models for such systems is given in Section VI.

## 1. The Growth Path and the Growth Path Envelope

Focus attention upon a single $\beta$ particle as it changes during a microstructural transition. The sequence of metric states through which this particle passes as the transformation proceeds is defined as the *growth path* for the particle. In systems in which the particle shape is time independent, the growth path may be fully represented by the variation of any convenient particle dimension with time. For example, if the particles were spherical, the variation of the radius of the sphere during the microstructural change is sufficient to completely specify the geometric state of the particle at any instant. If the particles were platelets of fixed thickness, the variation of the radius of the plate with time is sufficient to specify the growth path. Other examples might be the variation of the length of a rod of fixed radius with time, or the edge length of a polyhedral ideomorph.

If the shape of the particle changes during the process, then it is necessary to represent the growth path in three or more dimensions. For example, consider a platelet whose edge is moving with a constant steady state velocity, while its thickness grows parabolically. The growth path for this system is a space curve in radius-thickness-time space, and is illustrated in Fig. 3a. The growth path for this system might also be described alternatively in terms of separate relationships between the radius and the thickness and the radius and time. In this case, the growth path for the particle would be graphically similar to Fig. 3b, with an implied auxiliary relation between radius and thickness, so that the specification of the radius at any stage of the micro-structural evolution would also specify the thickness, and hence the geometric state of the platelet.

Particles that nucleate at the same time may, in general, be expected to follow somewhat different growth paths, because the system may not be expected to be homogeneous. Thus, for example, the behavior of a given particle may depend upon the configuration and proximity of adjacent particles, i.e., the growth path may be expected to depend upon the position of the $\beta$ particle in the sample. This situation is illustrated schematically in

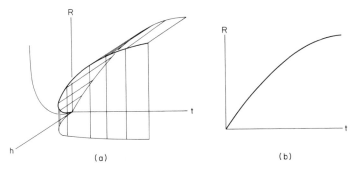

*Fig. 3.* Growth paths for a particle for which the radius and thickness grow at different rates. *a*, Path in radius; *R*, thickness; *h*, time; *t*, space; (b) same path with independently specified *R-h* relation.

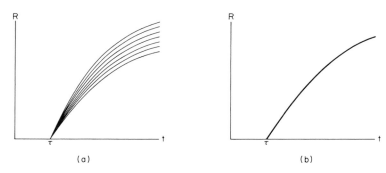

**Fig. 4.** Schematic growth paths showing (a) the spectrum of growth paths originating in any time interval, and (b) the average growth path representing this spectrum.

Fig. 4a. The problem of characterizing the distribution of growth paths emanating from any nucleation period has not been treated. In the present theoretical development, it is assumed that all of the particles that nucleate at a given time proceed along the same path (Fig. 4b), which may be regarded as an average for the distribution of paths shown in Fig. 4a.

Particles that nucleate at different times may be expected to follow different average growth paths; this assertion is shown in Fig. 5.

The continuum of curves describing the geometric evolution of all particles in the system is defined as the *growth path envelope* for the system (Fig. 5). In the development that follows, it is necessary to assume that the individual growth paths in the envelope do not cross each other. If this assumption were violated, then particles that reach crossover points as they evolve would be indistinguishable, so that an unambiguous path back to the nucleation event could not be traced. With this assumption it is possible to experimentally

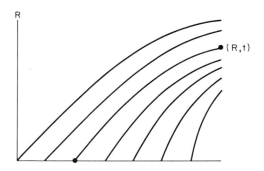

**Fig. 5.** Schematic illustration of a growth path envelope for a nucleation and growth process.

determine the entire growth path envelope for a system from a measurement of the evolution of particle size distribution during the microstructural transition involved. Conversely, given a model of the structural change, formulated in terms of the time dependence of the rate of formation of new $\beta$ particles and the growth path envelope, it is possible to predict the evolution of the particle size distribution with time for the system. The necessary mathematical apparatus for these analyses is presented in the next section.

## 2. THE FUNDAMENTAL RELATION GOVERNING SIZE DISTRIBUTION EVOLUTION

Let the size distribution of particles at any time $t$ be expressed in terms of the fractional frequency function $f(R,t)$, such that $f(R,t)dR$ is the fraction of particles in the system in the size range $R$ to $R+dR$ at time $t$. The corresponding cumulative distribution function $F(R,t)$ is defined in the usual way as the fraction of particles smaller than $R$ at time $t$:

$$F(R,t) = \int_0^R f(R,t)dR \qquad (30)$$

It will also be useful to define a *number frequency function* $n_v(R,t)$, such that $n_v(R,t)dR$ is the number of particles per unit volume of microstructure that exist at time $t$ in the size range $R$ to $R+dR$. Thus, $n_v(R,t)$ is $N_V(t)f(R,t)$.

The quantity $[1-F(R,t)]$ is the fraction of particles larger than size $R$. Let $N_V(t)$ be the total number of particles in unit volume of structure at time $t$. Then, the number of particles in unit volume which are larger than size $R$ is

$$N_{V>}(R,t) \equiv N_V(t)[1-F(R,t)] \qquad (31)$$

The quantity $N_{V>}(R,t)$ is tentatively termed the "$N$-$V$-greater-function." This function plays a central role in the experimental determination of the growth path envelope from measurements of the particle size distribution.

Assume that the growth path envelope exists for the system, and that $t$ is the time at which the system is being observed. Focus attention upon those particles that have attained the geometric state specified by their size $R$ at time $t$. These particles all nucleated at some time $\tau(R,t)$, which could in principle be determined by tracing their growth path back until it intersects the time axis in Fig. 5. Thus, there is associated with the particles in any given size class at the time of observation of the structure, a time $\tau$, which is the time at which these particles nucleated. These three variables $\tau$, $R$, and $t$ are related by the integral equation

$$R = \int_\tau^t G(R,T)dT \qquad (32)$$

where $G(R, T)$ is the growth rate $(dR/dt)$ for a particle of size $R$ at time $T$.[†] In the general case, where $G$ is a function of size and time, the integration of Eq. (32) is straightforward only if a separation of variables may be accomplished; otherwise, numerical techniques may be required. However, even where Eq. (32) may not be integrated in a closed form, the functional relation $\tau(R, t)$ exists, so long as the growth path envelope exists.

The fundamental relationship governing the evolution of particle size distributions is derived from the following conservation statement: "Those particles which are larger than size $R$ at some observation time must have nucleated before $\tau(R, t)$." This assertion is easily verified by inspecting the growth path envelope (Fig. 5). Mathematically,

$$N_{V>}(R, t) = N_V(t)[1 - F(R, t)] = \int_0^{\tau(R, t)} \overset{\circ}{N}(T)\, dT$$

where $\overset{\circ}{N}$ is the rate of formation of particles at time $T$, i.e., the nucleation rate. To convert this equation to a more convenient form, operate on both sides with $(\partial/\partial R)_t$. This yields (see Appendix C)

$$n_v(R, t) = \frac{\overset{\circ}{N}[\tau(R, t)]}{G[R, \tau(R, t)]} \tag{33}$$

Thus, under fairly general circumstances, the evolution of the particle size distribution, expressed as the time variation of the number frequency for the structure, is equal to the ratio of a nucleation rate to a growth rate.

It is essential to emphasize the particular values of nucleation rate and growth rate that are required for application of this equation for a given position on the growth path envelope specified by the number pair $(R, t)$. (1) The nucleation rate that is inserted in Eq. (33) is that which was in effect at the time $\tau$ of nucleation of the particles under study. (2) The growth rate is that which is obtained by inserting the function $\tau(R, t)$ for the time variable in the expression for the growth rate that describes the growth path envelope for the system. These requirements will be amplified with examples of specific models presented in Section V, p. 272.

## V. Microstructural Changes in Geometrically Simple Systems

An analysis of the dynamics of microstructural change for a given system is considered to be complete in the present context when the rates of topological change and the distribution of interface velocities have been determined for

---

[†]Note that the $T$ in the integrand is a "dummy variable" having units of time, which disappears when the limits of the integral are substituted.

the process. In systems in which the second phase particles exist as isolated, simply connected volumes of simple geometry, it is possible to deduce the variation of the nucleation rate with time and the growth path envelope from an experimental determination of the evolution of particle size distribution with time, applying the techniques of quantitative microscopy. Alternatively, an analysis of the time variation of the global properties may provide the essential insights to topological change and the interface velocity distribution. If the system is microstructurally more complex, e.g., the particles are of tortuous shape and multiply connected, then the best available approach to a dynamical description is to resort to models for the geometric evolution that can be deduced or fitted to the experimental observations of the topological and metric properties. The development of techniques for analyzing the former kind of system are presented in this section; modeling procedures for complex systems are discussed in Section VI, pp. 277–279.

## A. Nucleation and Growth Processes

By definition, a nucleation and growth process is one in which volumes of a newly forming phase form spontaneously at positions distributed throughout the structure, and these volumes expand by the motion of the bounding $\alpha\beta$ interface. They may be examined by applying the growth path envelope analysis to the evolution of size distribution, or by studying the time variation of the global properties.

### 1. Size Distribution Analysis

The most detailed description of the geometric state of a simple two-phase system is contained in the determination of the size distribution of particles. It may be helpful to accompany such a determination with quantitative measurements of the metric properties determined by the direct counting methods in order to verify the validity of the model particle shape used, but the size distribution, though not in itself complete, provides the most detail. The size distribution may be reported as a fractional or number frequency function, or as a cumulative distribution function. In the present application it will be found most useful to plot the size distribution in terms of the $N_{V>}(R, t)$ function defined in Eq. (31). It is interesting to note that in the simplest method for measuring size distribution, the $N_{V>}$ function is most directly related to the measurements obtained on the microsection (Underwood, 1968; DeHoff and Bousquet, 1970).

A schematic illustration of the time variation of the $N_{V>}$ function during a nucleation and growth process is shown in Fig. 6a. Note that it must be a

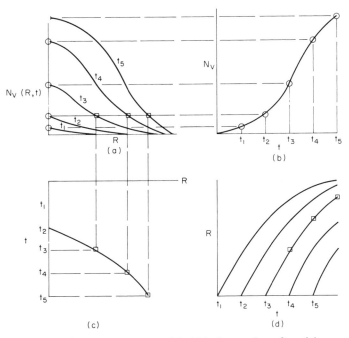

**Fig. 6.** A sequence of N-V-greater curves (a) yields the number of particles as a function of time (b), individual growth paths (c) and the growth path envelope (d).

monotonically decreasing function, which intersects the ordinate at $N_V(t)$, the total number of particles in the structure at the time of observation, and intersects the abscissa at the maximum particle size $R_{max}$. The nucleation rate and the growth path envelope may be estimated graphically from this plot by applying the following straightforward analysis.

A plot of the number of particles as a function of time may be constructed by reading the intercepts on the ordinate for each of the times for which the size distribution is determined (Fig. 6b). The slope of such a plot at any time $(dN_V/dt)$ is by definition the nucleation rate at time $\tau$.

Consider the particles that lie along a given growth path in Fig. 5, i.e., particles that nucleated at the same time $\tau$. In order to experimentally determine the growth path enveolpe from a measurement of the evolution of the particle size distribution, it is necessary to be able to identify, in successive observations of the size distribution, those particles that lie along the same growth path. From Fig. 5 it is clear that the condition that identifies such particles is that, at every time of observation, the number of particles which is larger than their current size is a constant. Thus, a constant value of the N-V-greater function links particles on the same growth path at different observation times. In order to determine a specific growth path from the experimentally measured evolution of the size distribution, it is only necessary to plot the size distribution functions as $N$-$V$-greater functions as in Fig. 6a, and draw a horizontal line on the plot. The growth path corresponding to a particular $N_{V>}$ value may then be plotted by reading successive $R, t$ combinations along this line (Fig. 6c). The full envelope of such curves may then be generated by choosing other values for $N_{V>}$. The envelope derived from Fig. 6a is shown in Fig. 6d.

## 2. GLOBAL PROPERTIES

Estimates of the nucleation rate require the determination of the number of particles per unit volume in the structure as a function of time. The only direct method available for such a determination involves serial sectioning, since number is a topological property. Indirect methods for estimating number all involve simplifying assumptions about particle shape or size of both (DeHoff and Rhines, 1961; Hilliard, 1967). The best of these indirect methods require the same information as is necessary to estimate the particle size distribution. Thus, the procedure described in the previous section, which also yields the growth path envelope, is perhaps the best method for estimating the nucleation rate.

The basic dynamical equations (8)–(15) provide the basis for characterizing the growth rate. A specific test for determining whether the interface velocity is everywhere uniform has already been discussed. This condition will characterize the microstructural change if it is found that $\overline{H}_G = \overline{H}$ and/or $\bar{v}_S = \bar{v}_H$. Cahn (1967) has pointed out that if this is not true for a given system, then the magnitude and direction of departure from equivalance of these properties gives some insight into the distribution of interface velocities in the system. Thus, the condition $\overline{H}_G > \overline{H}$ implies that a correlation may exist between the local interface velocities and local interface curvature such that the interface velocity increases as the curvature is increased. The greater the difference between $\overline{H}_G$ and $\overline{H}$, the more strongly is the local velocity dependent upon the local curvature. Conversely, the condition $\overline{H}_G < \overline{H}$ implies that the local interface velocity is decreased in regions where the curvature is high.

The growth path envelope analysis provides a simple flexible basis for modeling nucleation and growth processes and for calculating the path and kinetics of change of the global properties. It is only necessary to assume a form for the time dependence of the nucleation rate and a growth path envelope to construct such models, since the global properties may be readily calculated if the size distribution is known:

$$V_V = \int_0^{R_{\max}} V(R)\, n_v(R, t)\, dR \tag{34}$$

$$S_V = \int_0^{R_{max}} S(R) n_v(R, t) \, dR \tag{35}$$

$$M_V = \int_0^{R_{max}} M(R) n_v(R, t) \, dR \tag{36}$$

where the functions $V$, $S$, and $M$ are respectively the volume, surface area, and total curvature of a particle of size $R$.

As an illustrative example of this modeling approach, consider the following simple case: spherical particles, nucleating at a constant rate, and growing according to the rate law:[†]

$$v = a/R \to R_{max} = (2at)^{1/2}$$

This condition of "size dependent, time independent" growth is representative of a system in which each particle grows in proportion to its current size, with no influence from surrounding particles. The growth of each particle thus depends only upon how long it has been in existence (time since nucleation), and not upon how long the overall microstructural change has been in process. For this model, applying Eq. (33), the size distribution function is

$$n_v(R, t) = N/G = N/(a/R) = (N/a) R$$

When this result is substituted into Eqs. (34)–(36), together with appropriate functions for $V$, $S$, and $M$ for spheres, integration yields the following time dependences for the global properties:

$$V_V = \tfrac{4}{15}\pi (N/a) R_{max}^5$$
$$S_V = \pi (N/a) R_{max}^4$$
$$M_V = \tfrac{4}{3}\pi (N/a) R_{max}^3$$

the following velocity averages:

$$\bar{v}_S = \tfrac{4}{3}(a/R_{max})$$
$$\bar{v}_H = \tfrac{3}{2}(a/R_{max})$$

and the following curvature averages:

$$\bar{H} = \tfrac{4}{3}(1/R_{max})$$
$$\bar{H}_G = 3(1/R_{max})$$

[†]The example chosen describes diffusion-controlled growth if the parameter $a$ is an appropriate combination of interface compositions and the diffusion coefficient.

where $R_{\max} = (2at)^{1/2}$. Note that this model describes a case in which the interface velocity increases with the local curvature, so that the predicted inequalities $\bar{H}_G > \bar{H}$ and $\bar{v}_H > \bar{v}_S$ are satisfied. It is evident that other, more complex and detailed models could easily be constructed, and time dependence of all of the global properties and averages obtained by this straightforward set of calculations.

## B. Redistribution Processes

Let a redistribution process be defined as one for which the number of particles decreases with time, while the average particle size increases. While it is often true that the total volume fraction of the particles in the structure is fixed, as in overaging or carbide coarsening, this is not a necessary restriction.

### 1. SIZE DISTRIBUTION ANALYSES

Such processes are characterized by a size distribution evolution described by $N_{V>}$ curves in which the intercept of the ordinate decreases with time, while that on the abscissa increases. Thus, successive N-V-greater curves cross one another (Fig. 7a). The rate of topological change is expressed as a rate of annihilation of particles in the structure. As in the nucleation and growth case, this rate may be determined by plotting the ordinate intercepts as a function of time, and taking the slope of the curve at any time (Fig. 7b).

The construction of the growth path envelope from a time sequence of size distributions is identical to that for nucleation and growth process, and is illustrated in Fig. 7c and d. In this case, the fact that the N-V-greater curves cross requires that small particles shrink while large particles grow, as is typical of coarsening processes.

Once the growth path envelope has been determined experimentally, very detailed tests of physically based predictions of the size and time dependence of the growth rate may be performed. For an example of an application of this approach to the analysis of the rate controlling process in the coarsening of carbides in plain carbon steel, see DeHoff (1971).

### 2. GLOBAL PROPERTIES

Because redistribution processes involve a relatively subtle rearrangement of the structure, changes in the global properties are not as informative as they are for nucleation and growth processes. For example, in most cases the volume fraction, which is the most important global parameter in nucleation and growth studies, does not change with time. This implies that the surface area average interface velocity $\bar{v}_s$ which is proportional to the rate of change of

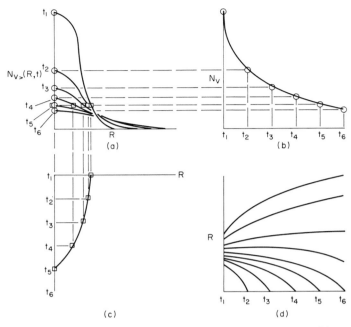

**Fig. 7.** Schematic illustration of relations between the $N$-$V$-greater curve (a), annihilation rate (b), growth path (c) and growth path envelope (d) for a redistribution process.

volume fraction [Eq. (11)] is zero for such processes. Also, Eq. (14) shows that $H_G$ is infinite for $V_V$ time invariant. Thus, the conditions $\bar{H}_G > \bar{H}$ and $\bar{v}_H > \bar{v}_s$ are true in general for microstructural change at constant volume fraction. As a consequence, these tests do not provide useful information for analyzing redistribution processes.

Models for growth during redistribution may be tested by applying the dynamical equations (9) and (10) in some cases. For example, one theory of coarsening controlled by diffusion between the particles gives for the rate of growth of a particle of size $R$

$$v = k[\bar{H} - (1/R)] \tag{37}$$

Substitution of this model into Eq. (9) and integration gives

$$dS_V/dt = k[\bar{H}M_V - 4\pi N_V] \tag{38}$$

for the rate of surface area change. Thus, measurements of the global properties $S_V$, $M_V$, and $N_V$ as functions of time during coarsening could be used to test the physical model expressed in Eq. (37). However, even in this case, a determination of $N_V$ is required. Accordingly, it is concluded that the growth path analysis of the evolution of particle size distribution as described in Fig. 7 is the most powerful approach to the analysis of redistribution processes.

### C. Grain Boundary Reactions

If the formation of a second phase is confined entirely to the grain boundaries in the parent phase, then it is useful to reformulate the description of the process in terms of a two-dimensional grain boundary transformation. This formulation will be straightforward if the parent $\alpha\alpha$ grain boundaries do not move during the process; if simultaneous grain growth occurs, the situation is very complex. If the $\alpha\alpha$ grain boundaries do not move, then the formation of the $\beta$ phase may be characterized as a nucleation and growth process in which the rate of nucleation is expressed in terms of the number of new particles formed per second, per square centimeter of grain boundary area, and the rate of growth is governed by the distribution of velocities upon the $\alpha\alpha\beta$ triple lines that bound the $\beta$-occupied areas in the grain boundary. The fraction of the $\alpha\alpha$ grain boundary area occupied by $\beta$ regions is a measure of the extent of the reaction, and the length of $\alpha\alpha\beta$ triple line measures the extent of the boundary between the growing and parent regions. Equations (21)–(29) may be applied to such an analysis.

As an alternative approach, it is possible to determine the size distribution of the two-dimensional $\beta$ regions in the $\alpha\alpha$ grain boundaries, if they are simply connected and isolated. This requires the measurement of the distribution of lengths of distances between $\alpha\alpha\beta$ triple points on a section (DeHoff, 1962; DeHoff and Bousquet, 1970). If the evolution of size distribution is measured, the nucleation rate and growth path envelope may be determined by a procedure which is precisely analogous to that outlined in Section V, A, 1.

Models for grain boundary reactions may be constructed by applying Eq. (33) as outlined in Section V, A, 2. As a simple example, a grain boundary reaction may be modeled as an aggregate of isolated circles, forming with a constant nucleation rate per square centimeter of grain boundary area $N$ and growing with a constant velocity $v$. The size distribution in this two-dimensional phase transformation is, according to Eq. (33),

$$n_A(R, t) = N/v \tag{39}$$

The global properties of interest are

$$S_S{}^\beta = \int_0^{R_{max}} \pi R^2 \frac{N}{v} dR = \frac{\pi}{3} \frac{N}{v} R_{max}^3$$

$$L_V^{\alpha\alpha\beta} = \int_0^{R_{max}} 2\pi R \frac{N}{v} dR = \pi \frac{N}{v} R_{max}^2$$

where $R_{max}$ is $vt$, and $S_S$ is the fraction of the original $\alpha\alpha$ grain boundary area that has become occupied by $\beta$ regions, i.e.,

$$S_S{}^\beta = 1 - (S_V^{\alpha\alpha}/S_{V_0}^{\alpha\alpha}) \tag{40}$$

Because the interface velocity is assumed everywhere constant in this model, the various averages of interface velocities [Eqs. (24) and (25)] and curvature averages [Eq. (26)] are equal. With the flexible growth path approach, more complex models could readily be generated, in which these properties would be different from one another, reflecting the distribution of interface velocities. The validity of the model could be examined by measuring the corresponding global properties as a function of time, fitting the nucleation and growth rates, and then performing appropriate statistical tests.

## VI. Modeling Microstructural Change in Geometrically Complex Systems

In the present context a microstructure is said to be "geometrically complex" if the features under study are not simply connected, isolated regions. In a nucleation and growth process, this occurs after roughly 0.1 of the volume has transformed to the new phase. At this stage a significant number of growing grains will have encountered each other. This *impingement process* produces $\beta\beta$ grain boundaries and $\alpha\beta\beta$ triple lines. As the number of impingement events increases, these particles form connected chains, and eventually the entire phase may become connected, and quantitatively possess a high value of connectivity. Continued transformation ultimately isolates chains of the parent phase, which become broken into small simply connected regions that are consumed as the volume fraction of $\beta$ approaches 1. In the intermediate range of the transformation, the microstructural state is clearly "complex."

The sintering process is another example of a microstructural change that is "complex." Sintering begins in the middle of the sequence outlined in the preceding paragraph when the structure is multiply connected, and proceeds to the isolation of pores that vanish as full density is attained. For a detailed description of the quantitative path of microstructural change during sintering, see Kronsbein *et al.* (1965) and Aigeltinger and DeHoff (1971).

### A. Modeling the Impingement Process

A direct study of the impingement process that occurs during growth has not been attempted. The rate of nucleation of $\beta\beta$ grain boundaries, which is the rate of occurrence of impingement events, and the velocity of the $\alpha\beta\beta$ triple line thus formed, which is their rate of growth, could be characterized in terms of the path of change of these particular microstructural features. In existing theories the details of the impingement process are avoided, and a more or less phenomenological description has been developed; this phenomenological approach is presented in this section. It will be noted that this approach has the advantage that it simply provides a correction to properties of the structure that are calculated from models that ignore impingement of the particles in the growing phase. Thus, all of the flexibility for model construction presented in Section V may be applied to complex structures by operating upon the results with an impingement correction.

## 1. UNIFORM IMPINGEMENT: THE JOHNSON MEHL–AVRAMI–EQUATION

Imagine a structure calculated from some model based upon an assumed nucleation rate function and growth path envelope. If the methods developed in Section V are used for calculating the global properties of this structure, the corresponding properties in the real structure may be expected to be overestimated; the magnitude of the overestimation increases with the time of transformation. As a concrete example, the model discussed at the end of Section V, A, 2 predicts that the volume fraction will approach $\infty$ as $t \to \infty$, whereas values of $V_V$ greater than 1 do not have geometric significance. The overestimates result because no account is taken of impingement of the growing particles; in the model they simply ignore each other's presence, and grow right through one another. The resulting structure, in which the modeled particles of the growing phase freely overlap, is defined as the *extended structure* for the model.

To derive the volume fraction of the real structure from that of the extended structure, Johnson and Mehl (1939), Avrami (1939), and Kolmogorov (1937) independently proposed the following procedure. Consider an incremental change in time, which produces an incremental increase in the volume fraction of both the real and extended structures $dV_V$ and $dV_{V\,ex}$. In the extended structure this increment is a surface layer not necessarily of constant thickness, spead over all of the existing particle surfaces. Focus upon a small element of this incremental surface layer $\delta(dV_{V\,ex})$. The basic assumption in the phenomenological approach is that these elements of incremental volume are distributed *uniformly* in the volume of the extended structure. If this is granted, then the fraction of these elements that lie in regions in the extended structure that are not already occupied by the $\beta$ phase is equal to the fraction of the volume in the real structure, which has not yet transformed: $(1 - V_V)$. Only these elements contribute to increasing $V_V$ in the real structure. Thus, the change in $V_V$ in the real structure resulting from a change in $V_{V\,ex}$ in the extended structure is

$$dV_V = (1 - V_V)\,dV_{V\,ex} \tag{41}$$

This is the basic equation governing uniform impingement. Integration of this equation yields

$$V_V = 1 - \exp(-V_{V\,ex}) \qquad (42)$$

It is clear that this result has the proper general form required, since as $t \to \infty$ and $V_{V\,ex} \to \infty$, $V_V \to 1$. The value of $V_{V\,ex}$ may be calculated for any assumed nucleation rate and growth model by applying the methods developed in Section V.

Additional equations relating the other metric properties of the extended structure to those of the real structure are derived for the uniform impingement model in Appendix D. The results of these derivations are[†]

$$S_V = (1 - V_V)\,S_{V\,ex} \qquad (43)$$

$$M_V = (1 - V_V)\{M_{V\,ex} - [\bar{H}(ex)/\bar{H}_G(ex)]\,S_{V\,ex}^2\} \qquad (44)$$

As an example of the application of these equations, consider the model presented in Section V, A, 2. The results are properties of the extended structure for the case wherein impingement is important. Substitution of these results into Eq. (42) through (44) gives the time dependence of the metric properties of the impinged structure:

$$V_V = 1 - \exp\left[-\tfrac{4}{15}\pi(N/a)\,R_{max}^5\right] \qquad (45)$$

$$S_V = \exp\left[-\tfrac{4}{15}\pi(N/a)\,R_{max}^5\right]\pi(N/a)\,R_{max}^4 \qquad (46)$$

$$M_V = \exp\left[-\tfrac{4}{15}\pi\,R_{max}^5\right]\left[\tfrac{4}{3}\pi(N/a)\,R_{max}^3 - \tfrac{2}{9}(\pi^2\,N^2/a^2)\,R_{max}^8\right] \qquad (47)$$

where $R_{max}$, $a$, and $N$ have the significance described earlier. Best fit values of these parameters may be obtained by computer analysis, or, if the results are simple enough, by appropriate graphical analysis.

### 2. More General Phenomenological Impingement Equations

The introduction of an adjustable parameter (Hillert, 1959) into Eq. (41) introduces a flexibility into the phenomenological equation that permits the description of impingement in structures in which growing particles tend to cluster (early impingement) or order (late impingement) in the volume. Nucleation at grain boundaries undoubtedly produces a tendency for clustering. The precipitation of $\gamma'$ phase in nickel base superalloys (Ardell, 1970) is a most exaggerated case of ordering of the particles of the growing phase. These

---

[†]These results contain simplifying assumptions about the interface velocity averages; for details, see Appendix D.

effects may be accounted for in a phenomenological way by introducing the parameter $i$ in Eq. (41)

$$dV_V = (1 - V_V)^i \, dV_{V\,ex} \tag{48}$$

The range $i > 1$ implies clustering; $i < 1$ implies ordering of the particles. Integration gives

$$V_V = 1 - [1 - (1 - i) V_{V\,ex}]^{(1/1 - i)} \tag{49}$$

The corresponding equations for the other metric parameters are (see Appendix D):

$$S_V = (1 - V_V)^i \, S_{V\,ex} \tag{50}$$

$$M_V = (1 - V_V)^i \{ M_{V\,ex} - i [\bar{H}(ex)/\bar{H}_G(ex)] \, S_{V\,ex}^2 \} \tag{51}$$

In this approach, the uniform impingement equation is the specific example, $i = 1$.

## VII. Future Development of the Dynamics of Microstructural Change

The present status of the dynamics of microstructural change provides a variety of avenues of approach for analyzing solid state processes, ranging from the direct analysis of paths of topological and metric change as quantitatively determined for a specific system, to the prediction of such paths from physically based models of the mechanisms of nucleation and growth or redistribution of solutes. However, the apparatus must be considered to be in its infancy. Some of the relations lack real mathematical rigor at their foundation. In other cases, the limitations of application may be too carefully drawn, so that better mathematical foundations may yield a broader range of application. Some of the problems remaining to be solved, and the vagaries remaining to be clarified are discussed in this section.

### A. Scope of Application of the Dynamical Equations

In most of the derivations of the dynamical equations, which are summarized in Appendices A and B, it is necessary to assume that the structural features they describe are smooth and continuous. In fact, this is rarely true in applications to solid state structures. Most surfaces have triple lines and quadruple points imbedded in them; these interrupt the smoothness. Most triple lines have quadruple points imbedded, which represent discontinuities in the triple line. In order to apply the dynamical equations, it is essentially necessary to neglect any effects associated with such edges and cusps. Clearly some theoretical work is necessary to (1) establish the order of importance of the effect of neglecting discontinuities upon the dynamical equations, and (2) develop corrections that specifically take these discontinuities into account.

## B. The General Impingement Problem

Much of the difficulty of fitting observed behavior in complex systems to physical models probably arises from an inadequate description of the effects of impingement of the growing particles upon the paths and kinetics of metric change. The phenomenological approach outlined in Section VI suffers from an obvious lack of straighforward geometrical significance. Some direct experimental study of the impingement process in a range of systems is indispensible to the development of better characterizations of geometric evolution in complex systems. A growing interest in applying two-phase materials with volume fractions in the middle range makes the description of the formation of such structures technologically important. The analysis of the impingement process in terms of the nucleation and growth of $\beta\beta$ grain boundaries is promising, but has not been tried.

## C. Broadening the Growth Path Envelope Analysis

Two limitations were placed upon the development of the growth path envelope analytical apparatus developed in Section IV, pp. 265–268. It was necessary to assume that all particles that nucleate at the same time follow the same growth path, and that these growth paths do not cross. Further theoretical development of this approach might permit one or both of these restrictions to be relaxed.

It should be possible to construct models in which each time interval for nucleation spawns a set of particles that follow growth paths that follow some distribution of configurations, as shown in Fig. 4a. This distribution could be characterized by one or more physically determined parameters, which then become adjustable parameters that may be evaluated for any given physical system. The growth path model is highly flexible in its present state of development; such studies could also establish whether introducing the concept of a distribution of paths improves the geometrical characterization presently available.

Although it has not been established, it seems likely that if a growth path is generated from a mathematical model, the evolution of particle size distribution could be predicted even if the growth paths cross. The origins of individual particles are unambiguously established in this case because the physical model is known. The reverse problem, that of deducing the growth path envelope from a set of measurements of an evolving size distribution, is not straightforward if the real growth paths cross. In addition, the derivation of Eq. (33), which forms the basis for calculating metric properties from a growth path model, is not straightforward if the paths cross.

Both of these restrictions seem susceptible to further study by fairly direct approaches.

## D. Development of Quantitative Microscopy

The basic analytical tool for the dynamics of microstructural change, quantitative microscopy, continues to develop. Although the methods by which simple extensive properties can be estimated are now well established, it is by no means clear that all of the governing fundamental relationships have been discovered. Intensive study at the mathematical foundations, couched in integral geometry, exterior algebra, and group theory, is necessary before this tool can be declared to be fully developed.

## E. Application to Time-Independent Processes

Throughout this paper the dynamical relationships have been formulated in terms of the time dependence of each of the geometric properties considered, or in terms of velocities and rates. The fundamental displacement equations, and the growth path envelope concept, are formulated in terms of comparing microstructural states at two successive times of observation. The restriction to time as the cause or monitor of geometric change is completely unnecessary. For example, equations entirely analogous to those developed in this paper may be formulated for describing changes in structure accompanying changes in the applied stress, or changes in temperature as in athermal processes. Concepts analogous to the nucleation "rate" and growth "rate" are easy to visualize. The growth path envelope replaces the time axis with stress or temperature. Thus, the dynamical equations may eventually find application to the evolution of dislocation structure during work hardening, to the athermal formation of martensite, and so on.

## VIII. Summary

A broad yet detailed approach to analyzing microstructural processes has been presented in this article. This approach is based upon formalized concepts of the qualitative and quantitative microstructural state for a material system, and the path of microstructural change that specifies the process under study. Two approaches to analyzing the dynamics of microstructural change are presented. The first involves the determination of the time variation of the global topological and metric properties of the system, and analyzes the dynamics by applying displacement equations from integral geometry. The second is based upon the growth path envelope concept, and requires the determination of the evolution of the particle size distribution. These two approaches may be combined in modeling microstructural change, with the growth path envelope used to model the evolution of particle size distribution, which may in turn be applied to predict the time variation of the global properties.

Although selected specific applications have been presented to illustrate these approaches, it is clear that they provide a broad and flexible basis for describing the dynamics of microstructural change in unprecedented detail.

### Appendix A: Interface Displacement Equations

Consider a smooth, closed surface, which is a phase boundary in a two-phase structure. Changes in the geometric state of this structure are accomplished by the displacement of this interface. The displacement of the interface alters one or more of its total geometric properties, namely, the volume $V$ it encloses, the area $S$ of the surface, and the total curvature $M$ of the surface. Changes in these properties may be considered as the sum of the changes contributed by the motion of each surface element on the interface.

Relationships between changes in the total properties, and the properties themselves, are derived by first evaluating the changes that result from the displacement of a single, arbitrary element of the surface, and then integrating the result over the surface area.

#### A. Local Displacement Equations

Consider the element of area shown in Fig. A-1a. When this element is displaced (Fig. A-1b) it (a) sweeps through a volume element $\delta(dV)$, (b) changes its area $\delta(dS)$, and (c) changes its total curvature $\delta(dM)$.
It will be shown how these changes are related to the initial geometry of the surface element.

#### 1. THE VOLUME CHANGE

If the area of the element is $dS$, the volume it traverses when displaced an infinitesimal distance $dn$ is given by

$$\delta(dV) = dnds \qquad \text{(A-1)}$$

as is easily verified from Fig. A-1b.

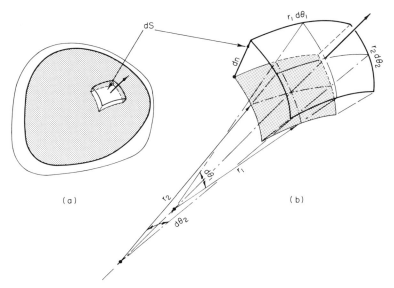

**Fig. A-1.** The expansion of a volume by the nonuniform motion of its bounding surface (a) produces changes in the local geometry of the surface (b).

## 2. THE SURFACE AREA CHANGE

Expressed in terms of the local radii of curvature, and the characterization of the element shown in Fig. A-1b, the surface area before displacement is

$$dS = r_1 d\theta_1 r_2 d\theta_2$$

The displacement increases the radii of curvature by $dn$, so that the surface area after displacement is

$$dS + \delta(dS) = (r_1 + dn) d\theta_1 (r_2 + dn) d\theta_2$$

The change in surface area resulting from the displacement is

$$\delta(dS) = (r_1 + r_2) dn d\theta_1 d\theta_2$$

$$= \frac{r_1 + r_2}{r_1 r_2} dn r_1 r_2 d\theta_1 d\theta_2$$

From the definition of local mean surface curvature

$$H \equiv \frac{1}{2}\left[\frac{1}{r_1} + \frac{1}{r_2}\right] = \frac{1}{2}\left[\frac{r_1 + r_2}{r_1 r_2}\right]$$

Thus, the change in surface area of the element due to its displacement is

$$\delta(dS) = 2dnHdS \tag{A-2}$$

## 3. THE TOTAL CURVATURE CHANGE

The total curvature has a value for the surface as a whole, i.e., it is a global property, as is surface area. The surface element shown in Fig. A-1 contributes an infinitesimal share to the total curvature of the surface, given by

$$dM = HdS = \frac{1}{2}\left[\frac{1}{r_1} + \frac{1}{r_2}\right] r_1 r_2 \, d\theta_1 \, d\theta_2$$

$$= \tfrac{1}{2}(r_1 + r_2) \, d\theta_1 \, d\theta_2$$

The displacement of the surface element increases the radii of curvature by an amount $dn$, so that after the displacement, the contribution of the element to the total curvature is

$$dM + \delta(dM) = \tfrac{1}{2}[(r_1 + dn) + (r_2 + dn)] \, d\theta_1 \, d\theta_2$$

The change in this contribution, resulting from the motion of the element, is thus

$$\delta(dM) = dn \, d\theta_1 \, d\theta_2$$

From the definition of the local value of the Gaussian curvature,

$$K \equiv \frac{1}{r_1} \frac{1}{r_2}$$

This change may be written

$$\delta(dM) = dn \frac{1}{r_1} \frac{1}{r_2} r_1 r_2 \, d\theta_1 \, d\theta_2 = dnKdS \tag{A-3}$$

The three equations (A-1)–(A-3), represent contributions to the changes in the total properties of the closed surface that result from the motion of the surface element $dS$.

## B. Global Displacement Equations

Changes in the total geometric properties of the surface may be obtained by summing the contributions of each surface element. In the present context, this summation amounts to a surface integral. For the volume change

$$dV = \int\int_S dndS \tag{A-4}$$

The change in total surface area,

$$dS = \int \int_S 2dnHdS \tag{A-5}$$

The change in total curvature of the surface,

$$dM = \int \int_S dnKdS \tag{A-6}$$

In each case, $dn$ is the local value of the displacement of the interface. The corresponding *rate of change* may be obtained by dividing both sides of these equations by $dt$, the time interval in which the displacement occurred. Further, by noting that $dn/dt$ is the local velocity $v$ of the interface at each surface element, Eqs. (A-4)–(A-6) become

$$\frac{dV}{dt} = \int \int_S vdS \tag{A-7}$$

$$\frac{dS}{dt} = 2 \int \int_S vHdS \tag{A-8}$$

$$\frac{dM}{dt} = \int \int_S vKdS \tag{A-9}$$

These three equations constitute the fundamental relationships between the rates of change of global properties, and the local geometric properties of the surface.

If the surface under examination contains sharp linear ridges or point cusps, then the description of the local surface element and its displacement illustrated in Fig. Al is no longer valid. Contributions to the volume change resulting from motion of such singularities are of second order and may be neglected. However, contributions to the change in surface area and total curvature are not negligible. Indeed, if the surface is a flat-faced polyhedron, for example, all of the changes in surface area and total curvature are attributable to motion of edges and corners. The proper inclusion of these features in the displacement equations requires additional study.

## Appendix B: Triple Line Displacement Equations

Figure B-1 illustrates a particle of the $\beta$ phase growing along an $\alpha\alpha$ grain boundary. In cases wherein the growing phase is confined to the grain boundaries of the matrix, it is useful to describe microstructural changes in terms of the formation and motion of the $\alpha\alpha\beta$ triple lines, which bound the $\beta$ phase in the αα interfaces. Equations similar to those derived in Appendix A may be written relating the displacement of this triple line to changes in the total geometric properties of the structure.

In deriving these equations, it is necessary to assume that the grain boundary in which the particle is growing is flat, and that the αβ triple lines that are moving are smooth and continuous, i.e., that they contain no quadruple points of the type ααβ or αββ. It is possible that these restrictions may ultimately be relaxed, but the rigorous treatment of the problem of growth on growing grain boundaries of networks of triple lines requires a good deal more study.

### A. Local Displacement Equations

Consider the element of length of αβ triple line shown in Fig. B-1. When the triple line moves along the αα boundary, this element (a) sweeps through an area $\delta(dS)$; and (b) changes its length $\delta(dL)$. Magnitudes of these changes are related to the initial geometry of the line element.

### 1. THE CHANGE IN SURFACE AREA

If the length of the line element in Fig. B-1 is $dL$, then its displacement a distance $dn$ sweeps through an area of αα boundary given by

$$-\delta(dS) = dndL \qquad (B-1)$$

The minus sign is included because the outward motion of $dL$ decreases the αα grain boundary area.

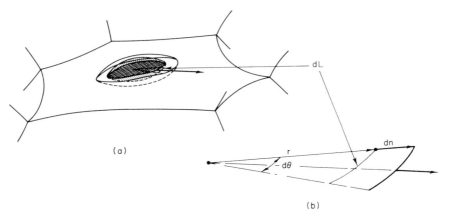

**Fig. B-1.** Growth of a particle in a grain boundary by motion of its bounding triple line (a) produces changes in the local geometry of the line (b).

## 2. THE CHANGE IN LINE LENGTH

The length of the element $dL$ before its displacement may be expressed in terms of its local curvature and the angle subtended by its normals, see Fig. B-1b:

$$dL = rd\theta$$

After displacement, its length is

$$dL + \delta(dL) = (r+dn)\,d\theta$$

since the radius of curvature changes an amount $dn$. Thus, the change in length is given by

$$\delta(dL) = (r+dn)\,d\theta - rd\theta = dnd\theta$$

$$\delta(dL) = dn(1/r)\,rd\theta \tag{B-2}$$

$$\delta(dL) = dnkdL$$

where $k$ is the local curvature of the triple line at $dL$.

### B. *Global Displacement Equations*

Changes in the global properties of the structure due to motion of all of the elements of the $\alpha\alpha\beta$ triple line is the sum of the changes associated with each line element, and may be expressed in terms of line integrals. For the change in surface area of $\alpha\alpha$ boundary, integration of equation (B-1) gives

$$-dS = \int_L dndL \tag{B-3}$$

The change in length of $\alpha\alpha\beta$ line arising from its motion is

$$dL = \int_L dnkdL \tag{B-4}$$

the corresponding rates of change of these global properties may be obtained by dividing both sides of equations (B-3) and (B-4) by $dt$, the time interval in which the displacement occurred, and noting that $dn/dt = v$:

$$-dS/dt = \int_L vdL \tag{B-5}$$

$$dL/dt = \int_L kvdL \tag{B-6}$$

These equations are the basic geometric relationships governing line displacements in two dimensions, and, as they are applied here, triple line displacements in grain boundary reactions.

## Appendix C: Fundamental Equation for the Evolution of Particle Distribution

In the growth path approach to describing microstructural evolution, the basic conservation statement, "Those particles that are larger than size $R$ at some observation time must have nucleated before $\tau(R, t)$" has been shown to have the following mathematical equivalent:

$$N_{V>}(R, t) = N_V(t)[1 - F(R, t)] = \int_0^{\tau(R, t)} \mathring{N}(T) \, dT \qquad \text{(C-1)}$$

where all terms are as defined in the text. Operate upon both sides of this equation with $(\partial/\partial R)_t$:

$$\left\{ \frac{\partial}{\partial R} [N_{V>}(R, t)] \right\}_t = \left\{ \frac{\partial}{\partial R} [N_V(t)(1 - F(R, t))] \right\}_t$$

$$= \left[ \frac{\partial}{\partial R} \int_0^{\tau(R, t)} \mathring{N}(T) \, dT \right]_t$$

The derivative of the left side of this equation may be evaluated directly, since the derivation of the cumulative frequency function $F(R, t)$ is the fractional frequency functions $f(R, t)$:

$$\left\{ \frac{\partial}{\partial R} [N_V(t)(1 - F(R, t))] \right\}_t = N_V(t) \left\{ \frac{\partial}{\partial R} [-F(R, t)] \right\}_t$$

$$= N_V(t)[-f(R, t)]$$

$$= -n_V(R, t) \qquad \text{(C-2)}$$

The operation on the right side of Eq. (C-2) is an application of Leibnitz' rule governing the derivative of a definite integral with respect to a variable contained in the limits of the integral. In the present application,

$$\left[ \frac{\partial}{\partial R} \int_0^{\tau(R, t)} \mathring{N}(T) \, dT \right]_t = \mathring{N}(\tau) \left( \frac{\partial \tau}{\partial R} \right)_t - \mathring{N}(0)(0) - \int_0^\tau \left[ \frac{\partial \mathring{N}(T)}{\partial R} \right]_t dR$$

$$= \mathring{N}(\tau) \left( \frac{\partial \tau}{\partial R} \right)_t \qquad \text{(C-3)}$$

It remains to evaluate $(\partial \tau / \partial R)_t$ in terms of the growth path envelope, as expressed analytically in Eq. (32). This may be accomplished by evaluating

$(\partial R/\partial \tau)_t$ in Eq. (32), i.e., by operating on both sides of Eq. (32) with $(\partial/\partial \tau)_t$. This is again an application of Leibnitz' rule, and gives

$$\left(\frac{\partial R}{\partial \tau}\right)_t = \left[\frac{\partial}{\partial \tau}\int_{\tau}^{t} G(R,t)\,dR\right]_t$$

$$= G(R,t)\left(\frac{\partial t}{\partial \tau}\right)_t - G[R,\tau(R,t)]\left(\frac{\partial \tau}{\partial \tau}\right)_t - \int_{\tau}^{t}\left[\frac{\partial G(R,t)}{\partial \tau}\right]_t dR$$

$$= -G[R,\tau(R,t)] \tag{C-4}$$

Combining Eq. (C-2–C-4) yields the equation governing size distribution evolution:

$$n_v(R,t) = \frac{\mathring{N}[\tau(R,t)]}{G[R,\tau(R,t)]} \tag{33}$$

The left side of this equation is the value of the number frequency function for particles of size $R$ at some time of observation $t$. The numerator of the right side is the value of the nucleation rate that was extant when these particles nucleated. The denominator is the value of the growth rate function $G(R,t)$ obtained by substituting the current size of the particles $R$ for the size parameter in the function, and their time of nucleation wherever in the function the time parameter appears.

## Appendix D: Impingement Equations in Nucleation and Growth Processes

### A. Uniform Impingement

The Johnson–Mehl–Avrami approach to describing discussed in Section VI, A, 1 leads to the basic phenomenological Eq. (41):

$$dV_V = (1-V_V)\,dV_{V\,ex} \tag{41}$$

This equation may be integrated to obtain a relation between the volume fractions of the real and extended structures:

$$\int_0^{V_V}\frac{dV_V}{1-V_V} = \int_0^{V_{V\,ex}}dV_{V\,ex}$$

$$-\ln(1-V_V)\,]_0^{V_V} = V_{V\,ex}]_0^{V_{V\,ex}}$$

$$\ln(1-V_V) = -V_{V\,ex}$$

$$V_V = 1 - \exp(-V_{V\,ex}) \tag{42}$$

Analogous equations for the other global properties may be obtained by applying the displacement equations derived in Appendix A to this equation. If both sides of Eq. (41) are divided by $dt$,

$$dV_V/dt = (1-V_V)(dV_{V\,ex}/dt)$$

Applying Eq. (11) to both derivatives gives

$$\bar{v}_s S_V = (1-V_V)\,\bar{v}_s(ex)\,S_{V\,ex}$$

If the volume elements in the extended structure are indeed uniformly distributed as was assumed in deriving Eq. (41), then it seems reasonable to assert that the growth rate averages $\bar{v}_s$ and $\bar{v}_s(ex)$ are equal, since one is a uniform sample of the other population. Thus

$$S_V = (1-V_V)\,S_{V\,ex} \tag{43}$$

which is Eq. (43) in the text. Taking the time derivative of this equation yields, in turn,

$$\frac{dS_V}{dt} = -\frac{dV_V}{dt}\,S_{V\,ex} + (1-V_V)\frac{dS_{V\,ex}}{dt}$$

The surface area derivatives may be evaluated by applying Eq. (12); the derivative of the volume fraction is another application of Eq. (11).

$$\bar{v}_H 2M_V = -\bar{v}_s S_V S_{V\,ex} + (1-V_V)\,\bar{v}_H(ex)\,2M_{V\,ex}$$

For the same reasons as outlined above, $\bar{v}_H$ and $\bar{v}_H(ex)$ may be assumed to be equal. Dividing by $2v_H$ gives

$$M_V = -\tfrac{1}{2}(\bar{v}_s/\bar{v}_H)\,S_V S_{V\,ex} + M_{V\,ex}(1-V_V)$$

Substituting for $S_V$ from Eq. (43), and noting that

$$\frac{\bar{v}_s}{\bar{v}_H} = \frac{\bar{v}_s(ex)}{\bar{v}_H(ex)} = 2\frac{\bar{H}(ex)}{\bar{H}_G(ex)}$$

yields

$$M_V = (1-V_V)[M_{V\,ex} - (\bar{H}(ex)/\bar{H}_G(ex))\,S_{V\,ex}^2] \tag{44}$$

Equations (42)–(44) form the basis for estimating properties of the real structure from a model for the microstructural process, under the assumption of "uniform impingement."

## B. Other Phenomenological Impingement Equations

A more general phenomenological description of impingement may be obtained by introducing an adjustable parameter into Eq. (41), as shown in Eq. (48):

$$dV_V = (1-V_V)^i\,dV_{V\,ex} \tag{48}$$

This equation may be integrated by separating the variables:

$$\int_0^{V_V} \frac{dV_V}{(1-V_V)^i} = \int_0^{V_{Vex}} dV_{Vex}$$

$$-(1-V_V)^{-i+1} \, ]_0^{V_V} = V_{Vex}]_0^{V_{Vex}}$$

$$(1-V_V)^{1-i} - 1 = -(1-i)V_{Vex}$$

$$V_V = 1 - [1-(1-i)V_{Vex}]^{(1/1-i)} \tag{49}$$

The global properties may be obtained for this impingement model in a manner analogous to that developed for uniform impingement. Dividing both sides of Eq. (48) by $dt$:

$$\frac{dV_V}{dt} = (1-V_V)^i \frac{dV_{Vex}}{dt}$$

Apply Eq. (11) to both sides:

$$\bar{v}_s S_V = (1-V_V)^i \bar{v}_s(ex) S_{Vex} \tag{D-1}$$

The assumption that the velocity averages are equal is no longer rigorously valid in this case, since it is not permissible to assert that $\bar{v}_s$ is the average of a uniform sample of interface velocities in the extended structure. However, it seems reasonable to assert that these averages will probably not be very different in most cases. With this assumption firmly in mind, Eq. (D-1) may be written:

$$S_V = (1-V_V)^i S_{Vex} \tag{50}$$

A relationship for the total curvature may be obtained by differentiating both sides of Eq. (50) with respect to $t$:

$$\frac{dS_V}{dt} = (1-V_V)^i \frac{dS_{Vex}}{dt} + i(1-V_V)^{(i-1)}\left[-\frac{dV_V}{dt}\right] S_{Vex}$$

Again inserting Eqs. (11) and (12), and assuming that the velocity averages are the same for the real and extended structures,

$$2M_V \bar{v}_H = (1-V_V)^i 2M_{Vex} \bar{v}_H(ex) - i(1-V_V)^{(i-1)}$$

$$\times [(1-V_V) S_{Vex} \bar{v}_s(ex)] S_{Vex}$$

$$M_V = (1-V_V)^i \{M_{Vex} - i[\bar{H}(ex)/\bar{H}_G(ex)] S_{Vex}^2\} \tag{51}$$

Equations (49) through (51) form the basis for modeling microstructural change with nonuniform impingement. The assumptions regarding the equivalence of the velocity averages for the real and extended structures should be carefully analyzed in any given application.

Acknowledgment

The author is indebted to the National Science Foundation, which supported the fundamental research upon which this paper is based.

# References

Aigeltinger, E. H., and DeHoff, R. T. (1971). *In* "Perspectives in Powder Metallurgy" (J. Hirschhorn, ed.), Wiley (Interscience), New York (in press).

Anderson, W. A., and Mehl, R. F. (1945). *Trans. AIME* **161**, 140.

Ardell, A. J. (1970). *Met. Trans.* **1**, 525.

Avrami, M. (1939). *J. Chem. Phys.* **7**, 1103.

Barrett, L. K., and Yust, C. K. (1969). ORNL Rep. No. 441.

Bierlein, T. K., Mastel, B., and Leggett, R. D. (1962). *Electron Microsc. Proc. Int. Congr. 5th, 1962* Vol. 1, Art. G-10.

Cahn, J. W. (1967). *Trans. AIME* **239**, 611.

DeHoff, R. T. (1962). *Trans. AIME* **224**, 474.

DeHoff, R. T. (1967). *Trans. AIME* **239**, 617.

DeHoff, R. T. (1968), *In* "Quantitative Microscopy" (R. T. DeHoff and F. N. Rhines, eds.), Chapter 10, p. 291. McGraw-Hill, New York.

DeHoff, R. T. (1971). "Dynamics of Microstructural Change," NSF Grant No. 16371, Annu. Rep.

DeHoff, R. T., and Bousquet, P. (1970). *J. Microsc.* **92**, Part 2, 119.

DeHoff, R. T., and Rhines, F. N. (1961). *Trans. AIME* **221**, 975.

DeHoff, R. T., and Rhines, F. N. (1968). *Eur. Symp. Powder Metallurgy, 2nd, 1968* p. 1.

Hillert, M. (1959). *Acta Met.* **7**, 653.

Hilliard, J. E. (1967). *In* "Stereology" (H. Elias, ed.), p. 195. Springer Publ., New York.

Hren, J. J., and Thomas, G. (1963). *Trans. AIME* **227**, 308.

Hsun Hu. (1963). *In* "Recovery and Recrystallization of Metals," p. 311. Wiley (Interscience), New York.

Hull, F. C., Coulton, R. A., Mehl, R. F. (1942). *Trans. AIME* **159**, 113.

Johnson, W. A., and Mehl, R. F. (1939). *Trans. AIME* **135**, 416.

Kolmogorov, A. N. (1937). *Bull. Acad. Sci. USSR Cl. Sci. Math. Natur.* No. 3.

Kronsbein, J., Buteau, L. J., and DeHoff, R. T. (1965). *Trans. AIME* **233,** 1961.

Pellesier, G. E., Hawkes, M. F., Johnson, W. A., and Mehl, R. F. (1942). *Trans. AIME* **30**, 1049.

Rhines, F. N., DeHoff, R. T., and Kronsbein, J. (1969). "A Topological Study of the Sintering Process," AEC Contr. No. AT-(40-1)-2581, Final Rep.

Speich, G. R., and Fisher, R. M. (1966). *In* "Recrystallization, Grain Growth and Textures," p. 563. Amer. Soc. Metals, Metals Park, Ohio.

Struik, D. J. (1950). "Lectures on Classical Differential Geometry." Addison-Wesley, Reading, Massachusetts.

Underwood, E. E. (1968). *In* "Quantitative Microscopy" (R. T. DeHoff and F. N. Rhines, eds.), Chapter 6, p. 149. McGraw-Hill, New York.

Vandermeer, R. A. (1970). *Met. Trans.* **1**, 819.

Vandermeer, R. A., and Gordon, P. (1959). *Trans. AIME*, **215**, 577.

Vedula, K. M., and Heckel, R. W. (1970). *Met. Trans.* **1**, 9.

# Studies in Chemical Vapor Deposition

R. W. HASKELL and J. G. BYRNE

*Division of Materials Science and Engineering*
*University of Utah*
*Salt Lake City, Utah*

## I. Introduction

The current article does not presume to review the immense and fast growing field of chemical vapor deposition (CVD); for that purpose the reader should refer to recent reviews such as by Holzl (1968), or the volume edited by Powell *et al.* (1966). Rather, the work to be described will be largely restricted to the author's own investigations in the areas of deposition kinetics, annealing behavior, and mechanical behavior of CVD tungsten. However, in order that this will not be completely foreign to those unacquainted with CVD, a very brief introduction to the general field will first be given.

CVD is a coating or deposition process which requires a chemical reaction at the surface of a heated substrate or deposition surface. This reaction feature distinguishes CVD from either physical vapor deposition, which is the condensation of elements or compounds from the vapor state, or from electrodeposition, in which individual ions discharge themselves on a conductive surface after transport through a liquid medium. All three molecular- or atomic-forming processes can produce deposits of controlled density, thickness, orientation, composition, or texture; however, CVD has the greatest throwing power or ability to deposit uniformly on relatively complex shapes and may be applied to the widest range of materials. A possible disadvantage is the usual requirement that the substrate be heated to relatively high temperatures.

Typically a volatile metal compound is reduced to form the desired metal species on the substrate. Obviously complicated shapes can be formed in this way. This is particularly advantageous when the metal involved lacks ductility and thus convenient formability by conventional means. It is generally true that metals which are readily electroplated are not those which are well suited to chemical vapor deposition and vice versa so that the two processes are complementary rather than competitive.

It is not always necessary to have a reducing agent in the feed gases to deposit a metal. Alternatively many metal bearing compounds may undergo a dissociation or pyrolysis reaction upon encountering a substrate. These compounds are usually organometallics or iodides. This is fairly convenient, however, difficulties exist in attaining high deposition rates. Another problem is attaining high purity levels with the organometallics because of carbon contamination.

The most common reduction reaction in CVD is the hydrogen reduction of volatile metal halides; for example $WF_6$. Such reactions are so common that CVD has sometimes been called halide metallurgy. An obvious problem here is the possible suceptibility of the feed system and mandrel or substrate material to halide corrosion. In the main, however, hydrogen reduction of volatile metal halides, especially refractory metal halides, has distinct advantages over either powder metallurgy, electrodeposition, or physical vapor deposition in terms of purity, stability, density, and ease of manufacture.

## II. Kinetics

### A. Introduction

Chemical vapor deposition involves many of the following processes: transport of the reactants to a substrate, adsorption, surface diffusion, reaction, and transport of the products from the susbstrate surface. These processes normally occur across gradients in temperature and composition. Also, during the course of these processes, intermediate and unknown species may be produced. For these reasons, the vapor deposition process is quite complex and is not usually amenable to exact analysis. Nevertheless, considerable effort has gone into analyzing CVD rate data. In the chemical vapor deposition of tungsten, which we shall be discussing, nearly all of the above considerations are important.

For the process of tungsten deposition, involving a $(WF_6, H_2)$ feed, we have an extremely versatile system. This process may be easily operated from any temperature between about 500° and 850° C, and at pressures between a few Torr and atmospheric. Typical data showing the variation of the rate with pressure, temperature, and flow rate, have been given by Holman and Huegel (1967a). It is generally found that the deposits deteriorate in quality with increasing rates so that practical deposition in this system is usually performed at rates less than 0.3 mils/min. Further reference to this data shows that such rates may be conveniently obtained at total pressures in the range of 300 to 500 Torr (using an aspirator) and at temperatures near 550° C. At pressures less than an atmosphere, measureable rates are not observable much below 500° C, hence, the cold gases $H_2$ and $WF_6$ may be brought together into the reaction chamber without having deposition on any surface except that which is intended. The only precaution in this is that the temperatures in the lines between the $WF_6$ tank and the reaction chamber do not drop below 19° C, otherwise $WF_6$ liquid may appear. Induction is a popular method for heating the substrate, for in this way, deposition is easily confined. Tube furnaces may also be used with deposition occurring on the inside of a tube. Product and unreacted $H_2$ and $WF_6$ gases are usually first brought through a scavenger furnace containing iron filings at about 850° C. This converts the remaining reactants to HF. The HF may then be caught in cold traps or passed over NaF pellets which react to form $NaHF_2$. (The NaF may be regenerated by heating the bifluoride to 300° C in a stream of nitrogen.)

## B. Transport Processes

Of the many processes operative in chemical vapor deposition, the one most studied and best understood is the diffusion of reactant and product gases to and from the substrate. In the deposition of tungsten by hydrogen reduction of the hexafluoride, we have a ternary gas phase system which is believed to satisfy the overall reaction:

$$WF_6 + 3H_2 = W + 6HF \tag{1}$$

Brecher (1970) has solved the set of Stefan–Maxwell multicomponent diffusion equations for this system. His solutions for the vapor phase mole fractions at the vapor–substrate interface, may be written in the following form:

$$X_1^s = X_1^0 + A_2(e^{r+\lambda} - 1) + A_3(e^{r+\lambda} - 1)$$
$$X_2^s = X_2^0 + B_2(e^{r+\lambda} - 1) + B_3(e^{r-\lambda} - 1) \tag{2}$$

where the subscripts $(1, 2, 3)$ refer to the species $(WF_6, H_2, HF)$ and the superscripts $(0, S)$ refer to the bulk and surface concentrations, respectively. The quantities $r_+$ and $r_-$ are the roots of the characteristic equation:

$$[r - (1 - 3R)][r - (3 - 5R_1)] - 3(R_1 - 1)(R - 1) = 0$$

where $R$ and $R_1$ are the diffusivity ratios $(D_{12}/D_{23})$ and $(D_{12}/D_{13})$, respectively. The quantity $\lambda$ is the ratio of the flux $(J)$ to the mass transport coefficient $(k_m)$. It is the latter quantity which is a measure of the flow properties of the system. From kinetic theory considerations, Brecher (1970) has determined that $R = 0.50$ and $R_1 = 3.00$. The coefficients $(A_2, A_3, B_2, B_3)$ are functions of $(R_1, R, X_1^0, X_2^0)$ and these equations may be found in Brecher's paper. The utility of considering solely diffusion in the gas phase is that one, in principle, may determine the maximum possible rate for the process for a given set of conditions. This maximum rate of deposition is determined when one or both of the surface concentrations $(X_1^s, X_2^s)$ vanish. This condition gives the curves shown in Fig. 1.

In practice, it is most difficult to determine the mass transport coefficient $k_m$, and we may use Fig. 1 and our input compositions and flux for a first estimate; however, a difficulty arises in flowing systems. Strictly speaking, Brecher's analysis applies to a system where the properties of the boundary layer do not change significantly over the substrate. Of course, in a flowing system, the boundary layer thickness changes with distance along the substrate and if the deposition efficiency is high, one may also have to account for a change in the bulk concentrations of the reactants over the substrate surface. In this case, the analysis becomes much more complicated, but one may still regard the representations in Fig. 1 as applying to a small region of the substrate.

If we consider the reaction of Eq. (1) under the conditions of low temperatures and high flow rates (i.e., large $k_m$), then $X_1^s \to X_1^0$ and $X_2^s \to X_2^0$. Taking an empirical reaction rate to be of the form:

$$J = k_r(X_1^s)^m(X_2^s)^n = J_k(X_1^s/X_1^0)^m(X_2^s/X_2^0)^n \tag{3}$$

we see that under these conditions $J \to J_K$, where $J_K(T, X_1^0, X_2^0)$ represents the maximum kinetic flux for these temperature and composition conditions.

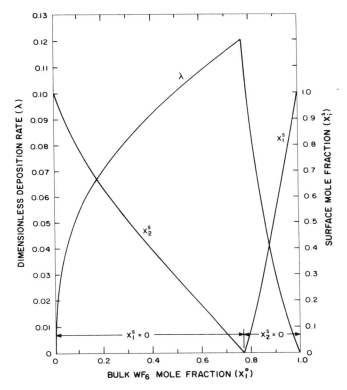

***Fig. 1.*** Plot of deposition rate and surface concentration versus bulk concentration.

Now as $T$ increases, $J_K$ increases through the forward reaction rate constant $k_r$, whereas the ratios $(X_1{}^s/X_1{}^0)$ and $(X_2{}^s/X_2{}^0)$ decrease since the ratio $(\lambda = J/k_m)$ is becoming larger in Eq. (2). Eventually a temperature is approached where one or the other of the surface concentrations becomes vanishingly small. An example of this process taken from our own data is shown in the data of Fig. 2. The temperature and flux values where the plot changes slope is an indication of the change in the rate controlling mechanism. Below this point, the principal resistance is due to some, or several, of the kinetic processes mentioned above. Above this point, the principal resistance is due to transport. Returning to Fig. 1, we may estimate an overall mass transport coefficient for the process by finding the $(\lambda = J/k_m)$ ratio corresponding to the bulk composition and solving for $k_m$ knowing $J$ (in the transport controlled region).

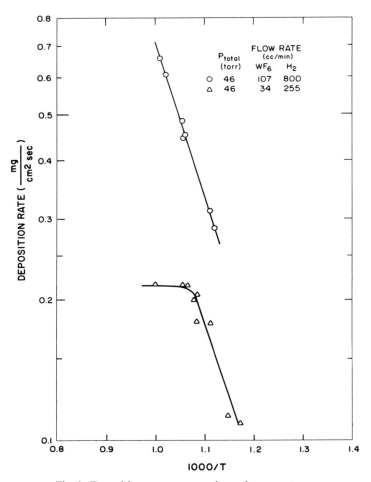

**Fig. 2.** Deposition rate versus reciprocal temperature.

## C. Reaction and Growth Mechanisms

Holman and Huegel (1967a) found for the region $J \to J_K$ that the apparent activation energy (based on radius change with time) was 16 kcal/mole. In our laboratory, we have found an activation energy of 15 kcal/mole (based on weight change versus time). Brecher (1970), in his analysis of the data of Berkeley et al. (1967), for a static system, used Eqs. (2) and (3) and from the temperature dependence of $k_r$ determined an activation energy of 27 kcal/mole. He was also able to determine the values of $m$ and $n$ to be $1$ and $3$, respectively. Berkeley obtained his rate data from short-time diameter change measurements in a static system. While Brecher was able to match this data quite well in the mass transport control region, the temperature variation of the rate is less reliable due to the scatter in Berkeley's data at the lower temperatures. It should also be mentioned that the activation energies measured by Holman and Huegel and at this laboratory must represent a lower limit since even the slope in the low temperature region contains a contribution due to mass transport.

This meager amount of information is representative of the state of our knowledge of most CVD systems. As is evident from the above, some estimates concerning mass transport rates may be given, but little information is available concerning kinetic parameters.

Going beyond the empirical rate expression of Eq. (3), we may consider what happens when the molecules $H_2$ and $WF_6$ encounter a tungsten substrate. Undoubtedly, the initial adsorption reaction is quite fast, so it may be inferred that this process is not rate controlling. Furthermore, the microstructure of the deposit is quite pertinent. The fact that long columnar grains are found in the growth direction implies that nucleation is negligible [Holman and Huegel (1967a) have enhanced nucleation by brushing the growing deposit], and there must be significant surface diffusion to certain sites where the

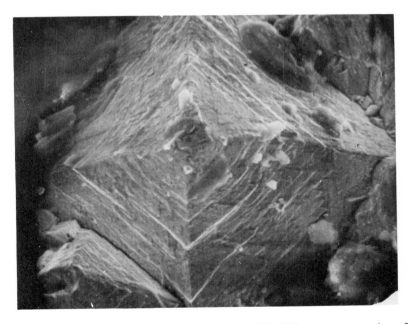

***Fig. 3.*** Scanning electron micrograph of a commercial CVD tungsten growth surface (855×). From Auck and Byrne (1971).

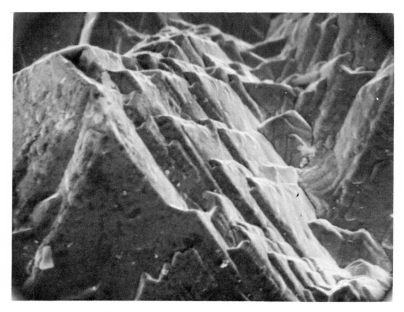

***Fig. 4.*** Scanning electron micrograph of growth ledges for CVD tungsten deposited at 500° C (900×). From Shim and Byrne (1971).

tungsten is incorporated into the lattice. Figure 3 is a scanning electron micrograph of the top of a growing columnar grain. This pyramidal structure is similar to the structures reported in the electrodeposition of copper single crystals (Sard and Weil, 1970). Figure 4 and Fig. 5a and b represent deposition at 500° and 700° C, respectively. It appears that the pyramids developed at 700° C have steeper sides and fewer facets than those developed at 500° C. Holman suggested that the pyramid faces in CVD tungsten (produced from an $H_2$–$WF_6$ feed) are {111}. This would agree with our earlier pole figure results which indicated a fiber axis which was largely ⟨100⟩. It is of interest to note that gases are generally preferentially adsorbed on {111} faces on tungsten (Haywood and Trapnell, 1964).

There appear to be many layers of steps on the pyramid faces. We believe these steps are nucleated at the grain boundaries of the columnar grains and then proceed to grow up the pyramid faces. Figure 6 is a two-dimensional schematic of this process. It should be noted that any proposed growth mechanism must preserve this columnar structure. There is evidently a matching problem when steps from different pyramid faces on the same columnar grain approach the top of the pyramid, and we notice that the tops of many grains contain small cavities as shown in Fig. 4.

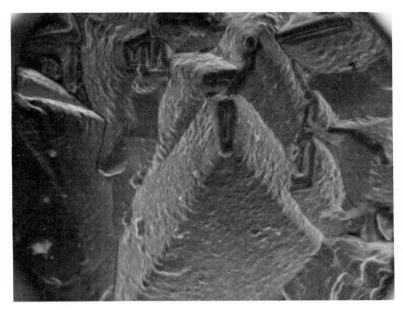

*Fig. 5.* Scanning electron micrograph representing deposition at 700° C (900×). From Shim and Byrne (1971).

Because of the significant surface diffusion involved in this picture of the growth process it seems unlikely that the $WF_6$ molecule undergoes complete dissociation or is stripped of all its flourine before reaching a growth step. If this did occur then atomic tungsten would have to diffuse to the growing steps. The activation energy for this process is about 70 kcal/mole (Barbour *et al.*, 1960) so we conclude that tungsten migrates to the growing steps as a fluoride molecule. The activation energy for this process may be of the order

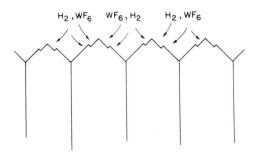

*Fig. 6.* Schematic diagram of growth process. From Auck and Byrne (1971).

of the heat of evaporation for $WF_6$ which is 6.35 kcal/mole (Kubaschewski and Evans, 1958). It is of interest to note that the activation energy for surface diffusion of hydrogen on tungsten is about 5 kcal/mole and that of oxygen is about 25 kcal/mole (Haywood and Trapnell, 1964). Oxygen is mentioned since it will be found that the heat of adsorption of fluorine is probably about the same as the heat of adsorption of oxygen, so as a first approximation we might take their activation energies for surface diffusion on tungsten to be the same.

When the tungsten fluoride complex reaches the growth step it must be reduced and then incorporated into the lattice. A pertinent observation here made in our laboratory is that if the CVD process is run with a $H_2:WF_6$ feed ratio of 1:1 instead of the usual ratio of 7:1 the rate is greatly reduced, as would be expected. If the ratio is then returned to 7:1 we observe that the system does not return to the increased plating rate. Apparently, we are observing a poisoning of the surface by fluorine. Evidently the pyramid surfaces and the regions of the growth steps become saturated with fluorine, which is strongly bound, and this makes it difficult for the hydrogen to become adsorbed and to take part in the reduction reaction. Indeed, fluorine films on tungsten have been reported to be stable up to 2000° C (Metlay and Kimball, 1948). A measure of the relative stabilities of fluorine and hydrogen films may be obtained by considering their heats of chemisorption on tungsten. The heat of chemisorption of fluorine may be estimated by assuming the chemisorption process to be equivalent to the gaseous reaction

$$(1/3) \, W_{(g)} + F_2 = (1/3) \, WF_6 \qquad (4)$$

As observed by Haywood and Trapnell (1964) this should be a good approximation provided the chemisorption process does not significantly perturb the metal–metal bonds of the substrate (note, the metal must be considered in the gaseous state). Taking the heat of sublimation of W to be 183 kcal/mole (Kubaschewski and Evans, 1958) and the heat of formation of $WF_6$ to be $-419$ kcal/mole (JANAF, 1960 to date), the heat of chemisorption is found to be 200 kcal/mole. This is about the same value as that measured for oxygen, and in the case of oxygen the calculated value (based on $WO_3$ formation) agrees quite well with the measured value. For hydrogen, the heat of chemisorption on a bare surface is found to be about 45 kcal/mole. As the surface coverage of hydrogen increases the heat of chemisorption decreases (Haywood and Trapnell, 1964). For hydrogen this decrease can amount to 27 kcal/mole for a completely covered surface, so the effect is quite significant. The high heat of chemisorption and the high activation energy for surface diffusion indicate that the processes involving fluorine govern the kinetics. Thus the rate-limiting steps may involve surface diffusion and/or reactions occurring at a growing step.

Haskell (1970) has considered the rate-controlling step to be the desorption reaction between adsorbed hydrogen and fluorine considered to be in equilibrium with the gaseous reactants $H_2$ and $WF_6$. The activation energy for this reaction is approximately one sixth of the enthalpy change for Eq. (1), i.e., one sixth of 30 kcal/mole (JANAF, 1960 to present) or 5 kcal/mole. If HF were desorbed as a dimer, the activation energy would be 10 kcal, which is closer to the activation energy observed for flowing systems. It is more important to note, however, that this proposed mechanism requires $m$ and $n$ in Eq. (3) to be $\frac{1}{6}$ and $\frac{1}{2}$, respectively. As noted earlier, Brecher (1970) found values of 1 and 3 for these variables. Both pairs of these values have the same ratio and imply that a maximum kinetic rate would occur at $X_1{}^0 = 0.25$. (It should be noted how this contrasts with the maximum rate for mass transport control shown in Fig. 1, where $X_1{}^0 = 0.75$.) In flowing systems, however, a maximum rate appears closer to a ratio of $7:1$ implying a maximum kinetic rate at $X_1{}^0 = 0.125$. This might be explained by hydrogen reacting as molecular hydrogen and fluorine reacting as atomic fluorine. It is known that several layers of adsorbed hydrogen can develop above the atomic, chemisorbed layer. Hence, the possibility of molecular hydrogen taking part in the reaction cannot be ruled out.

## D. Alloying and Thermodynamic Considerations

Tungsten–rhenium is the only alloying system which has received much attention in CVD tungsten work. This CVD system was first studied by Federer and Steele (1965) and later by Holman and Huegel (1967b). These latter workers made a careful study of the effect of the variables of pressure, temperature, gas composition, and flow rate on the deposit composition, microstructure, and mechanical properties. It should be appreciated that the production of alloys by chemical vapor deposition is beset with many problems. One of the most difficult is to find feed gases that will plate at similar temperatures. Furthermore, it is difficult to know what effect the metal bearing feed gases will have on one another's plating characteristics. Fortunately, there is a compound $ReF_6$ having properties very similar to $WF_6$ and the plating of W–Re alloys may be performed under the same temperature and pressure conditions as for pure tungsten plating. For this system it is of interest to consider the thermodynamics of the displacement reaction

$$ReF_6 + W = WF_6 + Re$$

At equilibrium, assuming the vapor to be ideal, we have for the relative ratios of Re to W in the deposit and in the vapor the relation

$$\left(\frac{X_{Re}}{X_W}\right) = \left(\frac{\gamma_W}{\gamma_{Re}}\right)(e^{-\Delta F^\circ/RT})\left(\frac{X^v_{ReF_6}}{X^v_{WF_6}}\right) \qquad (5)$$

where $X$ and $X^v$ refer to the mole fractions in the deposit and in the vapor, respectively, and $\gamma_{Re}$ and $\gamma_W$ refer to the activity coefficients of Re and W in the deposit, respectively. Also, $\Delta F^\circ \sim -40$ kcal/mole (Kubaschewski and Evans, 1958) is the standard free energy change for the displacement reaction. From these considerations Re would be expected to be found greatly enhanced in the deposited material as compared to the vapor. Indeed, the displacement reaction is so strongly directed to the right that one might expect that nearly each $ReF_6$ molecule adsorbed on the substrate is reduced by W. Of course this entire discussion is based on the assumption that the kinetics are favorable for such a reaction. The results of Holman and Huegel (1970) substantiate these thermodynamic considerations. They find that the amount of Re deposited is solely dependent on the $ReF_6$ flow for the conditions of their ex-

*Fig. 7.* Dislocation segments in tungsten foil parallel to tube surface.

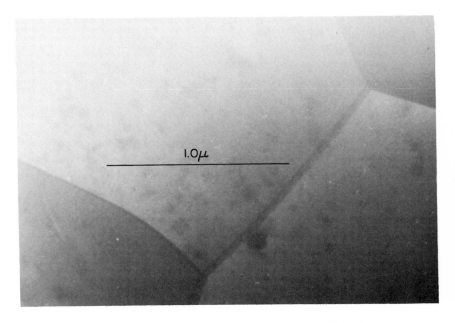

***Fig. 8.*** Transmission electron micrograph of tungsten–22% rhenium alloy.

periment. Such considerations as are made here should be important in considering other alloying systems.

Although the microstructure and properties of the tungsten–rhenium alloys are superior to those of pure tungsten, the amount of rhenium needed to refine the pure tungsten structure and the high cost of rhenium makes it desirable to find other alloying elements that might yield comparable properties.

## III. Structural and Annealing Studies

### A. Transmission Electron Microscopy

Only preliminary transmission electron microscopic observations are available to date, however, some interesting structures have been observed. The following electron micrographs pertain to material from San Fernando Laboratories. Figure 7 shows some dislocation segments in CVD tungsten. The plane of the foil is parallel to a cylindrical tube surface and should be approximately {100}. A number of electron micrographs were taken of the tungsten–22% rhenium alloy. Most triple point grain boundary intersections are at angles very close to the equilibrium 120° value. Very few dislocation images and no obvious segregation effects are seen. Figure 8 is typical of the appearance.

### B. Textures

An X-ray inverse pole figure technique (Barrett and Massalski, 1966) was employed to measure preferred orientations in each of the longitudinal, circumferential, and outside sections of CVD tubes. These specimens were spark cut, electropolished, and mounted together to form composite samples of each section as illustrated in Fig. 9. Zr filtered Mo $K_\alpha$ radiation was used

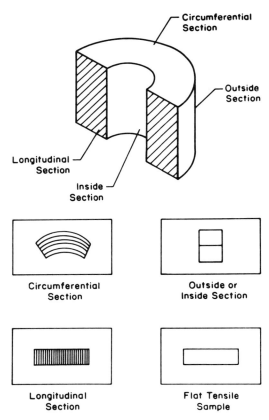

**Fig. 9.** Schematic illustration of method of specimen mounting for X-ray inverse pole figure studies of various tube sections. From Chun *et al.* (1971a).

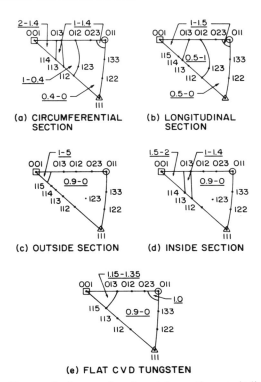

**Fig. 10.** Inverse X-ray pole figures of various tube sections as indicated. From Chun et al. (1971a).

for three scans of each composite sample. Each sample was moved about $\frac{1}{4}$ inch in the specimen holder between scans to allow the total sample surface to be covered by the X-ray beam.

Texture coefficients [T.C.$_{(hkl)}$], were obtained by substituting the measured integrated intensity of a given $(hkl)$ reflection into the following equation:

$$\text{T.C.}_{(hkl)} = \frac{I_{(hkl)}}{I^{\circ}_{(hkl)}} \Bigg/ \frac{1}{n} \sum_{0}^{n} \frac{I_{(hkk)}}{I^{\circ}_{(hkl)}} \tag{6}$$

where $I$ = measured integrated intensity of given $hkl$ reflection; $I^{\circ}$ = theoretical calculated intensity for the same $hkl$ reflection; $n$ = total number of reflections measured; and T.C. = texture coefficient or pole density for a given reflection, The T.C. values corresponding to the various diffracting planes can be plotted on a stereographic projection. The inverse pole figures obtained are shown in Fig. 10. Contour lines are drawn on the stereographic triangles to include regions of similar T.C. values.

The texture of a $\frac{7}{8}$ inch diameter tube was first considered. This is shown in Fig. 10a–d. It can be seen that the grains in the circumferential, longitudinal, outside, and inside sections contain mainly {100} planes and that the fiber axis of the tube is ⟨100⟩.

Inverse pole figures were also determined for flat pieces of CVD tungsten of the type used in the tensile tests to be reported in the next section. Figure 10e is the inverse pole figure describing this material. It is generally similar to the tubular material.

In Sections IV, D, and V an interesting consequence of textural changes due to substrate temperature changes will be discussed.

## C. Grain Growth

In discussing grain growth of typical CVD tungsten, one is confronted with the problem of characterizing changes in an array of linearly oriented columnar grains. We chose to first examine the inner and outer tube surfaces, i.e., to follow changes in the end views of the grains during annealing (Auck and Byrne, 1970). More recently, we have studied changes in columnar grain width when observing sections cut perpendicular to the tube axis.

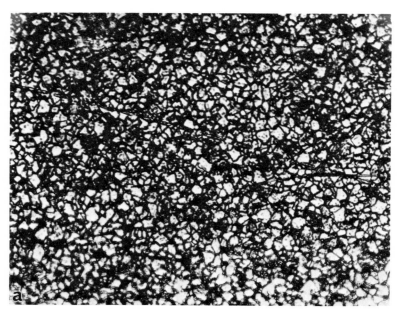

*Fig. 11.* [pp. 308–310]. Microstructure of commerical CVD tungsten. (a) inside section of tube (I.D.), 284×; (b) end of tube, close to I.D., 284×; (c) end of tube, close to O.D., 284×; (d) outside section of tube, (O.D.), 284×.

*Fig. 11* (b). For caption, see p. 308.

*Fig. 11* (c). For caption, see p. 308.

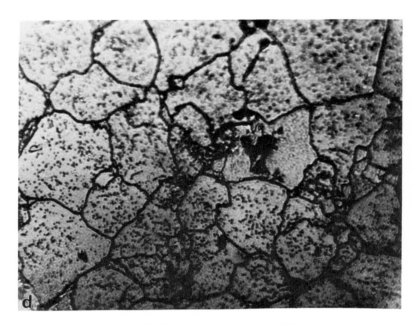

**Fig. 11** (d). For caption, see p. 308.

Figures 11a–d show the as-received microstructures of the inside tube surface, the tube end view close to the inside surface, the tube end view close to the outside surface, and the outside tube surface, respectively. To avoid the overly wide grain size distributions given by lineal analysis, we recently introduced a more realistic estimating method which we call cross-counting (Auck and Byrne, 1970). It consists of drawing two orthogonal axes through the approximate center of each of a large number of grains of all sizes. When the orthogonal grain diameters are measured, one obtains a much sharper grain size distribution than would result from lineal analysis. This is because there are no longer small apparent contributions such as a random line gives when it subtends the "corner" of a large grain.

Figure 12a is a plot showing the change in average grain diameter for 30 hour anneals at temperatures between 1300° and 2200° C, for both the inside and outside tube surfaces. Figure 12b shows that for 60 hour anneals the sharp rise begins at approximately 1700° C rather than 2000° C as in the previous figure. These data were plotted according to the relation

$$D^2 - D_0^2 = kt \qquad (7)$$

where $D_0$ is the initial grain size, $D$ is the instantaneous grain size, $k$ is a constant, and $t$ is the annealing time. It was found that data from outer surfaces gave negative slopes between 1290° and 1520° C and positive slopes between 1520° and 1870° C. This at first suggested that some nucleation-type phenomenon such as recrystallization had occurred; however, no obvious metallographic evidence of this was present. In addition, no significant X-ray line broadening was present in the initial material. The absence of microstrains as seen by X-ray precluded the existence of much stored energy which could have provided a driving force for recrystallization. Other arguments against the occurrence of recrystallization were the fact that new small grains were never found in a deformed matrix but rather among equiaxed grains; had recrystallization taken place, we would not have found the persistence of columnar grains in tube end view sections.

We feel that the negative slopes are explained by the existence of submicroscopic grains or nuclei, created during deposition, which simply began to grow at the first opportunity, i.e., during the lower temperature anneals. They then began to be detected but because they were so much smaller than the surrounding matrix grains, they caused the average grain size to decrease sharply. The microstructure of the tube end finally did loose its columnar

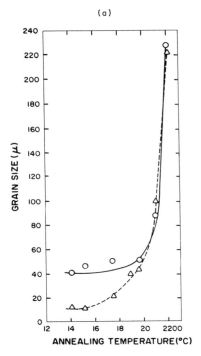

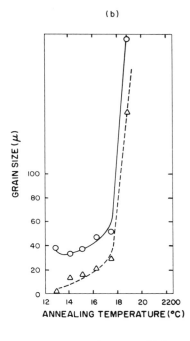

*Fig. 12.* Grain size changes for inner ($\triangle$) and outer (O) tube surfaces. (a) 30 hour anneals; (b) 60 hour anneals. From Auck and Byrne (1970).

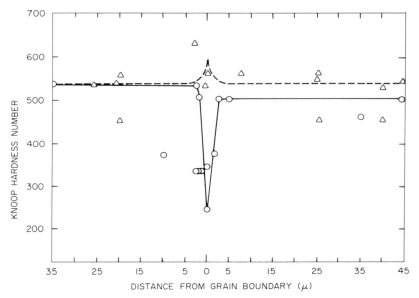

**Fig. 13.** Microhardness study of grain boundary after quenching (O) and after slow cooling (△), each from 2100° C. From Auck and Byrne (1971).

character after a 15 minute anneal at 3000° C. A more equiaxed structure developed, probably due to the growth of the small nuclei described above.

The slopes $k$ relating $D^2 - D_0^2$ and annealing time at $t$ equals 1 hour were used in an Arrhenius plot of the equation

$$k = k_0 \exp(-Q/RT) \tag{8}$$

The Arrhenius slope was used to calculate $Q$, the apparent activation energy for grain growth of 58.8 kcal/mole.

Some interesting effects of cooling rate on grain boundary hardness were noted in these same experiments. Figure 13 illustrates these effects for a specimen annealed at 2100° C for 13 hours. When the specimen was rapidly quenched the microhardness values followed the lower curve, indicating a boundary softer than the surrounding matrix. When a slow cool to room temperature (over a period of 12 hours) was used, the microhardness values gave the upper curve, which could either mean a slight boundary hardening or at least no boundary softening.

A possible explanation of the above results is the frequently reported formation of voids or bubbles on or near grain boundaries (Farrell *et al.*, 1967) especially if the specimen is under tensile stress (McCoy and Steigler, 1967).

We have observed large boundary cavities after anneals in the neighborhood of 2200° C (no applied stress). Whether these are due to fluorine or to vacancy condensation is not established; however, it seems reasonable that a quench would be more likely to preserve such voids at a boundary than would a 12 hour cool. In the latter, voids might be eliminated prior to microhardness testing. It is interesting to note that for the lower temperature anneals which gave negatively sloped log–log plots of $D^2 - D_0{}^2$ versus $kt$, no grain boundary softening effects were found, but rather slight increases of about 10% were found. This could be associated with the fact that the initial vacancy concentrations prior to quenching would have been much lower than in tungsten annealed at 2100° C. If fluorine is the source of these effects, it could merely be that fluorine bubbles do not segregate to or form at grain boundaries sufficiently to affect hardness, except during anneals at 2100° C or above. The need for additional work in this area is obvious.

Similar microhardness studies were performed on tungsten–22% rhenium alloy tube specimens quenched both from 2780° and 1750° C. For both temperatures quenching produced a slight grain boundary hardening rather than a softening. The reason may be that here solute atoms tie up most of the vacancies at the annealing temperature, preventing them from reaching the grain boundaries. Those vacancies which do reach boundaries may be accompanied by rhenium atoms, which could account for the slight hardening effect observed.

Optical microscopy of polished and etched outer surfaces of a tungsten–22% rhenium alloy tube revealed patterns which resemble the "chevron" patterns as reported by others (Holman and Huegel, 1967b). The current patterns were more like concentric pentagons than chevrons and had a spacing of approximately $1\mu$ between lamellae, as will be seen later in Fig. 18. Electron microprobe traces across such surfaces gave some indication of rhenium rich regions, spaced apart by approximately the distance which is observed metallographically. Rhenium $L_\alpha$ radiation and a LiF analyzing crystal were used. No corresponding dips in the tungsten probe profile were found when using tungsten $M_\alpha$ radiation and an ADP analyzing crystal. Since these first probe scans utilized a $1\mu$ beam diameter, the resolution was poor. Future attempts should be made on a probe with a finer electron beam; however, based on the etching behavior and the microprobe results it seems reasonable to believe that some segregation does exist. It will be recalled from Fig. 8 that our transmission electron microscope results on this material showed no evidence of solute segregation. This is not surprising, since even if such segregation does exist, TEM should not detect it in the present case since tungsten and rhenium, being adjacent elements in the periodic table and of about the same atom size, should not produce microstrains and consequent diffraction contrast when segregated.

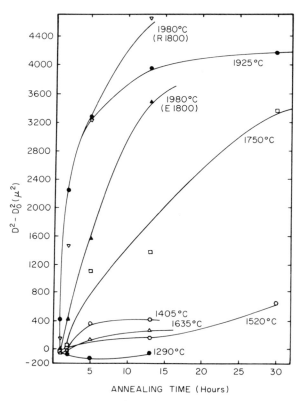

**Fig. 14.** Grain growth in terms of $D^2 - D_0^2$ versus annealing time for a tungsten–22% rhenium alloy between 1290° and 1980° C. From Auck and Byrne (1971).

In Fig. 14, for the tungsten–rhenium alloy, notice is called to the difference between the upper and lower curves for anneals at 1980° C. The data for the lower of the two was obtained after thermal grooving, and the data for the higher of the two was obtained after subsequent electropolishing and etching of the same surfaces. The difference is probably caused by the restriction of grain growth by external surfaces. Figure 15 shows tungsten–rhenium alloy grain growth in a higher temperature range. Finally Fig. 16 shows some additional pure tungsten grain growth data. It is of interest to note that if one compares Fig. 14 and 16 for an anneal of 30 hours at 1750° C, one finds a change in $D^2 - D_0^2$ in the tungsten–22% rhenium alloy of nearly twice that found in the tungsten (curve marked $\triangle$). This at first seems an anomaly. By way of explanation let us first point out that the apparent activation energy for grain growth in the alloy has been found to be 61.7 kcal/mole (Auck and

Byrne, 1971). This is not very different from the earlier mentioned value of 59 kcal/mole for CVD tungsten. It would then seem that something other than the presence of rhenium atoms is responsible. The remaining and obvious difference is that the grain morphology of the alloy is equiaxed, whereas that of the tungsten is columnar. This suggests that equiaxed grains can grow faster than columnar grains even when the former exist in the presence of a high concentration of solute atoms. This point will be pursued further when we consider lateral growth observations of columnar grains in Fig. 19.

First, however, some additional points should be mentioned in reference to Fig. 14. In that figure the 1290° C curve for the alloy shows an actual negative deviation. This can only mean that at 1290° C grains appear at short annealing times which are in fact smaller than the initial grain size. In an earlier report (Auck and Byrne, 1970) plots of log $(D^2 - D_0^2)$ versus log time

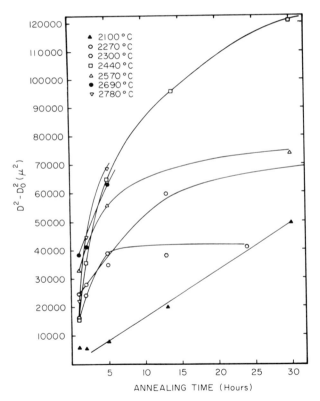

**Fig. 15.** Grain growth in terms of $D^2 - D_0^2$ versus annealing time for a tungsten–22% rhenium alloy between 2100° and 2780° C.

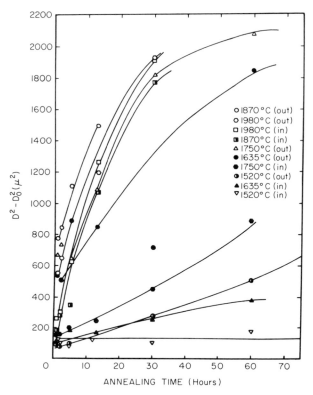

**Fig. 16.** Grain growth in terms of $D^2 - D_0^2$ versus annealing time for tungsten between 1520° and 1980° C.

showed negative slopes for the lower annealing temperatures for CVD tungsten. In those cases the negative slopes meant only that new grains were being counted which were smaller than their neighbors *at* the annealing temperature but not smaller than the original grains since such a negative difference could not appear on a log scale. Thus, the recrystallization possibility is perhaps not as remote in the alloy material as it was in the tungsten. If it occurs in the alloy, it may be connected with the nucleation of a new phase. If we compare Fig. 17 which represents the alloy after a 60 hour anneal at 1290° C (the temperature with the negative deviation in Fig. 14), with Fig. 18, which represents the as-received structure of the alloy, we see that indeed the grain size *is* smaller after annealing.

Let us return now to the question of the lateral growth of columnar grains. Isothermal annealing experiments at ten temperatures between 1290° and

***Fig. 17.*** Tungsten–22% rhenium microstructure representing 60 hours at 1290° C, 450×.

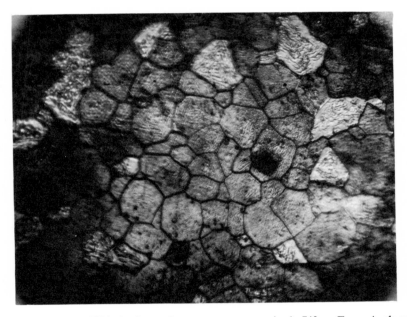

***Fig. 18.*** Tungsten–22% rhenium microstructure, as received, 743×. From Auck and Byrne (1970).

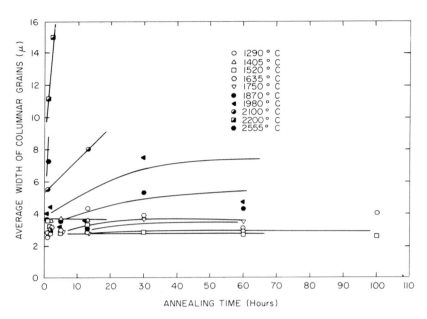

**Fig. 19.** Average columnar grain width as a function of annealing time for tungsten at various annealing temperatures between 1290° and 2555° C. From Auck and Byrne (1971).

2555° C were performed on tungsten tube end sections to examine the motion of the long, relatively straight columnar grain boundaries. Figure 19 is a plot of average columnar grain width versus annealing time. It is evident that very little boundary motion occurs at temperatures below 1870° C. Above that temperature the growth rate increases rapidly. Values on the axis of the ordinates cannot be compared directly to earlier $D^2 - D_0{}^2$ plots for end views of the same columnar grains. There is considerable inaccuracy inherent in defining the width of a columnar grain since first, these grains do not always go all the way through the tube thickness and second, they vary in width from inner to outer diameter of a tube. The small changes observed in comparison with end views do suggest, however, as did the earlier mentioned comparison at 1750° C between tungsten (columnar) and alloy (equiaxed) growth rates, that columnar boundaries do in fact migrate more slowly than the boundaries of more equiaxed grains.

---

**Fig. 20.** Schematic diagrams of (upper) high temperature hoop stress testing apparatus assembly and (lower) components of high temperature hoop stress testing apparatus. From Chun *et al.* (1971a).

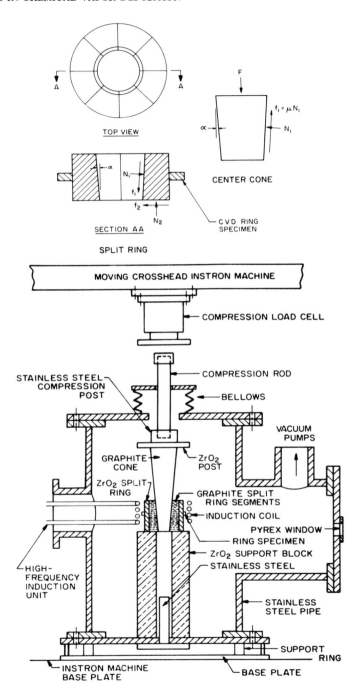

TOP VIEW

CENTER CONE

$f_1 = \mu N_1$

$N_1$

SECTION AA

CVD RING SPECIMEN

SPLIT RING

MOVING CROSSHEAD INSTRON MACHINE

COMPRESSION LOAD CELL

COMPRESSION ROD

STAINLESS STEEL COMPRESSION POST

BELLOWS

VACUUM PUMPS

GRAPHITE CONE

$ZrO_2$ POST

$ZrO_2$ SPLIT RING

GRAPHITE SPLIT RING SEGMENTS

INDUCTION COIL

PYREX WINDOW

RING SPECIMEN

$ZrO_2$ SUPPORT BLOCK

STAINLESS STEEL

HIGH-FREQUENCY INDUCTION UNIT

STAINLESS STEEL PIPE

SUPPORT RING

INSTRON MACHINE BASE PLATE

BASE PLATE

## IV Mechanical Behavior

### A. Testing Apparatus

Figure 20a shows a schematic of the high temperature hoop stress apparatus. A downward force is applied to a tapered graphite cone which transmits force to eight equal segments that press out uniformly on the inner surface of the ring specimen, putting the ring into tension. These components are shown in more detail in Fig. 20b. Figure 21 shows a schematic of the high temperature tensile test apparatus. The transformation from hoop testing to tensile testing requires only that the support block and graphite cone be replaced by pull rods and specimen grips.

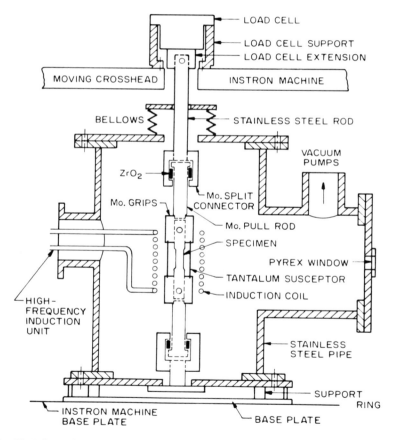

**Fig. 21.** Schematic diagram of high temperature tensile testing apparatus. From Chun *et al.* (1971a).

For descriptive purposes the apparatus is divided into the following parts: furnace chamber and heating system, load train, and temperature measurement.

The furnace chamber is basically a stainless steel pipe with thick stainless steel flanges. A stainless steel plate is sealed with an O-ring to the top flange of the chamber. Motion of the upper pull (or compression) rod is achieved through a stainless steel bellows and pull rod. The bottom flange of the chamber is O-ring sealed to a large stainless steel plate. A stainless steel pull rod is welded to the bottom plate as shown in Fig. 21. Two flanged openings are provided on opposite sides of the heating chamber. A Pyrex window facilitates observation (for temperature measurement with an optical pyrometer) as well as vacuum and/or gas atmosphere introduction. The other flanged opening admits the induction coil. Water-cooled copper tubing is wound around the outside and on the top of the furnace chamber.

A tantalum susceptor, heated by induction, radiates heat to the specimen. The power supply is a 20 kW Lepel induction generator. A 4-turn copper tubing coil with an inner diameter of 3 inches and a length of 1 inch was used for tensile specimens. The seamless tantalum susceptor had a sighting hole to permit specimen temperature measurement by an optical pyrometer.

The entire hoop stress load train assembly is shown in Fig. 20a. The top of the stainless steel compression rod is threaded to the main compression rod. A $ZrO_2$ post is placed between the end of the compression rod and the top of the graphite cone to provide thermal insulation. Eight equal split ring graphite segments of a split $ZrO_2$ ring are placed between the graphite segments and the CVD ring specimen to prevent any possible carburization reaction between the graphite and the CVD tungsten. A $ZrO_2$ tube served as a support for the two split rings, the specimen, and the cone.

The entire tensile test load train assembly is illustrated in Fig. 21. The top and bottom stainless steel pull rods are threaded into molybdenum alloy couplings. The latter are linked to molybdenum alloy pin grips.

CVD tungsten tubes of 1.5 inches I.D. and 0.050 inch wall thickness and flat 0.040 inch thick strips of CVD tungsten were purchased from the San Fernando Laboratories of Fansteel Metallurgical Company. The deposition conditions and chemical composition are presented in Tables I and II.

Ring specimens $\frac{1}{4}$ inch high were cut using a lathe tool post grinder with a 0.020 inch thick silicon carbide cutoff wheel operating at approximately 7000 rpm. The CVD tungsten tube was held in a soft collet. Gage length and pin holes for flat tensile specimens were spark cut. Any slight damage to the specimen caused by spark cutting (and tool post cutting) was removed by electropolishing (20 gr NaOH + 980 ml distilled water). The electropolishing produced a reduction of thickness of approximately 0.002 inches.

With the load train and specimen installed, the furnace chamber was evacuated, helium gas was admitted to the chamber, the specimen was slowly

**TABLE I**

DEPOSITION CONDITIONS OF COMMERCIAL CVD TUNGSTEN

| | |
|---|---|
| Gas Flow (cc/min) | |
| WF$_6$ | 950 |
| H$_2$ | 3800 |
| Vacuum | $\frac{1}{2}$ Atm. |
| Preheated gas temperature | 675° C in Argon |
| Deposition time | Approx. 3 hours |
| Substrate | 304 Stainless Steel |

**TABLE II**

CHEMICAL ANALYSIS OF COMMERCIAL CVD TUNGSTEN

| Element | ppm |
|---|---|
| Carbon | 6.0 |
| Fluorine | 2.0 |
| Hydrogen | 0.3 |
| Oxygen | 1.0 |
| Nitrogen | < 1.0 |
| Aluminum | 4.0 |
| Copper | N/D[a] 3.5 |
| Iron | 10.0 |
| Nickel | N/D[a] 15.5 |
| Silicon | 10.0 |

[a] N/D means not detected at unit of sensitivity shown.

heated to and then held at the test temperature for a period of five minutes to establish thermal equilibrium. Hoop stress and tensile tests were performed with an Instron tensile machine at a cross-head speed of 0.01 inch/min.

## B. Derivation of the Hoop Stress Equation

If a downward force $F$ is applied to the graphite cone (see Fig. 20b) the conditions for equilibrium require:

$$F = A_1 N_1 (\sin \alpha + \mu_1 \cos \alpha) \qquad (9)$$

where $A_1$ is the area of the tapered surface in contact with the split graphite ring, $N_1$ is the normal pressure on the tapered surface of the cone, $\alpha$ is the angle of taper, and $\mu_1$ is the friction coefficient between the graphite cone and the graphite split ring.

Considering the split ring; the conditions for equilibrium require

$$A_2 N_2 = A_1 N_1 (\sin \alpha + \mu_1 \cos \alpha) \tag{10}$$

where $A_2$ is the total area of the bottom of the split ring and $N_2$ is the normal pressure on that area. Similarly, equilibrium requires that for the split ring, in the radial direction

$$A_r P + A_2 N_2 \mu_2 = A_1 N_1 \cos \alpha - A_1 N_1 \mu_1 \sin \alpha \tag{11}$$

where $A_r$ is the inner area of the ring specimen, $P$ is the pressure on the inside of the ring, and $\mu_2$ is the friction coefficient between the split ring and the support block.

Dividing Eq. (11) by Eq. (10), using the fact that from Eqs. (9) and (10) $F = A_2 N_2$, and assuming that $\mu_1$ equals $\mu_2$[†], $P$ is expressed by

$$P = \frac{F}{A_r} \left[ \frac{(1 - \mu^2) \cos \alpha - 2\mu \sin \alpha}{\sin \alpha + \mu \cos \alpha} \right] \tag{12}$$

If Eq. (12) is substituted into the hoop stress equation:

$$\tau_{\text{hoop}} = Pr/t \tag{13}$$

where $\tau$ is the circumferential stress, $r$ is the internal radius of the ring specimen, and $t$ is the ring thickness, then the hoop stress is expressed by

$$\tau_{\text{hoop}} = \frac{F[(1 - \mu^2) \cos \alpha - 2\mu \sin \alpha]}{2\pi h t [\sin \alpha + \cos \alpha]} \tag{14}$$

$h$ being the height of the ring. Equation (14) thus has two unknowns, $\tau_{\text{hoop}}$ and $\mu$ the friction coefficient. The friction coefficient can be obtained from the following relation, assuming the tensile yield stress equals the value of the hoop stress at yielding.

$$\mu = \frac{ \begin{array}{l} -[(\sigma/k) \cos \alpha + 2 \sin \alpha] \\ \quad + \{[2 \sin \alpha + (\sigma/k) \cos \alpha]^2 - 4 \cos \alpha [(\sigma/k) \sin \alpha - \cos \alpha]\}^{1/2} \end{array} }{2 \cos \alpha} \tag{15}$$

where $k$ is $F/2ht$ and $\sigma$ is the yield stress, separately obtained from a tensile test. It can be shown that the strain is expressed by:

$$\varepsilon = (C_f - C_i)/C_i = \Delta r/r_i$$

$$\varepsilon = \frac{(\text{vertical distance moved by cone}) \tan \alpha}{r_i} \tag{16}$$

[†]This assumption will be unnecessary in future work where the support block will also be graphite and the $ZrO_2$ split ring will not be in contact with the latter.

where $C_i$ is the initial circumference, $C_f$ is the final circumference, and $r_i$ is the initial radius of the ring specimen.

It may be helpful to point out that there is an upper bound on the thickness which appears in Eq. (13). Timoshenko (Timoshenko and Goodier, 1951) showed an equation for the circumferential stress $\sigma_\theta$ in an open-ended cylinder of inner radius $a$ and outer radius $b$, subjected to internal pressure $p_i$:

$$\sigma_\theta = \frac{a^2 p_i}{b^2 - a^2}\left(1 - \frac{b^2}{r^2}\right) \tag{17}$$

This stress is a maximum on the inner surface; however, if $b - 1.1a$, $\sigma_{\theta max} > \sigma_{\theta min}$ by only 10.5%. Thus, there is a small error if we assume $\sigma_\theta$ to be uniformly distributed over the thickness and given by

$$\sigma_\theta = p_i a/b - a \tag{18}$$

which is the same as our hoop stress equation (13). In the rings we have been testing $a$ is approximately 0.75 inch and $t$ is approximately 0.050 inch; therefore, $b/a = 1.07$, which is less than Timoshenko's recommended upper bound of 1.1 for the safe assumption of a uniform stress distribution.

It might be argued that friction between the segments and the inside of the test ring should produce a frictional stress which would lower the yield stress of the ring. It is known first, that such friction must be static if in fact it exists, because no relative motion is found between the segments and the ring. Second, a consideration of the sign of such a frictional stress in the ring shows that it would be compressive and maximum on the inner surface of the ring. Recall that Timoshenko's $\sigma_\theta$ (Timoshenko and Goodier, 1951) was, in the general case, a maximum on the inner surface and tensile in nature. Thus, even if such a friction exists at this location, its effect would be to make $\sigma_\theta$ even more uniform than envisioned by Timoshenko for the case of frictionless internal pressure on the interior of a thin walled cyliner. We have also considered the possibility of stress differences between supported and unsupported ring sections (Chun et al., 1971a). This effect does not influence Eq. (15).

The relation between hoop stress and tensile stress was first studied at room temperature for the metals aluminum, copper, and brass. The tensile and hoop stress specimens were produced from the same plate for each metal. The apparatus components in this preliminary room temperature prototype were of stainless steel. The results of tensile tests for these metals were self-consistant and reproducible. The average value of 0.2% offset yield stress for brass was 48 ksi, for aluminum 38 ksi, and for copper 17 ksi. These values were substituted into Eq. (15) to obtain the friction coefficients for hoop stress measurements at room temperature. The results of the effective friction coefficient were as follows: for copper it varied from 0.17 to 0.19, for aluminum from 0.12 to 0.14, and for brass from 0.11 to 0.12. Thus, each metal produced a very consistent and reproducible effective friction coefficient value. We do not think that this implies that the friction at room temperature is affected by the particular materials involved in the segment–metal ring pair, but rather that the effective friction coefficient of the apparatus is force sensitive. At any temperature the major sliding is between the cone and segments: for the prototype this would be stainless steel versus itself. The friction coefficient $\mu$ can sometimes be a function of normal force, contrary to our usual concept of this coefficient. For example, it has been shown (Bowden and Tabor, 1964) that the friction coefficient of diamond decreases as the load increases, approximately following a law $\mu = kW^{-\frac{1}{3}}$, where $W$ is the load. In the current case, $\mu$ dropped from the range of 0.06–0.19 to that of 0.104–0.12 as the yield stress (indicative of normal force on cone) increased from that of copper (17–18 ksi) to that of brass (46–50 ksi). If one uses the upper values from each range of $\mu$ and each range of yield stress and calculates the constant in the previous relation, a difference in $k$ values of only 12.7% is found which suggests that the stainless steel behavior is similar to diamond in the force sensitivity of its friction coefficient.

Since these results indicated the feasibility of hoop stress measurement at room temperature, the technique was extended to temperatures up to 1600° C for CVD tungsten.


### C. High Temperature Results

Grain size and texture differences between the flat and thin-wall tube CVD tungsten were carefully noted since these could give mechanical property differences. For flat material the average grain diameter at the first deposited surface was approximately $6.8 \times 10^{-4}$ inches and for tubular material the first deposited surface grain diameter was approximately $3.1 \times 10^{-4}$ inches. The tube outside surface grain diameter was almost the same as the last deposited surface of the flat specimens. Texture coefficient values for the various diffracting planes, T.C.$_{(hkl)}$, were described in Section III, B.

Figure 22 shows the temperature dependence of the 0.2% offset yield stress and the ultimate tensile stress of flat specimens. The temperature dependence of the elongation is shown in Fig. 23. The yield stress of 66 ksi at 1400° C and the elongation of 30% at 1400° C are in good agreement with Taylor's results (Taylor and Boone, 1964).

The 0.2% tensile yield stress values were substituted into Eq. (15) to obtain the friction coefficient $\mu$ for hoop stress tests at each temperature. The results are summarized in Table III. Figure 24 shows that $\mu$ decreases linearly with increasing temperature. Table IV shows that the strains calculated by Eq. (16) give very good agreement with the actual measurements.

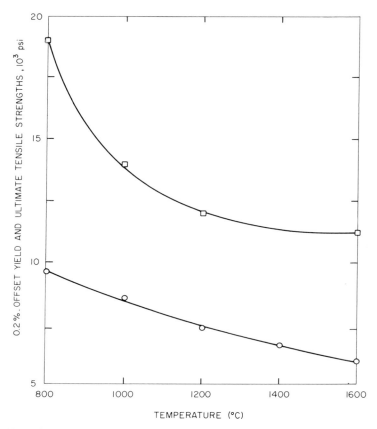

**Fig. 22.** Ultimate tensile strength (□) and yield strength (O) versus temperature for CVD tungsten. From Chun *et al.* (1971b).

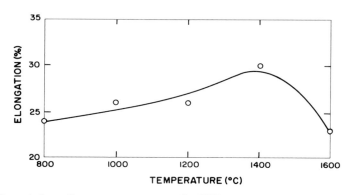

**Fig. 23.** Percent elongation versus temperature for CVD tungsten. From Chun *et al.* 1971b).

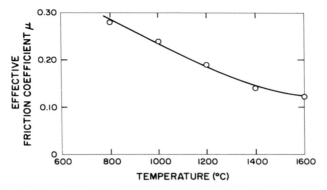

**Fig. 24.** Effective friction coefficient $\mu$ versus test temperature. From Chun *et al.* (1971a).

**TABLE III**

HOOP STRESS DATA

| Temperature (°C) | $\sigma$ (Yield stress ksi) tensile tests | $F$(lbs) | $K = F/2\pi ht$ | $\sigma/K$ | $\mu$ |
|---|---|---|---|---|---|
| 800  | 9.6 | 315 | 4013 | 2.39 | 0.28 |
| 1000 | 8.5 | 240 | 3058 | 2.78 | 0.24 |
| 1200 | 7.3 | 170 | 2166 | 3.37 | 0.19 |
| 1400 | 6.6 | 120 | 1529 | 4.32 | 0.14 |
| 1600 | 5.9 | 88  | 1276 | 4.62 | 0.12 |

**TABLE IV**

COMPARISON OF MEASURED AND CALCULATED VALUES OF HOOP STRAIN

| Temperature (°C) | Total measured strain (%) | Cross-head travel of instron (inches) | Strain calculated from Eq. (16) (%) |
|---|---|---|---|
| 800  | 3.7 | 0.31 | 3.6 |
| 1000 | 2.4 | 0.19 | 2.2 |
| 1200 | 4.4 | 0.38 | 4.4 |
| 1400 | 4.4 | 0.38 | 4.4 |
| 1600 | 4.0 | 0.35 | 4.0 |

Scanning electron microscopy has been applied to both ring and tensile samples on both the fracture surfaces and the electropolished specimen sides and edges where slip lines can be observed. In both ring and tensile specimens, the long direction of the columnar grains is through the thinnest dimension of the specimen. Thus, in both types of test the tensile direction is at 90° to the long dimension of the grains. Specimen side views give the end view and specimen edge views give the side view of the columnar grains. Two sequences of SEM photos are now shown to illustrate the kind of differences in fracture behavior, which rapidly become evident with this observation technique. The first sequence of SEM photos in Fig. 25 typifies the fairly flat and brittle

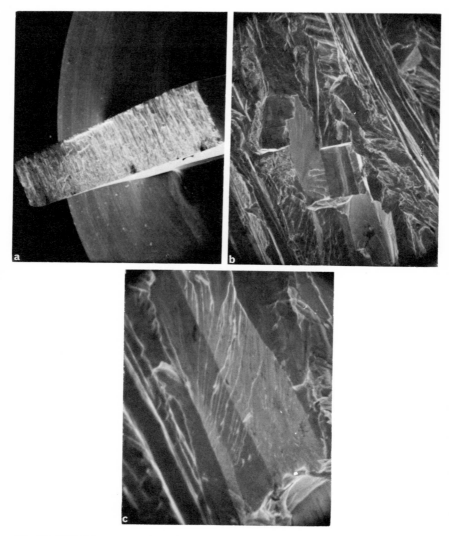

**Fig. 25.** 800° C fracture surface of a CVD tungsten ring observed with scanning electron microscopy: (a) 12×, (b) 120×, (c) 1200×. From Chun *et al.* (1971b).

fracture surface of a tungsten ring at 800° C. At 20 × one first sees the overall
flatness, then at 1000 × very brittle, transgranular fractures of long columnar
grains are seen in the fracture face. Zooming-in to 2000 × on this region reveals
the well-known "river patterns" on the exposed columnar grain surfaces.
These patterns are found on brittle cleavage surfaces.

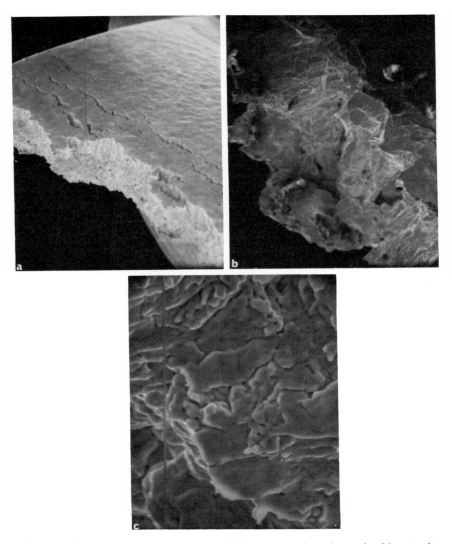

*Fig. 26.* 1400° C fracture surface of a CVD tungsten ring observed with scanning
electron microscopy: (a) 12 ×, (b) 120 ×, (c) 3100 ×. From Chun *et al.* (1971b).

Now consider the second sequence of SEM photos of Fig. 26. These represent a ring fracture at 1400° C. Note that at 20 × the fracture surface is much more fibrous than the 20 × picture for 800° C. This is a gross indication of greater ductility. Note also the many subsidiary side cracks which do not occur at 800° C. As we increase the magnification to 200 × and finally to 5000 ×, the ductile nature of the fracture becomes even more evident in comparison to the 800° C photos.

Between 1400° and 800° C the slip band spacing has no particular dependence on temperature, as can be seen in Table V. It may be interesting to view some other SEM photos which illustrate various features of the specimens tested in the above temperature range.

TABLE V

SCANNING ELECTRON MICROSCOPE OBSERVATIONS

| | Ring test | | | | Tension test | | | |
|---|---|---|---|---|---|---|---|---|
| | 1400°C | 1200°C | 1000°C | 800°C | 1400°C | 1200°C | 1000°C | 800°C |
| Slip band spacing (microns) | | | | | | | | |
| grain ends | [a] | [a] | 10 | [a] | 2–6 | [a] | 4–5 | 8–16 |
| grain sides | 2–17 | 4 | 2–6 | 2–16 | 1–7 | 3–22 | 2–5 | 3–7 |
| Subsidiary cracks (intergranular) | Many | Many | Some | None | Many | Many | Some | None |
| Description of fracture surface | Fibrous with voids | Less fibrous than 1400°C | Slip and river patterns across columnar grains | River patterns, very brittle trans- cryst. cleavages of columnar grains | Fibrous with voids | Less fibrous than 1400°C | Quite fibrous with voids | Some fibrosity |

[a] Not visible.

Figure 27 applies to 1400° C only. Figure 27a shows a side view of a ring near the fracture region. Note the exposed columnar grain with long dimension lying in the fracture surface. This has a similar appearance to the side view of a tensile sample tested at this temperature. Figure 27b shows slip bands in adjoining grains of a tensile sample edge view. Note that steps are produced along the grain boundary subsidiary crack at those positions where slip bands reach the crack.

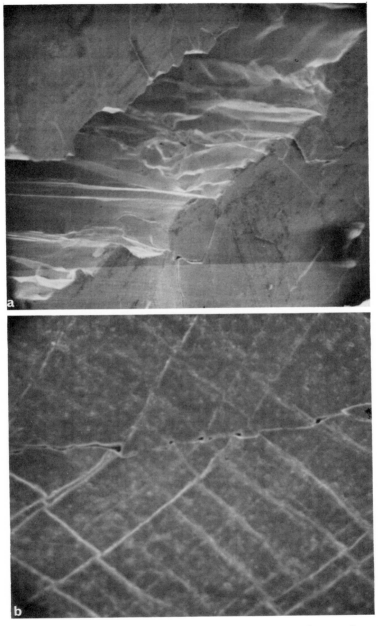

*Fig. 27.* CVD tungsten tested at 1400° C: (a) side of ring near fracture face, 400 × ; (b) slip bands in adjoining grains on edge of tensile sample, 1600 × . From Chun *et al.* (1971b).

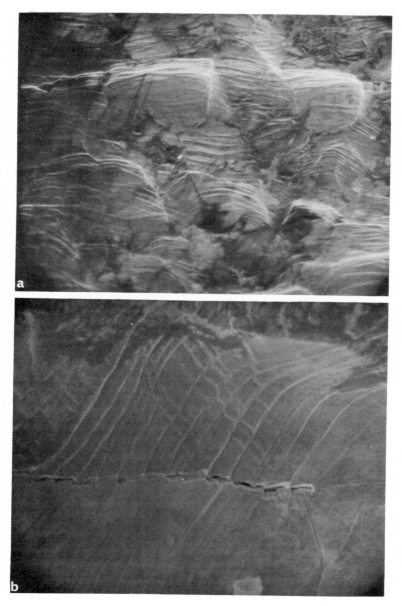

*Fig. 28.* CVD tungsten tested at 800° C: (a) tensile specimen side view showing slip bands on ends of columnar grains, 192×; (b) edge view of same specimen as in 15(a), 1120×.

Figure 28 represents a tensile specimen tested at 800° C. Figure 28a is a side view showing typical coarse slip bands on columnar grain ends. Figure 28b shows an edge view of the same specimen. The slip bands are finer than at 1400° C, and also form ledges where they reach the subsidiary (grain boundary) crack. It was normally observed (see Table V) that slip was finer on the sides than on the ends of the columnar grains. Another general observation is that only small deviations are found in the directions of slip bands from grain to grain. This merely suggests that a strong crystallographic fiber axis exists. The same conclusion has more recently been reached (Shim and Byrne, 1971) based on the appearance of cleavage faces in which river patterns were studied as they crossed columnar grain boundaries (see following section).

### D. Substrate Temperature Effects

On the premise that more defects are retained during chemical vapor deposition at lower substrate temperatures than at high temperatures (Farrell *et al.*, 1970), we have investigated the effect upon mechanical behavior of producing CVD tungsten at 500° and 700° C substrate temperatures. Hoop stress tests were performed on rings spark cut from tubes at each of these temperatures. These tests were conducted at temperatures of 240°, 520°, 800°, and 1000° C. We have found that at each test temperature the material deposited at 500° C has a yield stress higher by approximately 3 ksi than does the material deposited at 700° C. Both curves descend linearly in strength with increasing test temperature; the 500° C deposited material starting at 12.8 ksi and reaching about 11.2 ksi by 1000° C. The difference in kind, number, and arrangement of defects has not yet been examined, but soon will be with transmission electron microscopy. Perhaps an even more important "defect" in this regard is that at 500° C there is considerable "fanning out" of columnar grains from the substrate side as compared with 700° C at which temperature these grains run comparatively straight through the wall thickness of the deposit.

It has very recently been found (Shim, 1971) that the X-ray texture for 500° C deposition is fairly random in the first 0.010 inches from the substrate, but at greater distances it goes over to $\langle 100 \rangle$. For 700° C deposits the $\langle 100 \rangle$ fiber axis is strong even in the first 0.010 inches next to the substrate.

Our usual deposition pressure at both temperatures has been 500 Torr, however, one tube was also prepared at each temperature with a pressure of 300 Torr. This pressure difference produced virtually no change in the yield strength values at any test temperature.

Microhardness traverses have been run radially, through the tube thickness, on surfaces cut normal to the tube axis for each substrate temperature. It is found that the material produced at the 500° C substrate temperature is the harder of the two in agreement with the yield stress results. It is also interesting to note that each material has a microhardness which is highest at the substrate side and decreases gradually with distance toward the tube outer diameter. One must recall here that our deposition (unlike those of San Fernando Labs) is done on the inner surface of a copper substrate tube. Thus, when the copper tube is etched away, our finer grained surface is on the outer diameter of the tungsten tubes and the tube outer diameters are harder than the inner diameters. When the hoop stress test is performed, it is interesting to consider how crack propagation occurrs.

Scanning electron microscopy was used to examine these fracture surfaces. The direction of propagation is from inner to outer diameter. This is on the basis of river patterns which were observed to run approximately vertically, crossing columnar grain boundaries. The latter are evidently of low angle, tilt-twist type. The cracks thus appear to originate near the inner diameter at an edge, propagate vertically in the inner diameter region in a cleavage mode and then reach the outer diameter via a quasi-cleavage mode as the finer grained material is reached.

### E. Modulus Measurement

The dynamic Young's modulus was determined for CVD tungsten. Longitudinal sound waves were passed down a polycrystalline tungsten specimen from a quartz crystal in a direction perpendicular to the long axis of the columnar grains. The specimen was a long square prism, spark cut from a tube produced by San Fernando Laboratories. The substrate temperature had been 670° C; the other deposition conditions are given in Table I. This specimen was electropolished prior to measurement.

The length of the specimen was gradually decreased through the range in which a maximum voltage reasonance (1.843 inches and 49.857 kHz) was noted. The density was determined from a water displacement experiment and found to be 19.144 gms/cc. With this value of density Young's modulus was calculated from the well-known relation (Fine, 1952)

$$E = 4\rho f^2 l^2 \tag{19}$$

in which $l$ and $f$ are the values of length and frequency at maximum voltage resonance, respectively, and $\rho$ is the density. A value of $E = 60.4$ ksi resulted.

### F. Residual Stress

The experimental problem in measuring the surface residual stress of a CVD material by X-ray diffraction is complicated by the fact that as-deposited CVD surfaces are quite irregular. Any rough surface has projections which are not stressed in the same way as the bulk material and yet these projections contribute most strongly to diffraction. In tungsten this problem is aggravated by the high linear adsorption coefficient of tungsten for the most common X-ray wavelengths. Initially, we made numerous two-angle X-ray diffractometer residual stress measurements according to the well-known S.A.E. procedure (S.A.E., TR-182). The results on a given tube were found to fluctuate widely, sometimes by as much as 50 ksi, not only for the circumferential stress at various locations along a tube but even for locations around a given circumference.

In order to clarify the situation, a device was built which would permit a direct stress *versus* X-ray peak-shift calibration to be established. The peak shift referred to is that which is found only in the presence of residual stress when the angle of incidence of the incoming X-ray beam is varied.

A $\frac{3}{4}$ inch high ring of CVD tungsten was spark cut from a 1.5 inch diameter tube produced at San Fernando Laboratories. A longitudinal slot, $\frac{1}{32}$ inch wide, was then cut through one side of this ring to permit it to be flexed by the applied force. The wall thickness was $\frac{1}{16}$ inch. Strain gages were attached to the inner and outer ring surfaces at a location diametrically opposed to the slot. A correlation was determined between inner and outer surface strain readings for various values of applied force. By knowing the strain at the outer diameter for a given measured strain at the inner diameter we were able to remove the outer strain gage, focus the X-ray beam at that location, and then establish a correlation of applied strain and X-ray peak shift as the angle of incidence was changed for each of a series of applied strains. The stressing device permitted the ring specimen to be refocussed in the X-ray beam as its shape changed with each successive application of external force. In order to establish the instrumental error in the absence of any residual stress, a specimen of stress free tungsten powder was used. The (400) tungsten peak was used with copper radiation.

Let us consider the various contributions to the total long-range elastic stress $\sigma_T$ in a specimen. There could be an applied stress $\sigma_A$, a residual stress $\sigma_R$, and an instrumental stress or instrumental error $\sigma_I$. The theory (Cullity, 1956) shows that the stress measured by the two exposure diffractometer method is proportional to the difference in Bragg angle $\Delta 2\theta$ upon changing the angle of incidence by some angle $\psi$. Thus,

$$\sigma = K\Delta 2\theta \qquad (20)$$

Ordinarily $K$ is equal to

$$\frac{E \cot \theta}{2(1+v)\sin^2 \psi} \qquad (20)$$

where $E$ is Young's modulus, $\theta$ is the Bragg angle, and $v$ is Poisson's ratio. It is sometimes assumed that values of $E$ and $\nu$, measured in the ordinary way during a tensile test, may be used to calculate the value of $K$. Cullity (1956) points out several reasons why it is preferable to experimentally determine $K$, as has been done here, rather than calculate it.

Going back to the various components of stress, we can write for a given specimen

$$\sigma_{\text{total}} = \sigma_{\text{applied}} + \sigma_{\text{instrumental}} + \sigma_{\text{residual}} \qquad (21)$$

or

$$\sigma_T = K\Delta2\theta_A + \sigma_I + \sigma_R \qquad (22)$$

In practice the applied strain $\varepsilon_A$ was determined from a strain gage and converted to applied stress $\sigma_A$ through Hooke's law. These data were plotted against $\Delta2\theta_A$ as shown in Fig. 29. A number of two exposure measurements on stress free tungsten powder gave $\overline{\Delta2\theta_0}$, the instrumental error or instrumental stress found even for a stress free sample. In this case $\overline{\Delta2\theta_0}$ was positive, thus the original $\sigma_T$ curve of Fig. 29 is translated to the right by $\overline{\Delta2\theta_0}$ to make the instrumental correction. The value of the ordinate at $\Delta2\theta = 0$ for the latter curve is the value of the residual stress in the calibration ring. For simplicity we could further translate the latter curve to the origin to remove the effect of the residual stress. We then would have a true representation of $\sigma_A = K\Delta2\theta_A$ and the slope $K$ may be directly applied to subsequently measured values of $\Delta2\theta$ from other CVD tungsten tubes. The only distinction then will be that we will be measuring a $\sigma_{\text{residual}}$ since for an unconstrained CVD tube $\sigma_A$ is zero.

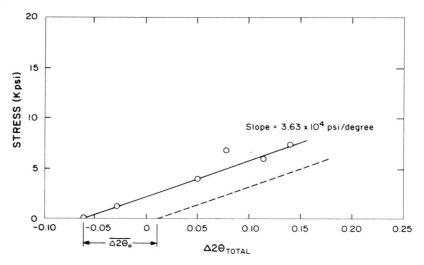

**Fig. 29.** Plot of stress versus change in Bragg angle for a tungsten calibration ring. ——Original total stress curve, – – – instrumentally corrected curve.

The calibration ring used had to be electropolished before reproducible data could be obtained. The slope of the lines in Fig. 29 is 36.3 ksi/degree. This value will be used on subsequent X-ray measurements of electropolished CVD tungsten rings. It must be recalled, however, that this $K$ will only be applicable to measurements made on the tungsten (400) peak with Cu$K\alpha$ radiation and $\psi$ values of 0° and 45°.

## V. Concluding Remarks

In the preceding pages a summary is presented of the current state of mainly our research on CVD tungsten. It would be appropriate to define some problems the solutions of which are felt to be most necessary for further progress in this field.

In the kinetics area there is an urgent need to better define the products of the reaction chamber. Undoubtedly in a flowing system the products are much more complex than indicated by the reaction shown in Eq. (1). A mass spectrometer study of these products would be very useful not only for those interested in tungsten deposition, but also for the fluorine chemist. Such a study might shed light on what happens when a $WF_6$ molecule encounters a tungsten substrate. Indeed this is an important question in many CVD systems; that is, what happens when a metal (or non-metal) halide encounters its own surface. A reference which may be pertinent here concerns adsorption studies made by Kemball (1950) who found that when $CCl_4$ was adsorbed on a Hg surface that a second-order chemisorption reaction took place. Apparently chlorine was being split out to form $C_2Cl_6$. Kemball stated that some higher mers might also be forming by the same process. Certainly adsorption studies of $WF_6$ on W and no doubt similar studies on other CVD systems could provide much basic information.

Concerning tungsten deposition, it is interesting to note the great differences in the effects, of fluorine and chlorine.

(1) Tungsten deposition using $WCl_6$ takes place at about 1000° C. The presence of hydrogen has little effect, and the reaction is apparently controlled by the pyrolysis of $WCl_6$. The microstructure is more refined and the orientation is such that {100} facets are observed (Glaski, 1970).

(2) Berkeley *et al.* (1967) found that additions of HCl to the $WF_6$–$H_2$ feed were far more inhibiting than HF additions. Indeed Brecher (1970) concluded that HF merely acted as a diluent in these experiments whereas HCl probably acted as a poison.

Perhaps a more practical need is to develop tungsten-base alloys other than the tungsten–rhenium system. As previously discussed, pure CVD tungsten has a very poor microstructure and is generally quite brittle at room temperature. Holman and Huegel (1967b) have shown how brushing during deposition greatly improves tungsten properties and also how alloying with rhenium is a great improvement. However, the addition via CVD of other alloying elements has not yet been studied.

Grain growth in CVD materials is intimately connected with the development of bubbles and their subsequent interaction with migrating grain boundaries. We have only noticed that for higher temperature anneals, plots of $D^2 - D_0^2$ versus annealing time decrease drastically in slope after several hours at temperature. Questions such as what the bubbles actually contain (usually thought to be $F_2$, but not as yet really proven) and how the bubbles affect the activation energy and the mechanism for grain boundary migration are at present unanswered. It has been shown (Willertz and Shewmon, 1970) that the nucleation of steps on the faceted surface of bubbles can satisfactorily account for observed bubble diffusion in gold and copper where the bubbles contain inert helium (due to irradiation). In the case of tungsten containing fluorine the solubility picture is quite different since fluorine does have a good affinity for tungsten. Further analysis and experimentation is planned in this area.

In the area of mechanical behavior perhaps the most interesting effect noted so far is the fact that the yield strength is raised considerably by lowering the substrate or deposition temperature. The increased nucleation rate at the lower temperature causes an almost random texture within 0.010 inches of the substrate. This has only just recently been proved by X-ray textural studies as a function of distance from the substrate for different deposition temperatures (Shim, 1971). At 500° C the randomness gives way within 0.010 to 0.020 inches to a strong $\langle 100 \rangle$ fiber axis. At 700° C the $\langle 100 \rangle$ texture is evident even within the first 0.010 inches from the substrate. The detailed mechanisms by which the more random texture gives rise to higher strength remain to be worked out, however, a crude analogy would be to compare the similarity of the disoriented material to a polycrystal, on the one hand, and the similarity of the oriented material with a single crystal, on the other, since only small misorientations are involved across the columnar boundaries of the latter material. Thus in the spirit of this analogy, the oriented (low temperature substrate) material would be stronger.

The high temperature hoop stress testing apparatus developed in this work could be used in quite another area, namely in the determination of high temperature friction cofficient data. This could be done by using the materials of interest in the cone and inner segment portions on the apparatus.

Finally it seems clear that although much work remains to be done on the fundamentals of CVD kinetics, annealing behavior and mechanical behavior, the times also demand more rapid application of what fundamentals are known, as well as much direct development work of a practical nature.

ACKNOWLEDGMENTS

This research has been sponsored by the Advanced Research Projects Agency. A number of co-workers have generously contributed data in advance of publication: Dr. John S. Chun on mechanical behavior, X-ray textures and scanning electron microscopy; H. S. Shim on substrate effects on mechanical behavior, textures, and morphology; Y. T. Auck on annealing behavior and transmission electron microscopy; H. Huggins on kinetics; J. S. Lo on residual stress, modulus measurements, and kinetics. Drs. P. S. Nicholson, A. Sosin, and J. S. Chun contributed to the design of the mechanical testing apparatus. The authors are very grateful to Drs. W. Holman and F. Huegel of Lawrence Radiation Laboratories, Dr. L. Brecher of Westinghouse, and F. Glaski of San Fernando Laboratories for many helpful discussions over the past two years.

# References

Auck, Y. T., and Byrne, J. G. (1970). *In* "Second International Conference on Chemical Vapor Deposition" (J. M. Blocher, Jr. and J. C. Withers, eds.), p. 241. Electrochem. Soc., New York.

Auck, Y. T., and Byrne, J. G. (1971). *AIME, 1971,* Atlanta, Ga., in press.

Barbour, J. P., Charbonnier, F. M., Dolan, W. W., Dyke, W. D., Martin, E. E., Trolan, J. K. (1960). *Phys. Rev.* **117**, 1452.

Barrett, C. S., and Massalski, T. B. (1966). "Structure of Metals," p. 203. Mc-Graw-Hill, New York.

Berkeley, J. F., Brenner, A., and Reid, W. E. (1967). *J. Electrochem. Soc.* **114**, 561.

Bowden, F. P., and Tabor, D. (1964). "The Friction and Lubrication of Solids," p. 350. Oxford Univ. Press (Clarendon), London and New York.

Brecher, L. E. (1970). *In* "Second International Conference on Chemical Vapor Deposition" (J. M. Blocher, Jr., and J. C. Withers, eds.), p. 37 Electrochem. Soc. New York.

Chun, J. S., Nicholson, P. S., Sosin, A., and Byrne, J. G. (1971a). *Electrochem. Soc.* (to be published). *J. Electrochem. Soc.* Vol. 118, pp. 1492–1498.

Chun, J. S., Shim, H. S., and Byrne, J. G. (1971b). Presented at AIME, Atlanta, Ga. (in press).

Cullity, B. D. (1956). "Elements of X-Ray Diffraction." Addison-Wesley, Reading, Massachusetts.

Farrell, K., Houston, J. T., and Schaffhauser, A. C. (1967). *Proc. Conf. Chem. Vapor Deposition Refract. Metals, Alloys, Compounds, 1967* pp. 363–390.

Farrell, K., Federer, J. I., Schaffhauser, A. C., and Robinson, W. C., Jr. (1970). *In* "Second International Conference on Chemical Vapor Deposition" (J. M. Blocher, Jr. and J. C. Withers, eds.), p. 263. Electrochem. Soc., New York.

Federer, J. I., and Steele, R. M. (1965). *Nature* **205**, 4971, 587.

Fine, M. E. (1952). *Amer. Soc. Test. Mater. Spec. Tech. Publ.* **129**, 43.

Glaski, F. (1970). Private communication.

Haskell, R. W. (1970). *In* "Second International Conference on Chemical Vapor Deposition" (J. M. Blocher, Jr. and J. C. Withers, eds.), p. 63. Electrochem. Soc., New York.

Haywood, D. D., and Trapnell, B. M. (1964). "Chemisorption." Butterworth, London.

Holman, W. R., and Huegel, F. J. (1967a). *Proc. Conf. Chem. Vapor Deposition, Refract. Metals, Alloys, Compounds, 1967,* pp. 127–148.

Holman, W. R., and Huegel, F. J. (1967b). *Proc. Conf. Chem. Vapor Deposition, Refract. Metals, Alloys, Compounds, 1967,* pp. 427–442.

Holman, W. R., and Huegel, F. J. (1970). *In* "Second International Conference on Chemical Vapor Deposition" (J. M. Blocher, Jr. and J. C. Withers, eds.) p. 171. Electrochem. Soc., New York.

Holzl, R. A. (1968). "Techniques of Materials Preparation and Handling," Vol. 1, Part 3. Wiley (Interscience), New York.

JANAF. (1960 to date). Thermochemical Data. Thermal Res. Lab., Dow Chem. Co., Midland, Michigan.

Kemball, C. (1950). *Proc. Roy. Soc., Ser. A* **201**, 377.

Kubaschewski, O., and Evans, E. L. (1958). "Metallurgical Thermochemistry." Pergamon, Oxford.

McCoy, H. E., and Stiegler, J. O. (1967). *Proc. Conf. Chem. Vapor Deposition, Refract. Metals, Alloys, and Compounds, 1967* pp. 391–425.

Metlay, M., and Kimball, G. E. (1948). *J. Chem. Phys.* **16**, 779.

Powell, C. F., Oxley, J. H., and Blocher, J. M., Jr. (1966). "Vapor Deposition." Wiley, New York.

S.A.E. "Measurement of Stress by X-Ray," Tech. Rep. TR-182. Soc. Automotive Eng., Inc., 485 Lexington Avenue, New York.

Sard, R., and Weil, R. (1970). *Electrochim. Acta* **15**, 1977.

Shim, H. S. (1971). Ph.D. Thesis, University of Utah.

Shim, H. S., and Byrne, J. G. (1971). *Int. Congr. Crystal Growth, 3rd, Marseille*, in press.

Taylor, J. L., and Boone, D. H. (1964). *J. Less-Common Metals* **6**, 157.

Timoshenko, S., and Goodier, J. N. (1951). "Theory of Elasticity." McGraw-Hill, New York.

Willertz, L. E., and Shewmon, P. G. (1970). *Met. Trans.* **1**, (12), 2217.

# Author Index

Numbers in italics refer to the pages on which the complete references are listed.

Abraham, F. F., 19, *77*
Ahmad, M. S., 216, 218, 220, 221, 237, *246*
Aigeltinger, E. H., 250, 252, 253, 254, 276, *292*
Anderson, W. A., 253, 258, *292*
Ardell, A. J., 278, *292*
Argon, A. S., 82, 87, 88, 89, 91, 92, 93, 95, 99, 100, 102, 108, *114*
Armenakas, A. E., 93, 94, 95, 110, *114*
Armstrong, R. W., 180, 181, *244*
Arsenault, R. J., 180, 181, 215, 216, 222, 237, 239, *244, 245*
Auck, Y. T., 299, 301, 308, 310, 311, 312, 314, 315, 317, 318, *339*
Avrami, M., 277, *292*

Bailey, D., 108, *114*
Barbour, J. P., 301, *339*
Barr, R. G., 227, *245*
Barrett, C. S., 306, *339*
Barrett, L. K., 250, 252, *292*
Bechtold, J. H., 180, 181, *244*
Begley, R. T., 180, 181, *244*
Behrens, E., 116, *176*
Berg, C. A., 109, *114*
Berkeley, J. F., 298, 337, *339*
Beshers, D., 182, *244*
Bierlein, T. K., 256, *292*
Biot, M., 106, *114*
Blickwede, D. J., 206, 207, *245*
Blocher, J. M., Jr., 293, *340*
Bolling, G. F., 3, 69, 73, *77*
Bomford, M. J., 108, *114*
Boon, M. H., 189, 195, 196, 216, 233, 234, 236, 237, 239, *245*

Boone, D. H., 325, *340*
Born, M., 116, 117, 119, *176*
Bousquet, P., 254, 269, 275, *292*
Bowden, F. P., 325, *339*
Bradshaw, F. J., 37, 39, 40, *77*
Brauner, J., 204, *245*
Brecher, L. E., 296, 298, 303, 337, *339*
Brenner, A., 298, 337, *339*
Broutman, L. J., 93, 108, *114*
Bunn, P. M., 216, 217, 218, 219, *244*
Burton, W. K., 43, *77*
Buteau, L. J., 250, 252, 253, 254, 276, *292*
Byrne, J. G., 299, 300, 301, 306, 307, 308, 310, 311, 312, 314, 315, 317, 318, 320, 324, 326, 327, 328, 329, 331, 333, *339, 340*

Cabrera, N., 43, *77*
Cahn, J. W., 43, 44, 54, 55, 59, 61, 63, 64, *77*, 251, 252, 261, 271, *292*
Chadwick, G. A., 3, *77*
Charbonnier, F. M., 301, *339*
Childs, W. J., 37, 39, 40, *77*
Chipman, J., 196, *244*
Chun, J. S., 306, 307, 318, 320, 324, 326, 327, 328, 329, 331, *339*
Clauer, A. H., 226, *245*
Cohen, M. H., 56, *77*
Coleman, B. D., 94, *114*
Collette, G., 197, 198, *244*
Corten, H. T., 80, *114*
Coulton, R. A., 253, 258, *292*
Couper, G. J., 202, 203, 228, *244*
Cullity, B. D., 335, 336, *339*
Cummings, D. G., 216, 217, 218, 219, *244*

*341*

# Subject Index

Page numbers in **bold type** denote the beginning of a chapter about the entry.